WILDLIFE HABITAT MANAGEMENT

Concepts and Applications in Forestry

Ermine in a Wisconsin forest by Michele Woodford (used with her permission).

WILDLIFE HABITAT MANAGEMENT

Concepts and Applications in Forestry

Brenda C. McComb

University of Massachusetts
Amherst, U.S.A.

CRC Press
Taylor & Francis Group
Boca Raton   London   New York

CRC Press is an imprint of the
Taylor & Francis Group, an **informa** business

Cover Photo: Ermine in a Wisconsin forest, courtesy of Michele Woodford (with permission).

Learning Resources
Centre

1293030X

CRC Press
Taylor & Francis Group
6000 Broken Sound Parkway NW, Suite 300
Boca Raton, FL 33487-2742

© 2008 by Taylor & Francis Group, LLC
CRC Press is an imprint of Taylor & Francis Group, an Informa business

No claim to original U.S. Government works
Printed in the United States of America on acid-free paper
10 9 8 7 6 5 4 3 2 1

International Standard Book Number-13: 978-0-8493-7489-0 (Hardcover)

Library of Congress Cataloging-in-Publication Data

McComb, Brenda C.
    Wildlife habitat management : concepts and applications in forestry / Brenda C. McComb.
       p. cm.
    Includes bibliographical references.
    ISBN 978-0-8493-7489-0
    1. Wildlife habitat improvement. 2. Forest management. 3. Wildlife management. I. Title.

QL82.M33 2007
639.9'2--dc22
                                                                  2007000602

Visit the Taylor & Francis Web site at
http://www.taylorandfrancis.com

and the CRC Press Web site at
http://www.crcpress.com

Table of Contents

Preface

This book is the result of over 30 years' work in academic and research organizations in which foresters, biologists, and scholars from other disciplines collaborated, fought, argued, and occasionally agreed about how forests should be managed. That 30-year period spanned a time that began as a shift from focus on game management to nongame species management to biodiversity conservation. Especially on public lands forestry evolved from commodity-focused even-aged management to green tree retention to ecological forestry. On private forest lands, green certification emerged as a nongovernmental licensing program that recognized commodity production and biodiversity conservation. Some scientists and managers involved in disciplines of wildlife management, forestry, sociology, biology, entomology, zoology, and botany as separate disciplines merged them into a common framework for conservation of biodiversity, which became known as conservation biology. In the 1980s and 1990s, ecological preservation often became pitted against economic growth. Passion for sustaining traditional ways of life opposed passion for sustaining biodiversity. Arguments were intense, but people changed. Eventually, neither philosophy emerged dominant, but rather realization grew that stable economies, viewing humans as part of ecosystems, and sustainability of ecological services were becoming more representative of societal values. Resource professionals, at times reluctantly, also began approaching natural resource management problems in this interdisciplinary light. My own philosophy changed during that time as well. I always felt that the common ground between wildlife biologists and foresters was greater than the chasm of differences, and I tried to represent one of many bridges between the disciplines in my classes, my research, and my interactions with other professionals.

The writing of this book stemmed from both a practical need and an emotional desire. Practically, I wanted to be able to use a book in my forestry and wildlife habitat management classes, and hopefully others would use it in theirs. There are excellent texts available, but I had long felt that I could make a contribution by integrating silvicultural and forest-planning principles with principles of habitat ecology and conservation biology. In addition, I wanted a forum where representatives of both disciplines could read my views and use them as a starting point for continued discussions, debates, arguments, and perhaps even agreement. The emotional motivations for writing this book are my two children. Although neither shows any inclination toward pursuing a career in natural resources, I would like to feel that I have done everything I could to leave a living inheritance to them as rich in the diversity of life as the world that I lived in. And that the foundation is laid for their generation to do the same for the ones that follow.

Many people contributed to the material in this book. People whom I considered mentors and who shaped my thinking tremendously at various points in my career are Drs. Malcolm Coulter, Mitch Ferrell, Bob McDowell, Bob Noble, Bart Thielges, and Logan Norris. In particular, I would like to thank the reviewers of various chapters. Several teaching assistants for my courses read and commented on the early drafts of all chapters and to them I am deeply in debt: Misty Cannon, Jesse Caputo, Stephanie Hart, Lori Keyes, and Holly Ober. Reviewers of individual chapters made excellent suggestions: Sal Chinnici, Sally Duncan, Cheryl Freisen, Joan Hagar, Bob Lackey, Josh Lawler, Karl Martin, Bob Mitchell, Tom Spies, Nobuya Suzuki, John Tappeiner, Denis White, and Ben Zuckerberg. Over 100 students in my WFCON 564 and FS/FW 453 classes used the early drafts of the chapters and provided excellent comments. Joann Smith, silviculturist on the Kisatchie

National Forest in Louisiana, generously provided stand exam data, which were used as the basis for an illustration of achieving a desired future condition.

Photos were generously provided by Susan Campbell, Joan Hagar, Mike Jones, Karl Martin, Bruce McCune, Kevin McGarigal, Jim Petranka, Dave Vesely, and Michele Woodford, as well as federal agencies including U.S. Geological Survey, U.S. Forest Service, National Park Service, and U.S. Fish and Wildlife Service. Numerous publishers and journals generously allowed me to use previously published figures and text, and they are cited herein.

I thank the Department of Forest Science at Oregon State University, and particularly Dr. Tom Adams for providing me with office space and computer support while I was on sabbatical leave to work on this book. The University of Massachusetts–Amherst provided me with a sabbatical leave to focus on finishing this book. John Sulzycki and Pat Roberson from Taylor & Francis publishers provided outstanding editorial support.

Finally, I thank my family and the friends who have supported me through this project, all my many other projects, and the various trials and tribulations of life that led me to this place and time in my life. Special thanks to Bill Swiggard who has supported me in more ways than I can count.

The Author

Brenda C. McComb is a professor in the Department of Natural Resources Conservation, University of Massachusetts–Amherst. She is author of over 130 technical papers dealing with forest and wildlife ecology, habitat relationships, and habitat management. She was born and raised in Connecticut at a time and place when the rural setting provided opportunities to roam forests and fields. She received a BS in natural resources conservation from the University of Connecticut, a MS in wildlife management from the University of Connecticut, and a PhD in forestry from Louisiana State University. She has served on the faculty at the University of Kentucky and Oregon State University. She was the head of the department of natural resources conservation at the University of Massachusetts–Amherst for 7 years, and served as the chief of the Watershed Ecology Branch in Corvallis for U.S. Environmental Protection Agency for 1 year. She is currently on the editorial board for *Conservation Biology*, and has been a member of The Wildlife Society and the Society of American Foresters for over 25 years. Her current work addresses interdisciplinary approaches to management of multiownership landscapes in Pacific northwest forests and agricultural areas.

1 Introduction

Aldo Leopold is generally accepted as providing the philosophical basis for wildlife management in the United States. Leopold was trained and employed initially as a forester, and in the early years the academic and disciplinary home of wildlife management was aligned with forestry. But over time, because of the need to be recognized as a discipline in its own right, the forestry profession diverged from wildlife biology and management, and in recent years the views of some wildlife professionals have been at odds with those expressed among some forestry professionals. This book is an attempt to bridge the disciplines of wildlife habitat management and forest management. It provides the conceptual bases for stand and landscape management to achieve habitat objectives for various species and communities; it also provides case studies from across the United States to illustrate how these concepts can be applied. When foresters are provided with an explanation of the concepts of habitat selection, habitat relationships, habitat elements, element dynamics in stands and landscapes, habitat permeability, connectivity, and exogenous pressures (climate change, invasive species, and development), they can understand how these factors influence decisions made during stand and forest management. Further, biologists are provided with explanations of stand and forest landscape dynamics, silvicultural approaches to providing habitat elements, and harvest planning. Case studies in each section of the book provide examples of how these concepts can be applied to achieve habitat goals at stand, landscape, and regional spatial scales. Finally, the information culminates in stand prescription development and forest planning — the key prerequisites to sustainable management practices. In addition, this planning process must include the concerns and objectives of various stakeholders. Foresters and wildlife biologists *must* work together, cooperatively, with those concerned to ensure that management approaches are adaptable to the inevitable social changes and competing demands for ecosystem services and aesthetic qualities of forests while also ensuring that current decisions are not likely to forgo future options.

WHAT IS HABITAT?

Despite the need to work together to achieve mutual goals across forested landscapes, the language of the disciplines itself can interfere with success. Throughout this text, I will try to define terms that could be confusing or misinterpreted between the disciplines. For instance, a brief search of the Web using the terms "forest wildlife habitat" produced 7.7 million hits, including the following quotes from resource professionals: "In recent years, an increasing number of landowners have realized the economic importance of timber management as a way to enhance wildlife habitat." "Several practices have damaged the wildlife habitat, including habitat fragmentation; past roads, excessive logging and development all worked to fragment large areas of intact habitat." So, who is correct? All 7.7 million? Well, yes and no. We cannot understand how to manage forests to provide habitat for wildlife species, or more generally for biodiversity, unless we understand what habitat is and is not. But first, what is wildlife? It is important to first recognize that it is not a singular noun but a plural one. Wildlife encompasses many species. To think of wildlife as one thing is making the mistake of considering us (humans) and them (all other species) as two separate groups; we are all in this together, humans and other species. Biodiversity goes beyond the collection of animals that we often perceive as wildlife to include all forms of life — plants, animals, microbes, and all the bits of the earth that support them. Leopold (1949) suggested that ". . . to keep every cog and wheel is the first precaution of intelligent tinkering." Save the pieces. In our efforts to manage forests to meet wildlife and biodiversity goals, the pieces are the species and the resources those species need to survive and reproduce. So, let us think about what a species needs as habitat.

Habitat is the place where a population lives. It includes the physical and biological resources necessary to support a self-sustaining population. Each species and each population has its own habitat requirements (Krausman 1999). References to "wildlife habitat" are meaningless, unless a particular wildlife species is identified, because almost every place is habitat for some species. Krausman (1999), Hall et al. (1997), and Garshelis (2000) have made compelling arguments to clarify the confusion that results from using the term habitat to mean vegetation types or other classes of the environment that are not directly related to a species.

Although habitat has been defined in many ways, I define *habitat* as the set of resources necessary to support a population over space and through time. Hence, every species has its own habitat needs, and the term "wildlife habitat" has little real meaning. Further, this definition focuses on populations and not simply on individuals. *Populations* are self-sustaining assemblages of individuals of a species over space and through time. *Communities*, by contrast, are assemblages of populations over space and through time. This definition of habitat is consistent with the approach taken by Hall et al. (1997), but Garshelis (2000) makes the point that quite often foresters and wildlife biologists both will refer to vegetation types, or other discrete classes of the environment, as habitats. More accurately, these are *habitat types* in that some species can be associated with some vegetation types and not with others. But these associations occur only because some or all of the resources needed by the species occur in those types. Consequently, it is important to think about how a habitat functions to provide those resources to each species.

HABITAT FUNCTION

It is useful to think of habitat meeting not only an individual's needs but also a population's needs. It is, after all, the population that is to be sustained. Individuals, though clearly essential to population maintenance, just come and go in the process, just as you and I come and go in the process of maintaining a human population. A number of factors, especially energy, drive the success of a population. Although nutrients, water, and other factors clearly have a role in maintaining the fitness of individuals and populations (e.g., Jones 1992), the lion's share of the system is driven by energy. A population gains energy from food resources and conserves energy by exploiting cover resources (Figure 1.1).

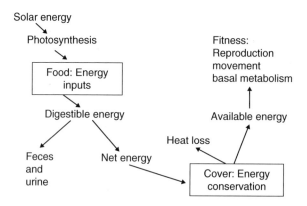

FIGURE 1.1 The concept of energy flow through individuals to influence individual and population fitness. (Based on Mautz, W.W. 1978. Redrafted from McComb, W.C. 2001. Management of within-stand features in forested habitats. Chapter 4. In Johnson, D.H., and T.A. O'Neill (managing editors), *Wildlife Habitat Relationships in Oregon and Washington*. OSU Press, Corvallis, OR.)

Energy is the currency for population sustainability. Give it more and the population will grow, give it less and the population will decline. When an individual goes energetically bankrupt, it dies. When a population goes energetically bankrupt, it goes extinct.

Food provides the source of energy (and nutrients) for individuals and populations. Food quality matters. Tree species vary in their ability to provide protein and carbohydrates to herbivores. Some parts of plants are more digestible than others, and some plant species are more digestible than others (Mautz et al. 1976). It is this digestible energy, and the net energy remaining after digestion and metabolism, that influences the fitness of individuals and populations. But fitness, the ability to survive and reproduce, also is influenced by cover quality.

It is advantageous for an animal to conserve any energy that it acquires. Mammals and birds that maintain a constant body temperature expend a large amount of energy to maintain that temperature. Cover provides a mechanism for conserving energy. The *thermal neutral zone* is the range of ambient temperatures in which an animal has to expend the least amount of energy maintaining a constant body temperature. Thermal cover places the animal closer to the thermal neutral zone. Energy expenditures to maintain body temperature are minimized in an animal's thermal neutral zone (Figure 1.2; see Mautz et al. 1992 for an example).

Any departure from the thermal neutral zone results in increased expenditure of energy, and so animals often select a habitat that reduces climatic extremes. There are upper and lower critical temperatures beyond which exposure for a prolonged period would be lethal. Cover from overheating is especially important to large animals with a low surface-area-to-body-mass ratio because they may find it particularly difficult to release excess heat unless water is available to aid in evaporative cooling. Cover from severe cold is especially important to a species with a high surface-area-to-body-mass ratio (e.g., small birds and mammals). Cover that allows an animal to stay within an acceptable range of temperatures (particularly those that approach the thermal neutral zone) is important for maintaining a positive balance of net energy and hence influences animal fitness. For instance, imagine yourself standing in a field wearing summer clothes in mid-January in Minnesota. Without any measurements, you know that you are expending a significant amount of energy to stay warm. Now, imagine you are in a field in Arizona in August at noon. You must expend energy to stay cool and not let your body temperature rise too high (e.g., heat stroke). In either case, moving into a building where the temperature is 18°C (65°F) allows you to spend less energy keeping your body at the appropriate temperature. Refer to Figure 1.2 and plot the metabolic rate for a small mammal or bird at a low temperature and then the metabolic rate for a temperature near the thermal neutral zone. The difference in metabolic rates along the y-axis is an index to the amount of energy

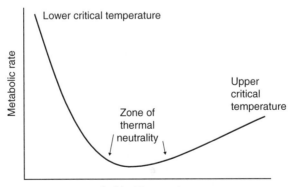

FIGURE 1.2   Relationship between metabolic rate and ambient temperature in a hypothetical mammal. (Based on Gordon, M.S. 1972, p. 325. Redrafted from McComb, W.C. (2001). Management of within-stand features in forested habitats. Chapter 4. In Johnson, D.H., and T.A. O'Neill (managing editors), *Wildlife Habitat Relationships in Oregon and Washington*. OSU Press, Corvallis, OR.)

that the individual can conserve by staying closer to the thermal neutral zone. For a small animal with a high metabolic rate and high ratio of surface area to body mass, that conserved energy can mean the difference between life and death on a cold winter night. But there are both behavioral and physiological adaptations that some species have in order to further conserve energy. Southern flying squirrels (see Appendix 1 for a list of scientific names of all plants and animals used in this book) and some species of cave-dwelling bats will often use communal roosts in winter to collectively maintain a lower surface-area-to-body-mass ratio. Flying squirrels pack many small bodies together to make one bigger, more energetically efficient big body by huddling (Merritt et al. 2001). Other species such as eastern chipmunks hibernate; striped skunks enter a state of torpor where metabolic rates are reduced and energy is conserved. Black-capped chickadees, a small 10 g bird that spends winters in very cold climates, will cache food, roost in cavities, and alter their metabolic rates seasonally to cope with temperature extremes (Cooper and Swanson 1994). So, the effects of conserving energy through use of thermal cover can be improved further by these physiological and behavioral mechanisms.

But the relationship portrayed in Figure 1.2 is different for species that do not maintain a constant body temperature. Most reptiles and amphibians (and some nestling birds) do not use large amounts of energy to maintain a constant body temperature. They are *ectotherms* — they receive most of their body heat from the surrounding environment, unlike *endotherms*, which generate their own body heat. For ectotherms, metabolic rates and food requirements vary as ambient temperature varies. The evolutionary advantage of such an approach is that these ectotherms require less food to survive, but they can be restricted from extreme environments that otherwise would be inhabitable by endotherms (some bird and mammal species). Hence, reptiles and amphibians often use cover to adjust the ambient temperature to allow them to survive, reproduce, and move in places and times in which they would otherwise be unable to (Forsman 2000). Consequently, cover is an important component of habitat for these species, to conserve energy as well as place them at a temperature at which they can be active.

Cover also can refer to the portion of habitat that an animal uses for nesting and escaping from predators. The most significant loss of energy by an animal is to have its energy converted into the energy of its predator! Hiding cover protects an animal from predation. Cottontail rabbits often spend resting hours in dense shrubby cover adjacent to grassy fields and meadows (Bond et al. 2001). The dense shrub cover protects them from predation by red-tailed hawks, whose body size and wing spread do not allow them to penetrate dense vegetation. Simple modifications to habitat such as allowing shrubs to proliferate along field edges can lead to increased survival and increased population growth for cottontails in this example.

Nesting cover provides the conditions necessary for raising young — appropriate temperature, and protection from predators and competitors. The effectiveness of nest box programs for wood ducks, eastern bluebirds, and other cavity-using species, demonstrates that manipulation of the quantity, quality, and availability of nesting-cover resources can be an effective management technique (McComb and Lindenmayer 1999). Forest managers can influence habitat of a species by altering food quality, quantity, and/or availability while also altering the quality, quantity, and availability of cover. This strategy can lead to drastic changes in habitat quality for the species.

Water is differentially important to animal species. Some species require free water or high humidity (mountain beaver, for example, have a primitive uretic system) (Schmidt-Nielsen and Pfeiffer 1970). Other species obtain most of their water from their food (e.g., pocket gophers). Some species use water as a form of cover to enhance evaporative cooling (e.g., elk) or to escape predators (e.g., white-tailed deer). Still others such as amphibians require free water or moist environments for reproduction.

The size of habitat is also an important determinant of its suitability for a species. A patch of habitat must be sufficiently large to provide energy inputs and energy conservation features to sustain a population. Habitat may occur in one large unit, but more commonly it is distributed in patches through other less suitable patches. If these habitat patches are too widely distributed, then the animal expends more energy moving among patches than what it receives from those patches. The amount of habitat, therefore, and its quality and distribution are interrelated. Increasing any or all of these attributes of habitat increases the net energy available to animals that use this energy to maintain body temperature, move to food and cover, and reproduce.

HABITAT FOR HUMANS

When trying to understand the concept of habitat it may be helpful to think of your habitat. You are an individual of a species, a mammal that maintains a constant body temperature, and you have your own set of resources that you need to survive and reproduce. So think about your food requirements. You eat a certain number of calories per day and this energy is converted to adenosine triphosphate (ATP) or stored for future use. You use this energy to maintain your body temperature and support your physical being, to move throughout the day from place to place, to raise your children, and to buy or harvest food. Generally, human food is rather digestible and high in energy, though in some societies digestible energy can limit not only human health but also survival.

Humans, like most other mammals, also use cover. We have homes where we attempt to keep the ambient temperature as close to our thermal neutral zone as possible. We raise our children there and we use these homes at times as a place of refuge during inclement weather or catastrophic disturbances. We need clean water in adequate quantities so as not to become dehydrated. All of these things must in close proximity so that we do not spend more energy acquiring resources than we receive from them. Think of nearly any other animal species and we can similarly define the food, cover, and water requirements for that species, as we can for our own. But for each species those requirements differ. No two species are likely to coexist and have the same habitat requirements for very long if resources are limited and species are competing for them. There are instances, however, when predation pressure on two prey species can allow them to coexist using the same resources, but if predation pressure changes or resource availability changes, then one species will likely out-compete the other (McPeek 1998). Because we humans are so adaptable and because we usurp energy and other resources that could be used by other species, we have a profound effect on the number of species and individuals with which we share this planet. "Saving all the pieces" comes at a price. And it is not a price that society is willing to pay in all instances. There are 6.2 billion people, 403 million tons of human biomass, which must be supported daily on this planet, and two more people are added every second. Saving all the pieces may be a noble goal, but human self-preservation and preservation of lifestyles can trump that goal quickly, unless we give more thought to our own habitat needs within the context of the needs of the other species with which we share this planet.

FORESTS AS HABITAT

How we manage forests to partition energy among various forms of life is the essence of the challenge facing foresters and wildlife biologists. It is a challenge because the rate of net primary production is fixed over large areas and times, there is a solar constant, and climate changes relatively slow (the current climate change crisis not withstanding). Further, although herbivores in forests exist in a sea of plant energy, little of it is available for those herbivores to use. Food quantity is often not as important as food quality in a forest (Mautz 1978). Most of the energy in a forest is in cellulose, the wood that society demands, and for many species this wood is not very digestible. Animals can only use the *digestible energy* in food (Figure 1.1), and so indigestible portions of food (e.g., cellulose, lignin, chiton, or bones) or compounds (tannins and other phenols) in the plants that inhibit digestion reduce food quality (Robbins et al. 1991). These indigestible portions of a forest can become available as energy to many species if they are made available through decomposing

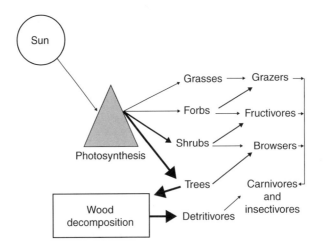

FIGURE 1.3   Energy pathways through a forest. Arrow sizes reflect relative differences in energy availability. (Based on Harris, L.D. 1984. *The Fragmented Forest: Island Biogeography Theory and the Preservation of Biotic Diversity*. University of Chicago Press, Chicago, IL, p. 211.)

organisms (Figure 1.3). Without the decomposers being available as digestible food for other species in forests, most of the energy would go unused. Indeed, without decomposers, animal diversity would be reduced to a relatively few species specialized in eating twigs, leaves, fruits, or nonwoody portions of the forest. The *decomposition pathway* is hence quite important in maintaining a diverse animal community in many forests. Further, providing plants with adequate sunlight and water to grow and produce flowers and fruits is the key for meeting the food needs of those species that rely on plant fruits for food.

HISTORICAL APPROACHES TO MANAGING FORESTS AS HABITAT

Management of forests as habitat for wildlife has been conducted for centuries in many cultures. Native Americans used fire to at least move animals during hunts if not to provide better forage for them (Boag 1992). But it was not until Leopold completed his *Game and Fish Handbook* for the Forest Service in 1915 and the subsequent publication of *Game Management* (Leopold 1933) that active management efforts began on many public and some private forest lands to promote selected species. There have been two common approaches to management of forests as habitat for species: management of individual species and management for biodiversity. Only recently has the focus of land managers shifted from utilitarian goals to the protection of rare species to conservation of entire ecosystems. Indeed the evolution of wildlife management as a discipline was driven largely by the philosophies underpinning forest management, and both disciplines have evolved in parallel with regard to the focus of their management efforts.

Both foresters and wildlife biologists agree that there is a way to manage species to provide the desired plants or animals in a nondeclining, sustained-yield manner. Harvest scheduling works for black oaks and black bears — we just need to provide the correct habitat and manage the density of the organisms to provide sufficient resources for the remaining individuals, and the system can go on and on and on. Right? Well, maybe, but by taking this approach there is the risk of losing other species that might be of value to society. So managers began to focus on indicator species (surrogates for other species) or guilds of species (groups of species with similar habitat needs) in order to meet the needs for many other species. These approaches also produced problems for exactly the reasons described in the previous section: each species has its own habitat requirements and each will respond differently to management activities (Mannan et al. 1984).

But some approach is needed to produce the desired species of plants and animals while minimizing the risk of losing some pieces that we may need later. Hence, the most recent approach to managing is to meet societal goals of aesthetics, game, biodiversity, recreation, and timber fall within the realm of *ecosystem management* — an approach designed to minimize risk to species and maximize the likelihood that the approach will be sustainable (Meffe et al. 2002). One basis for this approach recognizes that forest disturbances change the abundance of individuals in many populations, and those changes also influence the composition of plant and animal communities. Before technologically advanced humans began managing forests, natural disturbances caused the localized extinction of some species and opportunities for recolonization by others. Communities changed as forests regrew following these disturbances. Species tended to be adapted to the range of conditions that occurred under these conditions of natural disturbance. Understanding how species respond to these conditions and how management might replicate or depart from those conditions can be useful in understanding the effects of management on a suite of other resources (Landres et al. 1999). In addition, consider that the management of an individual species has consequences for other species in its community. Forest disturbances that benefit black-tailed deer, for example, probably would benefit creeping voles and orange-crowned warblers, two species found in early-successional forests, but not Douglas squirrels or pileated woodpeckers, two late-successional forest species.

Natural disturbances such as fires, insect defoliation, and hurricanes not withstanding, vegetation management by forestland managers is probably the greatest factor influencing the abundance and distribution of animals in our forests today. By understanding concepts of habitat function, population change, and habitat patterning, managers can make decisions that can find the appropriate societal balance among commodities, species, and ecosystems.

WHY MANAGE HABITAT?

We manage habitat for various reasons such as personal goals, corporate objectives, and legal requirements. Policies in the United States, such as the Endangered Species Act and National Forest Management Act, require people in various agencies to manage habitat. But why do we have these policies? Why should we spend time and money managing habitat for species that occur in our forests? Quite simply, we do or do not manage habitat because society either cares about these resources or does not, respectively. Wildlife are public resources that occur on both public and private lands. If society placed no value on a species or group of species, then we would not manage their habitat. Values that society places on animals evolve over time and from culture to culture. Take the beaver for example (Figure 1.4). Clearly, there are many reasons to manage habitat for beaver, though some segments of society would like to ensure that there are fewer animals and some would like more of them. To complicate matters further, oftentimes people with differing values are neighbors and the beavers do not care where the property line falls!

These values placed on a resource usually change slowly as other aspects of our society change. In some cultures, the species may be viewed as an important economic or otherwise subsistence resource that would be harvested and used for survival. As society becomes less reliant on or less engaged with native species, people may begin to place greater intrinsic value on them or fear them because they are unknown. Finally, the relative importance of a species may change markedly and rapidly as unexpected events occur, leading to rapid changes in societal values that have unanticipated impacts on our ability to manage natural resources. On September 11, 2001, I was on a west-bound United flight from Hartford to Denver, somewhere over Lake Erie at 9:15 AM eastern time. I was fortunate to have spent a week in Chicago rather than at other alternatives that morning. Those events changed our society's priorities suddenly. They certainly changed mine. Although it did not necessarily diminish the importance that people placed on environmental values, it raised human safety and welfare to a much higher priority. Human and financial resources once used to provide natural resource values for our society were diverted to these higher priorities. We saw a similar response following hurricanes Rita and Katrina in 2005. One can argue the political decisions were

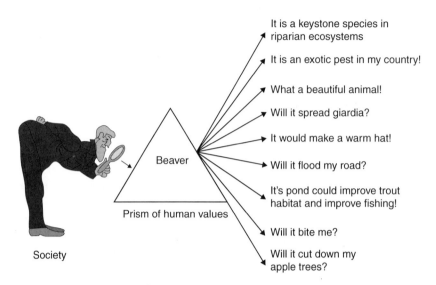

It is a keystone species in
riparian ecosystems

It is an exotic pest in my country!

What a beautiful animal!

Will it spread giardia?

It would make a warm hat!

Beaver

Will it flood my road?

It's pond could improve trout
habitat and improve fishing!

Prism of human values

Will it bite me?

Society

Will it cut down my
apple trees?

FIGURE 1.4 Society views natural resources, in this case beavers, through a prism of values. (Based on discussions with R.M. Muth, University of Massachusetts–Amherst.)

made at the time to achieve ideological as well as humanitarian goals, but changes did occur that impacted many aspects of our ability to meet natural resource and environmental quality goals for society. We are not alone in these struggles.

Overwhelming economic pressures are faced by many parts of the world. Huge loans have been provided by the International Monetary Fund (IMF) to countries such as Argentina. These debts to the IMF, combined with the overwhelming pressure to ensure that people survive on a limited and often declining natural resource base, significantly limits options to maintain environmental values. Forest reserves, popular approaches to biodiversity conservation in wealthy countries and recognized as important by developing countries, are usually an untenable option in much of the world unless significant foreign monetary support is provided. Even so, reserves become only one approach to protecting biodiversity. Indeed, the majority of the land and water resources that could support some components of natural systems are not within reserves and never will be. Social pressures force managers to consider options that are both economically feasible and ecologically sound. If large tracts of forest are managed in a manner that considers the structure and function of the habitat for valued species while still allowing some economic value to the landowner, then there is a greater likelihood that it will remain a forest or field. Once the value of a forest falls below that of other land uses, then there is a risk of conversion to a new use (e.g., industrial agriculture, grazing, or housing; Figure 1.5).

If the forest is managed to consider structure and function to valued organisms, then it may support these species, which otherwise would be found primarily in reserves, thereby complementing effectiveness of the reserve system. For instance, actions that maintain a forest rather than a pasture that likely would be overgrazed will decrease the probability that the site would be lost to desertification in the dry tropics. Active forest management to achieve multiple objectives such as grazing lands (Figure 1.6a), wood products (Figure 1.6b), and habitat for valued wildlife (Figure 1.6c) may be one step toward maintaining economic and ecological values.

We do not know with certainty how to manage all or even most forests to achieve multiple values. But we do understand vegetation dynamics, disturbance ecology, habitat selection, and population dynamics as well as the influence of local, regional, and global economies, cultural mores, and social value systems. If we use this information in a thoughtful manner, then we should be able to develop reasonable management plans to achieve multiple objectives. However, we will need to monitor

FIGURE 1.5 Once the value of land falls below the value associated with its current use, it becomes vulnerable to change to a new use, such as the conversion of this dry tropical forest in northern Argentina to an industrial soybean farm.

the effectiveness of the plans to ensure that we are meeting our goals. This *adaptive management* process (listen, synthesize, plan, implement, monitor, learn, listen, etc.) is an integral part of habitat management (Baskerville 1985). My objective is to provide the concepts, processes, and tools that you can use to develop resource management plans that will help you achieve landowner goals now and in the future.

CASE STUDY: THE FORESTS OF BRITISH COLUMBIA

Society directs the way that forests are managed. Over the past 100 years, we have seen dramatic changes in the principal values associated with forests in nearly every technologically advanced culture. Changes in values and beliefs associated with forest management have followed somewhat parallel courses in the United States, Canada, New Zealand, and Australia, though at different times. The following example is extracted from a paper by Kremaster and Bunnell (1998) and reflects those changing values associated with forest management in British Columbia (BC). Forests in BC extend from the subalpine region of the Canadian Rockies to the boreal forest in the north and south through temperate rainforests along the Pacific coast to pine forests of the interior. BC's forests cover an area twice the size of all of the New England states and the state of New York combined. The forest products industry has been and continues to be very important to the provincial and national economy.

The remaining text in this case study is paraphrased from Kremaster and Bunnell (1998) to illustrate these social changes. Until the 1940s, forests in BC were seen as inexhaustible suppliers of timber. It was not until after World War II that attention focused on sustainable forestry, and the public began to expect foresters to grow trees and continuously provide timber over the long term. Major concerns at that time were fire protection and the decline in timber volume as old growth

(a)   (b)

(c)

FIGURE 1.6 In northern Argentina an effort is being made at Salta Forestal to manage native dry tropical forests for (a) grazing, (b) wood products, and (c) habitat for various species of wildlife (in this case one of several species of armadillos found in the region) in a manner that helps to support local economies while conserving biodiversity.

was converted into managed stands. Foresters were expected to manage economically valuable tree species to ages well short of the potential life span of the species.

Wildlife concerns in the first half of the 1900s focused primarily on game species and fish. During the late 1960s and in the 1970s, foresters began to embrace the paradigm of "multiple use" or managing for many values on each piece of land. In BC, legislation was passed that recognized that many resources were provided by forests and all were important to its citizens. This idea of multiple use remained in place through the mid-1980s. Wildlife concerns expanded to include nongame species but attention was still focused mainly on game and fish species. Forest guidelines protected unstable soil and some streams, but growing trees and protecting live trees from insect and fire were the main concerns reflected in management guidelines.

During the late 1960s and in the 1970s, silvicultural systems focused on producing and recovering the maximum amount of economic fiber from the forest. Along the Pacific coast, clearcutting was the dominant practice, even as interest in forests expanded beyond trees alone (Figure 1.7). Typical rotations (the interval between harvests) increased the area of early stages of natural succession and truncated succession well short of tree ages in historical forests. Research began to document changes to wildlife species assemblages associated with various stages of forest development. It was recognized that managed and unmanaged forests change over time in response to natural and human-induced disturbances, and that different vertebrate species were more abundant in different stages of forest development.

Products and values desired from forests have continued to change. Although many people still embrace the notion of multiple use, there is general realization that a given piece of land, unless

FIGURE 1.7 Example of timber harvest in a watershed commonly seen in BC and Alaska in the 1970s.

enormously large, cannot provide all desired, and sometimes competing, resources. The current scientific perception is that all parts of the ecosystem are linked and activities that affect one aspect of the system will likely affect others. The current management focus is on managing ecosystems, sustaining biodiversity, and maintaining forests more like historical ones. These approaches are in response to current public concerns that include loss of species, productivity, future options, and economic opportunities. Sustaining biodiversity has become a fundamental goal. Forest practices and policies have continued to change. Scientists and managers have translated public concerns and their own improved understanding of forest systems into new approaches. Social concerns as well as long time periods and large areas were incorporated into the concept of "ecosystem management." Concerns about losing species and productivity impelled policy makers to create legislation (e.g., BC Forest Practices Code [FPC]), integrate recent scientific knowledge (e.g., Scientific Panel for Sustainable Forest Practices in Clayoquot Sound [CSP]), and initiate new approaches to planning (e.g., Innovative Forestry Practices Agreements). Legislation and planning try to include recent knowledge, but policy continues to precede reliable knowledge.

Legislation and planning processes have been enacted to translate public and scientific concerns into different forest practices. With the advent of the FPC in BC, foresters have a legislated responsibility for sustaining biological diversity compared to other natural-resource managers. Regulations governing agricultural and urban development do not reflect the same concern for maintaining ecosystems, even though these activities have had greater impact on biological diversity.

The FPC has encouraged less clearcutting and promoted a range of retention during even- and uneven-aged management. Retention of older stages of forest development, maintaining connectivity, and protecting buffers around several stream classes are now legislated. The levers used in FPC regulations reflect forest features that we believe are related to biodiversity and ecosystem productivity. Managers are limited to practical approaches, for example, remove snags or let them stand, or leave live trees to grow old or harvest them at an economic rotation. As a result, these levers include stand structures such as snags, downed wood, species mixtures, and large old trees, and forest-level measures such as seral stage distribution, amount of edge, forest interior, patch size, and corridors. Fortunately, these attributes link to public concerns and to species richness. To evaluate effectiveness of these approaches forest managers need to know "how much is enough of specific forest elements," "what spatial patterns are important," and "which species are likely to need more individualized approaches."

This example from BC has been repeated in many parts of the United States, Canada, New Zealand, Australia, and elsewhere. Clearly, social values have shifted to concerns regarding biodiversity conservation, and the management concepts and approaches used by biologists and foresters should reflect those concerns.

SUMMARY

Habitat is the set of resources necessary to support a population over space and through time. It is a species-specific concept and is different from habitat type, which is often a classification of the environment that may or may not be related to the resources necessary to maintain a population's fitness. Food represents the inputs of energy and nutrients. Digestible energy is not as available to many vertebrates in forests unless indigestible cellulose is broken down through decomposition. Energy is conserved by use of cover. Energy expenditure to maintain a constant body temperature is minimized in the thermal neutral zone; hence, thermal cover allows an animal to provide an ambient temperature closer to the thermal neutral zone. Escape and nesting cover protect an animal from risks of predation or competition.

Habitat was historically managed to increase populations of game species. The focus for habitat management has changed from primarily utilitarian in the mid-1940s in the United States to considering a broad suite of organisms, including endangered species and species enjoyed for their aesthetics. Biodiversity conservation has become the most recent social goal in forest management. Indeed, we manage forests to achieve human needs and desires. Wood products are clearly one reason for managing forests, but providing habitat for desired species, clean water, recreation, and rangeland resources are also of paramount importance. The challenge that private forest landowners face is that they are charged with meeting societal goals for a public resource (wildlife) on their private lands. To be effective at meeting both individual and public goals for forests, foresters must work collaboratively with wildlife biologists and other resource professionals to develop innovative approaches to forest management. That is the focus of this book.

REFERENCES

Baskerville, G. 1985. Adaptive management — wood availability and habitat availability. *For. Chron.* 61: 171–175.

Boag, P.G. 1992. *Environment and Experience: Settlement Culture in Nineteenth-Century Oregon.* University of California Press, Berkeley, CA.

Bond, B.T., B.D. Leopold, L.W. Burger, and D.L. Godwin. 2001. Movements and home range dynamics of cottontail rabbits in Mississippi. *J. Wildl. Manage.* 65: 1004–1013.

Cooper, S.J., and D.L. Swanson. 1994. Seasonal acclimatization of thermoregulation in the black-capped chickadee. *Condor* 96: 638–646.

Forsman, A. 2000. Some like it hot: intrapopulation variation in behavioral thermoregulation. *Evol. Ecol.* 14: 25–38.

Garshelis, D.L. 2000. Delusions in habitat evaluation: measuring use, selection, and importance. In Boitani, L., and T.K. Fuller (eds.), *Research Techniques in Animal Ecology: Controversies and Consequences.* Columbia University Press, New York.

Gordon, M.S. 1972. *Animal Physiology: Principles and Adaptations,* 2nd edn. MacMillan Co., New York, p. 591.

Hall, L.S., P.R. Krausman, and M.L. Morrison. 1997. The habitat concept and a plea for standard terminology. *Wildl. Soc. Bull.* 25: 173–182.

Harris, L.D. 1984. *The Fragmented Forest: Island Biogeography Theory and the Preservation of Biotic Diversity.* University of Chicago Press, Chicago, IL, p. 211.

Jones, R.L. 1992. Relationship of pheasant occurrence to barium in Illinois soils. *Environ. Geochem. Health* 14: 27–30.

Krausman, P.R. 1999. Some basic principles of habitat use. In Launcbaugh, K.L. et al. (eds.), *Grazing Behavior of Livestock and Wildlife*. Idaho Forest, Wildl. Range Exp. Sta. Bull. No. 70. University of Idaho, Moscow.

Kremaster, L.L., and F.L. Bunnell. 1998. Changing forests, shifting values and chronosequence research. *Northwest Sci.* 72: 9–17.

Landres, P.B., P. Morgan, and F.J. Swanson. 1999. Overview of the use of natural variability concepts in managing ecological systems. *Ecol. Appl.* 9: 1179–1188.

Leopold, A. 1933. *Game Management*. Charles Scribner's Sons, New York, p. 481.

Leopold, A. 1949. *A Sand County Almanac, and Sketches Here and There*. Oxford University Press, New York.

Mannan, R.W., M.L. Morrison, and E.C. Meslow. 1984. Comment: the use of guilds in forest bird management. *Wildl. Soc. Bull.* 12: 426–430.

Mautz, W.W. 1978. Nutrition and carrying capacity. In Schmidt, J.L., and D.L. Gilbert (eds.), *Big Game of North America, Ecology and Management*. Stackpole Books, Harrisburg, PA.

Mautz, W.W., H. Silver, J.B. Holter, H.H. Hayes, and W.E. Urban Jr. 1976. Digestibility and related nutritional data for seven northern deer browse species. *J. Wildl. Manage.* 40: 630–638.

Mautz, W.W., J. Kanter, and P.J. Pekins. 1992. Seasonal metabolic rhythms of captive female white-tailed deer: a reexamination. *J. Wildl. Manage.* 56: 656–661.

McComb, W.C. 2001. Management of within-stand features in forested habitats. In Johnson, D.H., and T.A. O'Neill (managing editors), *Wildlife Habitat Relationships in Oregon and Washington*. OSU Press, Corvallis, OR, Chapter 4.

McComb, W.C., and D. Lindenmayer. 1999. Dying, dead and down trees. In Hunter Jr., M.L. (ed.), *Maintaining Biodiversity in Forest Ecosystems*. Cambridge University Press, Cambridge, UK Chapter 10.

McPeek, M.A. 1998. The consequences of changing the top predator in a food web: a comparative experimental approach. *Ecol. Monogr.* 68: 1–23.

Meffe, G.K., L.A. Nielsen, R.L. Knight, and D.A. Schenborn. 2002. *Ecosystem Management: Adaptive, Community-Based Conservation*. Island Press, Washington, DC.

Merritt, J.F., D.A. Zegers, and L. Rose. 2001. Seasonal thermogenesis of southern flying squirrels (*Glaucomys volans*). *J. Mammal.* 82: 51–64.

Robbins, C.T., A.E. Hagerman, P.J. Austin, C. McArthur, and T.A. Hanley. 1991. Variation in mammalian physiological responses to a condensed tannin and its ecological implications. *J. Mammal.* 72: 480–486.

Schmidt-Nielsen, B., and E.W. Pfeiffer. 1970. Urea and urinary concentrating ability in the mountain beaver, *Aplodontia rufa. Am. J. Physiol.* 218: 1370–1375.

2 Vertebrate Habitat Selection

Managing forests to produce a desirable mix of forest resources, including timber products and wildlife species, requires an understanding of how animals respond to habitat in forests. Habitat provided within and among *stands* (units of homogeneous forest vegetation used as the basis for management) over a *landscape* (a complex mosaic of interacting patches including forest stands) can have significant effects on the abundance and distribution of animal species. Management strategies aimed at long-term population change are most likely to succeed if they alter habitat quantity, quality, and/or distribution. But knowing how species select habitat can provide clues as to what habitat elements to provide. *Habitat elements* are those bits and pieces of a forest important to many species, such as vertical structure, dead wood, tree size, plant species, and forage. We will discuss these in greater detail in the following chapter.

Habitat selection is a set of complex behaviors that a species has developed among individuals in a population to ensure fitness. These behaviors are often innate and have allowed populations to persist under the variable conditions that occur over time in forests (Wecker 1963). These behaviors also have allowed each species to select habitat in a manner that allows it to reduce competition for resources with other species. So, the evolutionary selection pressures on each species, both abiotic and biotic, have led species to develop different strategies for survival that link habitat selection and population dynamics. Some species are *habitat generalists*, and can use a broad suite of food and cover resources. These species tend to be highly adaptable and occur in a wide variety of environmental conditions. The deer mouse is a species that exemplifies this strategy, in that it can be found in all stages of forest development and in many *forest types* across the United States. Deer mice have high reproductive rates and can demographically take advantage of abrupt increases in food and cover resources (Figure 2.1). This species is also a primary food resource for many forest predators. Hence, providing habitat for deer mice in a forest is quite easy, although they do tend to be more abundant in early-successional forests than in late-successional forests.

Other species are *habitat specialists*. These species are adapted to survive in forests by capitalizing on the use of a narrow set of resources, resources that they are better adapted to their use than most other species. Consider where you might find spring salamanders in the eastern United States or torrent salamanders in the western United States. Both species occur in clear, cold headwater streams, and they tend to be most abundant where fish are excluded from the streams because fish are their predators. Both species are of interest to wildlife biologists because of the concern that forest management activities that reduce canopy cover and raise stream temperatures could threaten populations of these species (Lowe and Bolger 2002, Vesely and McComb 2002). Clearly though, habitat generalists and specialists are simply two ends of a spectrum of species' strategies for survival in forests faced with variable climates, soils, disturbances, competitors, and predators.

HIERARCHICAL SELECTION

Many studies have been conducted to assess habitat selection by forest wildlife species. The assumption made by biologists is that, if we can understand what characteristics of the environment are selected by a species, then we can infer what characteristics we may wish to provide during forest management to accommodate them in our stand or forest. There are some concerns surrounding this

FIGURE 2.1 Deer mice are habitat generalists that use a wide range of forest conditions. (Photo by Mike Jones, used with permission.)

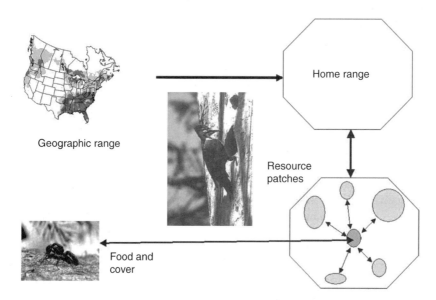

FIGURE 2.2 Hierarchical habitat selection as described by Johnson (1980). This generalized concept is illustrated using pileated woodpeckers as an example. (Range map from USGS Biological Resources Division, used with permission. Pileated woodpecker photo from Washington Department of Fisheries and Wildlife.)

assumption that we shall explore later, but the vast majority of information that we have available to manage habitat for species comes from these habitat selection studies. Consequently, we need to understand which level of habitat selection we can influence for a species through our forest management activities. Johnson (1980) suggested that many species select habitat at four levels and called these levels first-, second-, third-, and fourth-order selection (Figure 2.2).

First-order selection is selection of the geographic range. The geographic range defines, quite literally, where in the world this species can be found. In our example from Figure 2.2, pileated woodpeckers are found in forests throughout eastern and western North America. Now consider two extremes. In Figure 2.3, I have provided geographic range maps for two species, Weller's salamander, found in spruce forests above 1500 m (5000 ft) in the southern Appalachians, and black-capped chickadees found throughout the northern United States and southern Canada. The geographic range for humans is global, with occasional excursions to other planetary bodies! Of course, the geographic range for a few other species is also global: cockroaches, Norway rats, and other human *commensals* (species that typically are associated with humans). So, why is it that some species occur around the world and others are restricted to a few mountains in the Appalachians? Climate and history have had some effect on the distribution of some species. There is a complex set of geographic distributions for slimy salamanders in the southeastern United States that likely results from past glaciation that has "packed" species into the southern Appalachians and allowed species with limited capabilities to disperse from one valley to another (Figure 2.4). Similarly, other species such as the Siskyou mountain salamander may have been more widely distributed during periods when the northwest

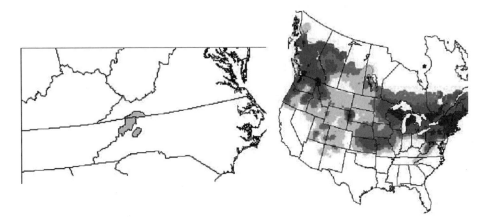

FIGURE 2.3   Range maps for a geographically restricted species, Weller's salamander (left) and a cosmopolitan species, black-capped chickadee (right). (Maps from USGS Biological Resources Division, used with permission.)

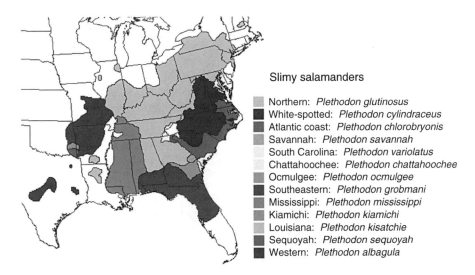

Slimy salamanders

Northern: *Plethodon glutinosus*
White-spotted: *Plethodon cylindraceus*
Atlantic coast: *Plethodon chlorobryonis*
Savannah: *Plethodon savannah*
South Carolina: *Plethodon variolatus*
Chattahoochee: *Plethodon chattahoochee*
Ocmulgee: *Plethodon ocmulgee*
Southeastern: *Plethodon grobmani*
Mississippi: *Plethodon mississippi*
Kiamichi: *Plethodon kiamichi*
Louisiana: *Plethodon kisatchie*
Sequoyah: *Plethodon sequoyah*
Western: *Plethodon albagula*

FIGURE 2.4   Distribution of a complex of slimy salamander species in the southern United States. (Map from USGS Biological Resources Division, used with permission.)

was cooler and wetter. This species became more and more restricted as the climate has changed, and now occurs only in a small region of southern Oregon and northern California.

In another example of the role of barriers as a mechanism for limiting the geographic distribution of organisms, consider that the Columbia River is the dividing line separating the geographic distributions of the western red-backed vole in Oregon and the Gapper's red-backed vole in Washington. One can only guess how this might all change when one or more of them hitches a ride in a recreational vehicle across the Dalles bridge! Humans, of course, have been important mechanisms for dispersing species into places that are climatically acceptable for a species, but barriers had kept species separated until humans moved them. The list of examples is growing rapidly, but includes such well-known ones as the European starling, tree-of-Heaven, and gypsy moth. Humans are breaking down barriers and allowing opportunities for exotic species to become invasive. Implications for native flora and fauna can be huge and the geographic ranges of some native species can be significantly altered as these invasive species proliferate. The influence of hemlock wooly adelgid on eastern hemlock mortality has led forest managers to extensively salvage dead hemlocks (Howard et al. 2000). This mortality and forest management has led biologists to worry about declines in the distribution and abundance of black-throated green warblers and other hemlock-associated species (Yamasaki et al. 2000). Invasive species can also influence forest wildlife populations by predation. A species of marsupial known as the woylie was once widespread over western Australia, but by 1980 had been reduced to three small populations due to the expansion of introduced red foxes (Figure 2.5). With recent widespread control of foxes using warfarin poisons (woylies are not affected by this poison because it occurs naturally in shrubs in their environment), populations have once again begun to increase.

Species geographic ranges have also been influenced by invasive competitors. Recently, barred owls have been found within the geographic range of northern spotted owls in the Pacific northwest of the United States. There is increasing evidence that northern spotted owls are declining in abundance in the presence of barred owls (Peterson and Robbins 2003) and there is clear evidence that the two species have hybridized (Hamer et al. 1994).

Just as humans have been the cause of changes in geographic ranges through species introductions, they also have been responsible for recovering species from areas where they were extirpated. Translocation efforts and reestablishment efforts have been successful in species recovery (Haight et al. 2000). For example, the red-cockaded woodpecker is a threatened species that occurs in forests

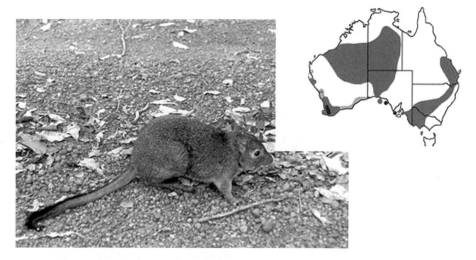

FIGURE 2.5 Past (gray) and current (black) distribution of the woylie, a woodland marsupial. Introduced foxes reduced the geographic range of the midsized marsupial. (Map from the Department of Environment and Conservation, western Australia. With permission.)

of the southeastern United States. This species requires old, living pines with heart rot in which to nest. As trees grow and forests age, areas of suitable habitat can be recruited. Rudolph et al. (1992) demonstrated that reestablishment of this species in this newly recruited habitat is possible. There are numerous other examples for game species such as wild turkeys.

Consider the importance of populations of a species at the center vs. the periphery of its geographic range. Populations at the periphery may be in lower-quality habitat if either biotic or abiotic factors are limiting its distribution. But recall that environments are not static, they are constantly changing. Climate changes; earthquakes change the topography; some species arrive while others leave. It is those populations at the periphery of their geographic range that are on the front line of these changes. Hence, although it may be tempting to think of these peripheral populations as somewhat expendable, they may be critical to population maintenance as large-scale changes in habitat availability occur. Given the rate at which climate change is occurring, these peripheral populations may be even more important over the next few hundred years.

Although Johnson (1980) does not describe *metapopulation* distribution as a selection level, it is important to realize that, within the geographic range, populations oftentimes are distributed among smaller, interacting populations that contribute to overall population persistence, or a metapopulation structure. Hence, these subpopulations may grow, go extinct, and be recolonized as habitat quality changes following forest disturbance and regrowth. The distribution of the subpopulations is important to consider during forest planning because if dispersal among subpopulations is restricted by forest management actions, then the subpopulations that might ordinarily be recolonized may be restricted from doing so.

Johnson (1980) described *second-order selection* as the establishment of a *home range*, an area that an individual or pair of individuals uses to acquire the resources that it needs to survive and reproduce. Not all species have established home ranges, but most do. Species that have nests, roosts, hibernacula, or other places central to its daily activities move in an area around that central place to acquire food, use cover, drink water, and raise young. Home ranges are not the same as territories. A *territory* is the space, usually around a nest, that an individual or pair defends from other individuals of the same species and occasionally other individuals of other species. Territories may be congruent with a home range, smaller (if just a nest site is defended), or may not be present at all. Many bird species, such as eastern bluebirds, defend a territory around a nest that includes the nest site and an area within which the pair finds food to feed their young. Other species such as fox squirrels defend a nest or den site when raising young, but have a home range that overlaps with other individuals' dens. Species such as flying squirrels seem not to establish territories and coexist with other individuals within their home range.

Home ranges vary in size with the body mass of the species (Figure 2.6). Species with a larger body mass need more energy to support that mass. Herbivores tend to have smaller home ranges than carnivores of the same size, because energy available to herbivores is more abundant, and with each increase in trophic level there is a decrease in energy availability. A *trophic level* is the feeding position in a food web — primary producers are typically plants, primary consumers are herbivores, secondary consumers are carnivores, and tertiary consumers are carnivores that eat carnivores. Hence, there is an energy or biomass pyramid, with more biomass in producers than in herbivores and more biomass in herbivores than in carnivores.

Home range sizes also vary among individuals within a species. As food resources are less abundant or more widely distributed, home range sizes increase. But within a species the home range size has an upper limit that is governed by balancing energy input from food with energy loss by movement among food patches. For instance, Thompson and Colgan (1987) reported larger home ranges for American marten during years of low prey availability than in years of high prey availability.

Third-order selection is the use of patches within a home range where resources are available to meet an individual's needs. Biologists often can delineate a home range based on observed daily or seasonal movements of individuals going about their business of feeding, resting, and raising

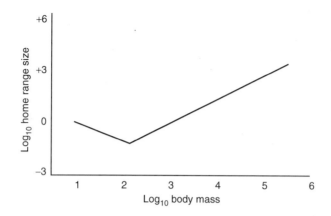

FIGURE 2.6   Relationship between body mass and home range size. Home range sizes are smallest for those species with body mass of ~100g and greater for sizes that depart for this size. (Based on data by Kelt, D.A., and D. Van Buren 1999, redrafted from Harestad, A.S., and F.L. Bunnell. 1979. *Ecology* 60: 389–402. With permission by the Ecological Society of America.)

young. But this area is not used in its entirety. Rather, there are some places within the home range that are used intensively and other parts of the home range that are rarely used (Samuel et al. 1985). Selection of these patches is assumed to represent the ability of the individual to effectively find and use resources that will allow it to survive and reproduce. But, as Garshelis (2000) makes clear, simply the amount of time or number of radio telemetry locations in a particular patch type does not necessarily reflect the importance or lack thereof to an individual. An individual may spend a small amount of time and be represented by relatively few locations in a particular patch type, but may receive important benefits from that patch type. For instance, you may spend 10% of your time in your kitchen and 30% of time in your bedroom, but the resources that you receive from your kitchen are as or more important than the rest that you receive in the bedroom. It is exactly those resources located in the patch types that are most important to maintaining an animal's fitness.

Fourth-order selection is the selection of specific food and cover resources acquired from the patches used by the individual within its home range. Given the choice among available foods, a species should most often select those foods that will confer the greatest energy or nutrients to the individual. Which food or nest site to select is often a trade-off among availability, digestibility, and risk of predation (Holmes and Schultz 1988). Factors that influence the selection of specific food and cover resources most often tend to be related to energetic gains and costs, but there are exceptions. The need for certain nutrients at certain times of the year can have little to do with energetics and much to do with survival and fitness. For instance, band-tailed pigeons seek a sodium source at mineral springs to supplement their diet during the nesting season (Sanders and Jarvis 2000).

Collectively, these levels of habitat selection influence the fitness of individuals, populations, and species. Habitat quality is dependent not only on the food and cover resources in the stand or forest but also on the number of individuals in that stand. Many individuals in one stand means that there are fewer resources per individual. Habitat quality and habitat selection is density dependent. Indeed, even if a patch has excellent food and cover quality, but a fixed quantity, too many individuals in the patch can cause some to leave to find other habitat patches of lower quality but with fewer individuals.

DENSITY-DEPENDENT HABITAT SELECTION

Fretwell and Lucas (1969) provided the conceptual basis for understanding density-dependent habitat selection. Consider a fixed level of resource availability in two patches, with resource availability in one patch higher than that in another (Figure 2.7). As the population density in patch 1 increases,

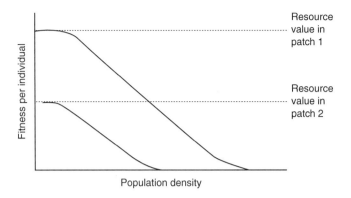

FIGURE 2.7   In the above diagram, the top horizontal line represents a fixed amount of resources in patch 1; the lower horizontal line represents the fixed amount of resources in patch 2. The resulting selection of habitat is dependent on the density of individuals in each patch. (Redrafted from Fretwell, S.D., and H.L. Lucas Jr. 1969. *Acta Biotheoretica* 19: 16–36. With permission from Springer Press.)

the resources available per individual and hence fitness per individual declines. Eventually, at a high enough population density, the fitness per individual declines to a point where each individual is afforded a level of resources that would be less than what they received if they moved to patch 2, a patch with lower total resource availability than patch 1. Hence, the selection of patch 1 and ultimately patch 2 by some individuals is influenced by the populations in each patch.

Under this approach each individual is free to choose the patch that will provide the greatest energy or other required resources. This concept is called the *ideal free distribution*. But in many populations, especially those that have a dominance hierarchy or are territorial, some individuals are less likely to move to patch 2 and some more so. Consider the case in which the species occupying patch 1 initially is territorial and, as more individuals are added to the population, each defends a specific territory. Territories help to ensure that individual fitness will not decline and that each individual or pair in the patch remains fit. Eventually, the patch will be saturated with territories and another territory cannot be packed into patch 1. If the territorial individuals are successful in patch 1 and they are successful at defending their territory from interlopers, then they are ensuring their fitness. Of course, the cost of territoriality is the energy expended defending it. Subordinate individuals in the population, those unable to displace an individual already on a territory, are relegated to patch 2. This situation represents an *ideal despotic distribution*, where individual fitness is maintained in the highest-quality patches at lower-than-expected densities through territoriality. If the individuals in the high-quality patch are fit enough to support a stable or growing population, then they occupy a *source habitat* — one with high individual fitness. Those individuals forced to move to a lower-quality patch may experience reduced opportunities for reproduction or survival. If these populations can only be maintained by immigration, then they occupy a *sink habitat*. Further, as resources are already restricted in the sink habitat, increasing density may cause individuals to abandon territorial behaviors, not reproduce successfully, or face a higher risk of mortality. Indeed, in some sink habitats, densities of subordinate individuals can be much higher than densities of individuals in source patches. Consequently, using density as an index to habitat quality may be inaccurate. Animal fitness is a better indication of habitat quality than animal numbers. Reproductive success, survival, and body mass are often used as indicators of animal fitness (Van Horne 1983).

RELATIONSHIP BETWEEN HABITAT QUALITY AND DEMOGRAPHICS

Foresters manipulate stand density to ensure that the trees that will eventually be harvested have sufficient resources to grow rapidly, produce seeds, and survive to maturity. Biologists do much the same thing when managing habitat for animals. Populations can be manipulated by modifying habitat and thereby influencing possibilities for survival and reproduction, the two primary indices to fitness. The linkages between animal demography and habitat are complex, but some understanding of these relations is necessary for successful habitat management. Each species has its own potential for population increase and this potential is described as the *intrinsic rate of natural increase*. There is a solar constant, so energy available to plants and animals is limited. Given adequate food, cover, and water, populations will grow. But consider what happens as the density of individuals increases. Food becomes scarcer or of poorer quality as the population grows. Cover is occupied by more individuals and so the risk of disease and parasitism increases. Intraspecific (among individuals of the same species) competition for resources causes some subordinate individuals to use suboptimal patches. As food, cover, or other resources become limited, the population growth rate decreases, because either mortality increases or reproduction decreases, or both. This process is termed *logistic growth* (Figure 2.8). If we assume that resources are constant, then the population reaches a point where births equal deaths and growth becomes zero. This point is termed the *carrying capacity* of the habitat for the population.

But resources are not constant; they change daily, seasonally, and annually. Birth rates, death rates, and movement rates are variable over both space and time as habitat changes through forest disturbances and succession. Carrying capacity, consequently, is always changing. The concept of a *dynamic carrying capacity* is useful to land managers, because it provides the link between the dynamics of forests, habitat quality, and population growth. Manipulating habitat to change carrying capacity is a particularly effective approach to long-term manipulation of wildlife populations.

But populations do not always reach carrying capacity in relation to habitat quality. Some species, such as voles, snowshoe hares, and ruffed grouse, follow a "boom and bust" population pattern. Populations grow for about 3 to 6 years and then rapidly decline for another 3 to 6 years. High-quality habitat usually increases the highs and decreases the lows of a population cycle, but habitat probably does not directly mediate these cycles, because they occur throughout much of the geographic range of the species (Keith and Windberg 1978).

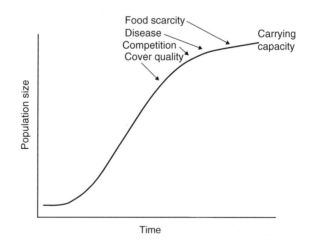

FIGURE 2.8 Population growth over time assuming fixed resource availability results in a carrying capacity where births balance with deaths and the population remains somewhat stable. In actuality, carrying capacity is quite dynamic as resources change over time and space. (From McComb, W.C. 2001. Management of within-stand features in forested habitats. In Johnson, D.H., and T.A. O'Neill (managing editors), *Wildlife Habitat Relationships in Oregon and Washington*, Chapter 4. OSU Press, Corvallis, OR. With permission from Oregon State University Press.)

POPULATION FITNESS

Individuals are fit when they have a high probability of surviving and reproducing successfully. Population fitness is high when the population is increasing or at least not declining. Individuals with high fitness can occur in populations with low fitness and vice versa. Since it is populations that are sustainable over the long-term and not individuals, we need reliable indicators of habitat quality using population fitness. Habitat quality refers to the ability of a locality to provide for the long-term persistence of a population over time. Biologists tend to measure habitat quality based on vital rates of the population. If a population is reproducing at an optimum rate and survival of young and adults is high, the habitat is considered to be of high quality. *Vital rates* are the demographic parameters that drive population change, primarily birth and death rates.

The rate at which animals reproduce is a basic component of population dynamics. Two measures of reproductive fitness are natality and fecundity. *Natality* refers to number of young individuals born or hatched per unit time. *Fecundity* is the number of young produced per female over a given time period and relates population fitness to the average fitness per female. Usually 1 year is the time period considered, but for smaller animals, especially those that may breed several times a year, a shorter time period may be selected. Thus, if a population of 1000 female bears produced 200 young in a year, the birth rate or fecundity, would be $200/1000 = 0.2$.

A number of factors affect a population's birth rate. Animals that are young or in poor nutritional condition usually have fewer young ones and/or breed less often. Age at first reproduction is also an important factor in determining birth rate. Large, long-lived animals typically do not become sexually mature until they are several years of age. A vole might become sexually mature and breed for the first time at 18 days of age. An Asian elephant on the other hand will typically be 9 to 12 years old when it first breeds. The birth interval is also important in determining birth rates. A vole might produce a litter of young every 30 days during the breeding season, but a grizzly bear may only reproduce every 3 or 4 years. The average number of young produced is of obvious importance in a population's birth rate. Some animals such as fish or amphibians produce hundreds or thousands of eggs (not all of them hatch and few survive), while many wildlife species only have one or two young at a time (e.g., barred owls). Potential population growth rates are related to fecundity rates. A doubling in the fecundity rate will more than double the population growth rate.

Mortality rate is another indicator of population fitness. Mortality rate is measured as the number of animals that die per unit of time (usually 1 year) divided by the number of animals alive at the beginning of the time period. Thus, if 1000 fawns are born in June and 400 are alive the next June, then the mortality rate is 600 (the number that died)$/1000 = 0.6$ or 60%. Survival and longevity are two other population parameters related to mortality. *Survival* is the number of animals that live through a time period and is the converse of mortality. Thus, if the mortality rate is 0.8 or 80% per year, then survival would be 0.2 or 20% per year. *Longevity* is the age at death of the average animal in a population.

Mortality rates are usually age- and often sex-specific, which means that animals of different ages or sexes die at different rates. In many species, the young and old animals die at faster rates than the mid-age animals. Often, males have higher mortality rates than females because of activities associated with territorial or mating behavior.

Different species have different *survivorship functions* related to their life-history traits. A Type I *survivorship curve* would be typical of animals that have relatively high survivorship until later in life, when they become subject to age-related mortality (Figure 2.9). Typically, these are animals with a high degree of parental care. Many larger mammals, such as whales, bears, and elephants, might have Type I survivorship curves. Some animals have fairly constant survivorship (Type II). Some birds and most reptiles and amphibians probably fit this pattern, although our knowledge of survivorship in birds is not very complete because they are difficult to study. A Type III survivorship curve would be typical of animals with little or no parental care and/or vulnerable young; mortality is high in the young-age classes and low in older animals. Insects and fish often have Type III survivorship curves.

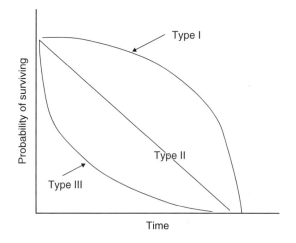

FIGURE 2.9   Survivorship curves for three example species. Type I species have high juvenile survival rates while Type III species have low juvenile survival rates.

MEASURING HABITAT SELECTION

Clearly we should use estimators of fitness as a measure of habitat quality and selection. Rarely have biologists taken that approach largely because measures of fitness are expensive and difficult in many species. Usually, occurrence or abundance is used, or some indirect indicator of fitness such as body mass and evidence of breeding may be used. Despite not measuring fitness attributes directly for many species, years of habitat selection research have produced repeatable patterns of use and selection for many species.

Assessing habitat selection is scale dependent. Most information available on habitat relationships of species in forests comes from one of two approaches: stand-based assessments or species-based assessments. In stand-based assessments, comparisons of abundance or occurrence are made between stands of different structure or following different treatments. This information can be very useful if the animal response is matched to the scale of the stands. For instance, we would not use the number of northern goshawks detected in thinned and unthinned stands that are 10 to 15 ha in size (second-order selection) as a response variable because the home range of one goshawk could encompass many stands. Rather, we might ask if stands were selected by goshawks for foraging (third-order selection) within their home range or if particular nest structures were used in these stands (fourth-order selection). Alternatively, we might place radio transmitters on goshawks and analyze the stand types used in comparison with their availability in each home range. Both approaches provide useful information and have potential weaknesses.

When considering how animals use habitat in forests, it is important to differentiate between use and selection. Animals can be found in various types of forest conditions. We occasionally find a species that is typically found in early-successional stages occurring in old-growth forests and vice versa. Dispersing spotted owls may occasionally be found in urban settings. It is important to know why they were found in these settings. Were they forced there by more dominant (and fit) individuals? Did random dispersal bring them there by chance? Are they surviving there? Reproducing? Observations of use can be important information, but they must be placed into the context of why the animal is found in these places.

Selection is the choice of one or more patch types among those that are available. For example, say you were able to make 100 unbiased observations of American marten, 80 in old forest and 20 in young forest. Within your study area, old forest comprises 40% of the area and young forest comprises 60%; therefore, if marten were using the area randomly (no selection), you would expect 60 observations in young forest and 40 in old forest. Hence, in this simple example, marten were using old forest out of proportion to its availability and could be said to "select old forest." Does that mean that young forest was avoided or unimportant? Not necessarily. If marten were eating raspberries in young forest during the summer, then they may not spend much time there, but that food resource was sought (not avoided) and could contribute to marten fitness. Again, it is important to know why marten were found in these forest types. Further, selection can only be assessed among the choices available to an organism. The organism may actually prefer some other condition that is not available. Given a choice of a beech–maple forest and a pine forest, a gray squirrel may demonstrate selection for the beech–maple forest, but it would prefer an oak forest if it were available.

Use–availability studies are further complicated, in that too often we classify forests by dominant tree species and/or age class and then see which classes are selected by a suite of species. The classes that were created often are done so based on human perceptions of differences (clearcuts, old-growth, hardwood, or conifer) and may be only marginally related to providing the habitat elements needed by the species being assessed. Take Swainson's thrush for example. Swainson's thrushes are associated with shrub cover where they nest and are found in woodlands, where shrub cover is dense. That there are hardwoods or conifers or pole-sized trees or old-growth trees in the overstory is somewhat irrelevant. But how often do we humans classify vegetation based on shrub cover beneath the overstory? And, of course, habitat is more than just vegetation. Soils, slope, aspect, etc., could all be mapped and classified, but it would need to be done differently for each species and that too rarely happens. Hence, the results of selection studies based on a priori classifications of forest condition unrelated to habitat elements important to the species of interest should be viewed with caution.

Ideally, experiments that manipulate resources and measure population vital rate responses are most reliable. For example, if an experiment were designed to test the effects of thinning on ruffed grouse, we would identify randomly located study sites and sample an aspect of fitness (e.g., survival rates and natality) for at least one full population cycle (approximately 10 years) prior to thinning. Study sites would have to be large to ensure that we could sample multiple individuals in each stand. Because the home range for a ruffed grouse is approximately 4 ha (10 acres), then stands might need to be 120 ha (300 acres) or more in size. Once the pretreatment data have been collected, then we could thin a randomly selected group of stands and monitor the same vital rates on the thinned stands as well as on untreated controls for another 10 years. Such an approach may be ideal, but in most circumstances it is impractical due to expense and logistics and for some species, it may be illegal.

Critical habitat is defined as specific areas that are essential to the conservation of a federally listed species under the U.S. Endangered Species Act and which may require special management considerations or protection. If the best available information would suggest that thinning would be detrimental to the species, then the aforementioned experiment may simply not be allowed in the United States.

In addition, experiments as described here present difficulties when assessing species that show high affinity for an area, also called *site fidelity*. Pairs may return year after year to the same location despite drastic changes in the habitat around them. Effects of the treatment may only be apparent once these pairs are gone, because new breeders may not be recruited to this site as it no longer has the cues they look for in a breeding area.

PROXIMATE AND ULTIMATE CUES TO HABITAT QUALITY

Use–availability studies often result in evidence for selection of certain habitat types, tree species, or vegetation structures. These structures are often related to the availability of resources that an animal needs for survival, but not always. The *ultimate* food and cover resources that each species needs are often found by the species using *proximate cues*. Migratory birds are a good example. As they move from breeding areas to wintering areas, they must make choices about where to rest or settle such that food and cover will likely be available for them. In these situations, vegetation structure seems to be a key proximate cue to these choices (Cody 1985).

The *structure* of a forest provides a cue to an animal that certain insect or plant food resources might be available, or that nest sites might be available. These cues may cause animals to establish a territory before (e.g., early spring) the ultimate resources (e.g., foliage-dwelling insects) are even available. Animals use visual, aural, and/or olfactory cues when establishing a territory or home range. Managers can identify the habitat elements that may be proximate cues to habitat selection and ensure that these habitat elements are present for those species that are desired in a stand or forest. Mangers manipulate aspects of the stand such as stocking levels, tree density, and tree size, but need to consider other specific habitat elements that may or may not be related to traditional stand management for timber production. It may be important to grow a large tree, but if the ultimate resource associated with the large tree (proximate cue) are bark-dwelling insects in deeply dissected bark, then simply having large trees may not yield higher quality habitat. Consider a human example. Humans use proximate cues every day. When we are hungry and need food fast (quality may be a separate issue), we do not go into every building and hunt around for a hamburger, but instead look for a proximate cue, for example, golden arches. If the place with the golden arches is out of burgers, then a typically reliable cue did not yield the desired resources, and you spent time and energy for nothing. Providing only proximate cues without considering ultimate resources is no different.

CASE STUDY: AMERICAN MARTEN HABITAT SELECTION

American marten are mustelids, members of the weasel family. Their geographic range extends across North America in boreal forests. Females have a home range of approximately 2.3 km^2 and males, being larger, have larger home ranges. There has been growing concern that this small carnivore may be adversely affected by intensive forest management in coniferous forests. This case study is based on a study by Potvin et al. (2000), who examined marten habitat selection at stand and landscape scales in intensively managed spruce forests in Quebec. About 10,000 km^2 of Canada's forests are clearcut each year. A consistent finding among many studies conducted throughout the geographic range of marten is that having >20 to 30% of an area recently cut leads to declines in marten abundance. That does not necessarily mean that clearcuts are not used by marten. Indeed, some types of food, especially berries and other sugar-rich fruits, may be more available in openings, but if openings cover too large an area, then prey (primarily Gapper's red-backed voles) are not sufficiently abundant at other times of the year.

Potvin et al. (2000) attached radio transmitters to 33 marten and they estimated the winter home ranges for each marten. An example is shown in Figure 2.10. Several factors become clear from this study. First, marten did use regenerating stands, but much less than would be expected by chance alone. Second, most locations were in the surrounding older forest. Stands of deciduous and mixed deciduous–conifer >30 years of age were selected out of proportion to availability. Conifer forests >30 years of age were used in proportion to availability. Recent cuts <20 years old where young trees had grown to a point where the crowns closed to form a continuous canopy were also used in proportion to availability, but recent cuts that still had an open canopy were used less than expected by chance.

The landscape level analyses showed similar patterns. Home ranges contained more area in forests >30 years of age and less than expected of young, open forest. Indeed, marten with smaller home ranges had less young, open forest than marten with large home ranges suggest that marten can find more necessary resources in a smaller area when there is less open, young forest. Based on these results and results of studies from Maine and Utah, it seems that marten cannot tolerate >30% of a home range in recent clearcuts, but that once young stands form a continuous canopy, they will use the area. Hence, forest planners developing a harvest schedule can use this information to guide where and when clearcutting could occur to minimize effects adverse on marten.

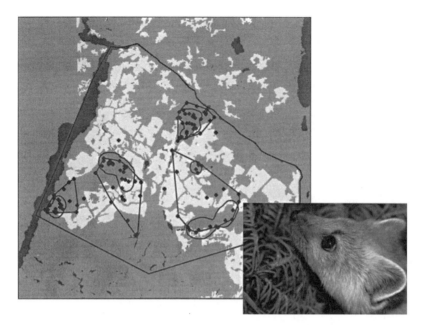

FIGURE 2.10 Radio locations, core areas and 95% polygon home ranges of American marten in Quebec boreal forests. White areas have been recently cut and shaded areas are older forest. (Map from Potvin, F. et al. (2000) *Conserv. Biol.* 14: 844–857. With permission by Blackwell Publishers; Marten photo by Mike Jones and is used with his permission.)

SUMMARY

Habitat is selected by many vertebrates at four levels: geographic range, home range, patches within the home range, and the ultimate resources needed for survival. Such selection is assumed to represent a complex set of behaviors that species have evolved to yield high population fitness despite environmental variability. Selection of habitat can be influenced by other species such as competitors and predators. Habitat selection is also density dependent, with the choice of habitat patches influenced by the effects of the population on individual fitness. In territorial species, subordinate individuals may be forced into sink habitat where survival and reproduction rates may be lower than in source habitat occupied by dominant individuals. Although we often gather information on habitat selection using use–availability studies, these results must be interpreted with caution unless we understand why species are using certain conditions. Lack of selection does not necessarily imply avoidance. Although experimental approaches that document effects of forest management on animal fitness are ideal, they are often impractical. Hence, forest wildlife biologists are usually faced with using information from associational studies to identify the proximate cues to habitat selection provided during forest management.

REFERENCES

Cody, M.L. 1985. *Habitat Selection in Birds*. Academic Press, New York.

Fretwell, S.D., and H.L. Lucas Jr. 1969. On territorial behavior and other factors influencing habitat distribution in birds. I. Theoretical development. *Acta Biotheoretica* 19: 16–36.

Garshelis, D.L. 2000. Delusions in habitat evaluation: measuring use, selection, and importance. In Boitani, L., and T.K. Fuller (eds.), *Research Techniques in Animal Ecology: Controversies and Consequences*. Columbia University Press, New York.

Haight, R.G., K. Ralls, and A.M. Starfield. 2000. Designing species translocation strategies when population growth and future funding are uncertain. *Conserv. Biol.* 14: 1298–1307.

Hamer, T.E., E.D. Forsman, A.D. Fuchs, and M.L. Walters. 1994. Hybridization between barred and spotted owls. *Auk* 111: 487–492.

Harestad, A.S., and F.L. Bunnell. 1979. Home range and body weight — a reevaluation. *Ecology* 60: 389–402.

Holmes, R.T., and J.C. Schultz. 1988. Food availability for forest birds: effects of prey distribution and abundance on bird foraging. *Can. J. Zool.* 66: 720–728.

Howard, T., P. Sendak, and C. Codrescu. 2000. Eastern hemlock: a market perspective. Pages 161–166 In K.A. McManus, K.S. Shields, and D.R. Souto (eds.), *Proceedings: Symposium on Sustainable Management of Hemlock Ecosystems in Eastern North America.* USDA For. Serv. Gen. Tech. Rep. 267.

Johnson, D.H. 1980. The comparison of usage and availability measurements for evaluating resource preference. *Ecology* 61: 65–71.

Keith, L., and L.A. Windberg. 1978. A demographic analysis of the snowshoe hare cycle. *Wildl. Monogr.* 58: 70.

Kelt, D.A., and D. Van Buren. 1999. Energetic constraints and the relationship between body size and home range area in mammals. *Ecology* 80: 337–340.

Lowe, W.H. and D.T. Bolger. 2002. Local and landscape-scale predictors of salamander abundance in New Hampshire headwater streams. *Conserv. Biol.* 16: 183–193.

McComb, W.C. 2001. Management of within-stand features in forested habitats. In Johnson, D.H., and T.A. O'Neill (managing editors), *Wildlife Habitat Relationships in Oregon and Washington.* OSU Press, Corvallis, OR.

Peterson, A.T., and C.R. Robbins. 2003. Using ecological-niche modeling to predict barred owl invasions with implications for spotted owl conservation. *Conserv. Biol.* 17: 1161–1165.

Potvin, F., L. Belanger, and K. Lowell. 2000. Marten habitat selection in a clearcut boreal landscape. *Conserv. Biol.* 14: 844–857.

Rudolph, D.C., R.N. Conner, D.K. Carrie, and R.R. Schaefer. 1992. Experimental reintroduction of red-cockaded woodpeckers. *Auk* 109: 914–916.

Samuel, M.D., D.J. Pierce, and E.O. Garton. 1985. Identifying areas of concentrated use within the home range. *J. Anim. Ecol.* 54: 711–719.

Sanders, T.A., and R.L. Jarvis. 2000. Do band-tailed pigeons seek a calcium supplement at mineral sites? *Condor* 102: 855–863.

Thompson, I.D., and P.W. Colgan. 1987. Numerical responses of marten to a food shortage in north-central Ontario. *J. Wildl. Manage.* 51: 824–835.

Van Horne, B. 1983. Density as a misleading indicator of habitat quality. *J. Wildl. Manage.* 47: 893–901.

Vesely, D., and W.C. McComb. 2002. Terrestrial salamander abundance and amphibian species richness in headwater riparian buffer strips, Oregon Coast Range. *For. Sci.* 48: 291–298.

Wecker, S.C. 1963. The role of early experience in habitat selection by the prairie deer mouse, *Peromyscus maniculatus bairdi. Ecol. Monogr.* 33: 307–325.

Yamasaki, M., R.M. DeGraaf, and J.W. Lanier. 2000. Wildlife habitat associations in eastern hemlock — birds, smaller mammals and forest carnivores. Pages 135–143 In McManus, K.A., K.S. Shields, and D.R. Souto (eds.), *Proceedings of the Symposium on Sustainable Management of Hemlock Ecosystems in Eastern North America.* USDA For. Serv. Gen. Tech. Rep. 267.

3 Forest Structure and Composition

Historically, a forester looked at a forest, she may at first focus on the tree species mix, the tree size, tree density, and other clues about how the stand might be managed to achieve wood products or other goals. When a wildlife biologist looked at a forest, she may at first focus on evidence of deer browse, pellet groups on the forest floor, tracks in the mud, or nests in trees. To effectively manage habitat in a forest the forester and the biologist must both assess the sizes, numbers, and arrangement of a set of habitat elements, the building blocks for habitat within a stand or forest. *Habitat elements* are those pieces of the forest that in certain numbers and arrangements meet the food or cover resources for a species. If these are highly variable within and among stands, then the needs for many species can be met. If they are very uniform, then the needs of only a few species can be met. The challenge to the forester and biologist is to walk into a stand and see the same habitat elements. In so doing, the biologist can explain why more or fewer of any set of them are needed to meet a species goal. Similarly, the forester can explain how silviculture might be used to achieve that goal.

FOOD AND COVER IN A CELLULOSE-MANAGED SYSTEM

Timber management has, for many years, been focused on producing wood products (cellulose) from managed forests. To maximize cellulose production, foresters want to be sure that the growing space for trees in a stand is fully utilized. But maximizing cellulose production maximizes biomass in the stand in a form that is quite indigestible for most animal species, unless the trees are allowed to decompose, which is not likely when they are managed for timber. So either some growing space must be allocated to habitat elements, or the cellulose must be made available to more species through wood decomposition. Both choices result in a decrease in the production of wood for humans. Consequently, the decision to manage habitat elements must come with the understanding that providing some habitat elements in some stands or in some parts of some stands may come at a financial cost to the landowner. The manager must decide which habitat elements can be provided in a way that is compatible with the goals for managing the stand for cellulose, and which will come at a cost, and how much cost the landowner is willing to bear.

Generally we think of providing these elements by altering the structure and composition of a stand. *Structure* refers to the physical features of the environment such as vegetation, soils, and topography. The complexity of the structure serves as both proximate cues in habitat selection as well as ultimate resources for cover (e.g., nesting and resting sites). *Composition* refers to the species of plants, types of soils, and other features that contribute complexity, for instance, through plant species richness. It is the combination of vegetation structure and composition that managers can change through their management actions.

VERTICAL COMPLEXITY

Read any forest plan or silvicultural prescription and early on there will be reference to forest area; 10, 100, or 1000 ha. But to most animals, forests are not areas. Forests are volumes — they have three dimensions: length, width, and height. One characteristic of forest development that influences

FIGURE 3.1 Vertical structure of a forest provides niches for many species. This is an example from the Willamette National Forest in Oregon.

the diversity of animals within a stand is the distribution of the foliage vertically within the stand. Some species use the foliage as a source of food either directly through herbivory (e.g., red tree voles in the Pacific Northwest), or indirectly by feeding on foliage-dwelling invertebrates (e.g., red-eyed vireos in hardwood forests of the southern United States).

Indeed, many species distribute themselves vertically within a forest to take advantage of food and cover resources while reducing competition among species for these resources (Figure 3.1). In some stands the foliage is distributed only in one layer, such as a dense plantation where all of the live foliage is in one canopy layer with little foliage beneath it. These stands typically provide habitat for a narrow range of species. Mixed-species stands have several tree and shrub species and often with foliage distributed in several layers (understory, midstory, and overstory). These multilayer stands tend to support more species of vertebrates than a one-layer forest. Wet tropical forests and mixed mesophytic forests develop vertical complexity naturally because of the mixture of trees species that occur within them. The distribution of foliage can be influenced by the shade tolerance of the tree species in the stand. *Shade-tolerant* species are those that can survive under low-light conditions. Shade-tolerant species use a strategy of survival and slow growth under low light until an opening or gap is provided in the tree canopy. When that gap occurs from tree fall or tree death, then the shade-tolerant species in that gap grow more rapidly to occupy the opening. These species can occupy lower levels in the stand for many years and often have deep crowns. Sugar maples, hemlocks, and American beech are examples of species that use this strategy. *Shade-intolerant* species do not survive under low-light conditions, and grow well only under full sunlight. Shade-intolerant species tend to grow rapidly once the seed has germinated. Because leaves are intolerant of shade, they tend to have smaller crowns in dense stands. Aspens, gray birch, and willows are examples of shade-intolerant species. A plantation of shade-tolerant species will have somewhat greater vertical complexity than a plantation of shade-intolerant species at a similar density simply because of the depth of the crowns of the trees.

Silviculture can be used to modify the vertical complexity in a stand in several important ways. Foresters usually will approach management of a stand using either even-aged or uneven-aged strategies. An even-aged stand is one where most of the dominant trees, those trees comprising the uppermost canopy, are of a narrow range of ages. That is, they all began life at about the same time

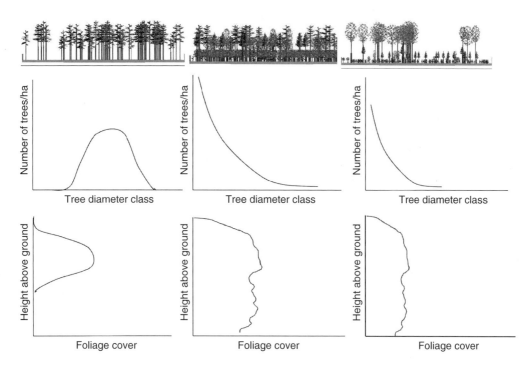

FIGURE 3.2 Even-aged, single species stands have a bell-shaped distribution of tree diameters, and one dominant vegetation layer (left). Even-aged mixed species stands have a skewed diameter distribution and a more complex vertical structure (middle). Uneven-aged stands have an inverse-J-shaped diameter distribution and a more complex vertical structure (right). Figures developed using Landscape Management Systems software (McCarter et al. 1998).

following a stand-replacement disturbance such as a fire, hurricane, or clearcut. If the trees are all the same age *and* of the same species, then the tree heights will be very similar and many of the tree diameters will also be similar (Figure 3.2). In plantations of one species, the distribution of tree diameters is often a bell-shaped curve. Foliage cover is also represented in one dominant layer and is a bell-shaped curve turned on its side (Figure 3.2).

Even-aged stands that contain a variety of species including ones that are shade intolerant and others that are shade tolerant often have more complex diameter distributions and vertical foliage structure (Figure 3.2). Slower-growing shade-tolerant species in the understory are "waiting" for the intolerant species in the overstory to die. Consequently, the range of tree diameters, heights, and distribution of foliage is more complex than in a single-species plantation.

Any disturbances, natural or silvicultural, that cause the diameter distribution of a stand to change from uniformity to uneven-sized will lead to more vertical complexity within the stand. In particular, silvicultural practices that create regeneration sites, where seeds can germinate and grow, can result in an uneven-aged stand — one with many tree ages and sizes (Figure 3.2). If these trees of many ages also represent a variety of tree species, then trees of a variety of diameters and height will be represented in the stand.

Vertical complexity is typically measured using an index to foliage height diversity (FHD). FHD is calculated in a manner similar to species diversity indices, by considering both the number of layers in a stand (comparable to number of species) and the percentage of cover by foliage in each layer (comparable to the number of individuals of each species). Hence, taller forests may have many layers and, if they have foliage cover in each layer, then FHD will be high (e.g., an old, tropical forest). Stands with short stature and all of the foliage in one or a few layers will have a low FHD (e.g., a young, single-species plantation). MacArthur and MacArthur (1961) reported that bird species diversity (BSD) was associated with FHD in the temperate forests of the eastern United States. Hence, there is a logical conclusion that increasing FHD could lead to increases in BSD within the stand, but the BSD–FHD relationship has not been found consistently in other forests, such as tropical forests, where competition among bird species may be important in structuring the community (Pearson 1975). Further, it is important to remember that not all species of animals benefit from vertically complex stands. Species such as northern goshawks are very well adapted to forage beneath the canopy of large even-aged stands that have sufficient flight space beneath the canopy for goshawks to forage successfully. In general, increasing the vertical complexity of a stand increases the potential number of niches for more species of birds (and possibly bats), but this generalization may not hold in some forest types and no single-stand condition is best for all species. Nonetheless, Flather et al. (1992) found that a combination of indices of vertical and horizontal complexity were reliable indicators of bird community integrity in the eastern United States.

HORIZONTAL PATCHINESS

The variability in tree size, species composition, dead wood, and other habitat elements is often related to the horizontal variability within a stand. Homogenous stands with evenly spaced trees and uniform canopies offer fewer niches, and hence animal diversity is often lower in these uniform stands. The size of openings and arrangement within a stand may positively affect some species and negatively affect others. Chapin et al. (1997) found that American martens' use of habitat was more influenced by the combination of vertical and horizontal complexity than by the species composition of the forest. Overwintering birds also seem to be associated with horizontal complexity in the southern United States (Zeller and Collazo 1995). Small openings or variability in tree spacing will likely have few adverse effects on most animal species and may benefit the maintenance and development of important ecological processes (Carey 2003). If openings or other discontinuities are larger than a species' home range, then we may begin to see species occupy the stand that otherwise would not occur there (DeGraaf and Yamasaki 2003), but there may be adverse effects of the openings on forest interior-associated species (Germaine et al. 1997). See Chapter 11 for more information on edge effects in forests.

FORAGE AVAILABILITY AND QUALITY

Forage for herbivores is influenced by many aspects of forest composition and structure. Conversely, the forest structure and composition can be significantly altered by herbivory. Vertebrate herbivores such as deer, elk, and moose typically *graze* more heavily on grasses (monocots) and forbs (herb-aceous dicot plants) than on woody plants during the growing season. Herbaceous plants are more easily digested than woody plants and can represent 50 to 80% of herbivore diets during the growing season. When cold weather or drought kills the upper portion of grasses and forbs, then vertebrate herbivores are forced to *browse* more heavily on woody plants. Browsing is concentrated on the new growth of the woody plants because that part is most easily digested. During the winter, browse can constitute over 90% of the diets of vertebrate herbivores, and for forest managers trying to regenerate forests, this level of browsing intensity can be a significant economic burden. Animals that try to survive a long winter on woody browse face significant stress. Nearly all browsing vertebrates lose body mass during the winter (Mautz et al. 1976). Indeed, browse quality is important in slowing the rate of starvation but not preventing it. Hence, providing high-quality grasses, forbs, and browse is important in the overwinter survival of many herbivores. Controlling the intensity of browse on tree seedlings is important in effectively regenerating forests.

During the growing season, stands that allow more sunlight to support grasses and forbs provide higher-quality forage for many species. These grasses and forbs are typically found in greatest abundance beneath or between overstory trees in open stands (e.g., savannahs) or following a large

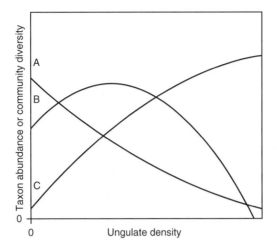

FIGURE 3.3 The change in taxa abundance or community diversity along an ungulate density gradient. Curve A is an idealized representation of taxa or communities that are adversely affected by browsing and Curve C represents taxa or communities that benefit. Curve B represents taxa or communities that benefit from intermediate ungulate densities. (From Rooney, T.P., and D.M. Waller. 2003. *For. Ecol. and Manage.* 181: 165–176. With permission of Elsevier Press.)

disturbance such as a fire or clearcut. The species composition of a stand can have profound effects on animal diversity and use by various species. The digestibility of twigs and leaves can be affected by tree species composition (Mautz et al. 1976).

Each species of herbivore is selective of the species of grasses, forbs, and woody browse that they choose to eat. In so doing, herbivores such as deer and elk can have profound effects on forest structure and composition (Rooney and Waller 2003; Figure 3.3). The potential effects of selective browsing include shifts in tree species composition of the forest understory (Strole and Anderson 1992). For instance, white-tailed deer in Illinois preferred to browse on white oak and shagbark hickory; sugar maple was browsed less than would be predicted from its abundance (Strole and Anderson 1992). Horsley et al. (2003) found that bramble abundance in several silvicultural treatments, the density of striped maple in clearcuts, and birch, American beech, and red maple in thinnings declined in abundance with increasing deer density. In Yellowstone National Park, elk preferentially browse on willows and aspens (Ripple et al. 2001). And moose in Newfoundland selected balsam fir, pin cherry, high-bush cranberry, and white birch over other plant species. Such selection may be influenced by the digestibility of the various plants, but not always. Mautz et al. (1976) compared digestibility among seven plant species eaten by white-tailed deer in the northeastern United States. They found higher levels of digestible energy in hobblebush, eastern hemlock, and balsam fir than in red maples, striped maple, mountain maple, or hazelnut. Despite these differences, white-tailed deer often feed heavily on maples in the winter in New England (Mautz et al. 1976); so factors other than simply digestible energy may be coming into play. Some plant species contain high levels of phenols, which reduce their digestibility for many herbivores (Friesen 1991, Sinclair and Smith 1984). Plants that produce high levels of phenols gain some protection against herbivory, but the coevolution of plants and herbivores has resulted in plant defense mechanisms that are less effective for herbivores such as deer. In the battle of coevolution, mule deer have evolved to produce saliva that contains a substance (prolene) that binds with the phenols and reduces their effectiveness (Austin et al. 1989, Robbins et al. 1987). Other chemicals such as lignin and cutin can influence digestibility as well. Plants with high content of lignin and cutin have lower digestibility than plants low in these compounds (Figure 3.4).

While some plants have chemical and physical defenses against herbivory (Farentinos et al. 1981), plants also respond to herbivory by altering growth rates. Among many grasses, forbs, and

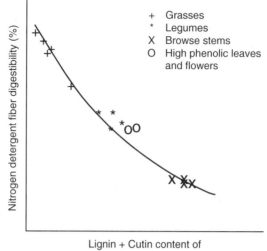

FIGURE 3.4 Plants with lower lignin and cutin have higher levels of digestibility than those with low levels for elk. (Redrafted from Hanley, T.A. et al. 1989. *Forest Habitats and the Nutritional Ecology of Sitka Black-Tailed Deer: A Research Synthesis with Implications for Forest Management.* USDA For. Serv. Gen. Tech. Rep. PNW-GTR-230.)

some shrubs, moderate levels of herbivory can actually stimulate growth above the levels of either undisturbed or heavily grazed or browsed plants (Belsky 1986, du Toit et al. 1990). It is widely assumed that browsed plants exhibit compensatory growth at the expense of reproduction and that herbivory, therefore, results in decreased seed production or smaller seed sizes (Belsky 1986).

Herbivores alter forest systems in ways other than consumption. They aid in the dissemination of seeds, and they may help maintain site quality. Some plants are well adapted to dispersal on animals (e.g., bedstraws). Other plant species (e.g., dogwoods and cherries) are well adapted to scarification that results from passing through animal digestive systems and "direct seeding" in a packet of fertilizer. Consequently, many fencerows are dominated by cherries, hawthorns, and dogwoods because birds often perch on fences after eating the fruits of these plant species.

Specialized herbivores, called mycophagists, that feed on fungi also play a role in ecological processes in forests. Through symbiotic relationships, mycorrhizal fungi aid vascular plants in the uptake of water and nutrients, and they can be particularly important to early plant growth and survival on harsh sites (Perry et al. 1989). These fungi produce fruits underground, unlike most other fungi, and they do not rely on aerial spore dispersal as do other fungi. They seem, instead, to be well adapted to animal dispersal. Some fungi known as truffles are important components of the diets of some small mammals, particularly red-backed voles in the United States and woylies in Australia (Maser et al. 1978, Taylor 1992; Figure 2.5). These animals eat fruits and ingest spores, which then pass through the digestive system in a few days and are deposited at a new site. A new fungal mat may then grow from this site and ensure the presence and widespread distribution of mycorrhizae in the soil (Cork and Kenagy 1989). Mixing organic matter in the soil by these burrowing animals also is likely to influence decomposition rates and influences soil processes (Maser et al. 1978).

The activities of some herbivores can have tremendous impact on habitat for other species (Naiman 1988). The activities of American beaver, for example, create early successional riparian forest patches and pools in the stream that can be important to other species. For example, Suzuki and McComb (2004) found very different amphibian and mammal communities associated with areas in the Oregon Coast Range that were impacted by beavers compared to similar areas where beavers did not build dams. Other examples include black bears that kill patches of trees in plantations, pocket

American chestnut

Northern red oak

White oak

Pignut hickory

American beech

Maple-leaf viburnum

Winterberry

FIGURE 3.5 Examples of hard and soft mast foods for vertebrates.

gophers that eliminate regeneration in patches, or elk herds that browse heavily next to riparian zones. All these activities create patchiness or heterogeneity in affected sites, and such patches can be important resource areas for other species.

FRUIT PRODUCTION

Fruits that are produced in forests by woody and nonwoody plants provide a key food resource for many species of animals. *Hard mast* means those hard fruits that are produced annually but tend to be highly variable in their production (Healy et al. 1999). Seed production is greatest in large, open grown trees. Generally those plants in full sunlight with large crowns produce larger mast crops more regularly, but year-to-year variability is high (Healy et al. 1999). Providing a variety of hard-mast-producing species in the stand may help to compensate for the variability in fruit production within any one species. For instance, the oaks in the United States are grouped into two subgenera, the white oaks (*Leucobalanus*) and red oaks (*Erythrobalanus*) (Figure 3.5). White oaks flower and are fertilized in the spring, the acorn matures in one growing season, and falls to the forest floor and germinates at the end of the growing season. Red oaks flower and are fertilized in the spring, but the acorn takes two growing seasons to mature before it falls to the forest floor. It then passes through a winter stratification period before germinating the following spring. Hence, red oak acorn production is delayed 1 year after fertilization compared with white oak acorn production. If both red oaks and white oaks occur in a stand, then a late frost that kills flowers in one spring may affect white oak acorn production that fall, but red oaks may still produce abundant acorns from flowers fertilized during the previous year. Similarly, providing a variety of other hard-mast-producing species such as hickories, beech, walnuts, and hazelnuts further reduces the risk of a complete mast failure in any one year. Unfortunately, one of the most reliable mast producers once dominant in eastern U.S. forests, American chestnut is now only a stump sprout in our forests because of the chestnut blight fungus.

Soft mast means soft fruits such as berries and drupes. These food sources are high in energy and used by many animal species. In a study in South Carolina, McCarty et al. (2002) found that 50% of fruits on 17 species of plants were eaten by vertebrates. Because different trees and shrubs flower and produce fruit at different times of the year, again a variety of species is important. For instance, serviceberries produce fruits early in the growing season, viburnums in late summer, and hollies retain fruit into the winter (Figure 3.5). These food sources may be particularly important

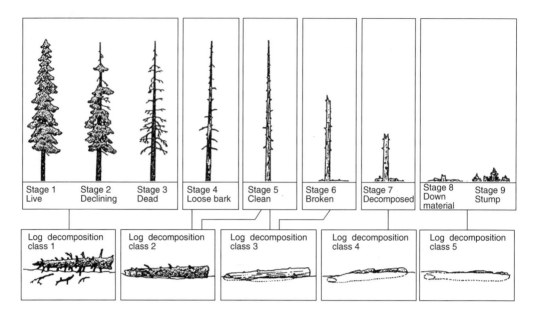

FIGURE 3.6 Stages of decay of trees and logs. As wood decays, the types of food and cover changes for various species. (From Maser et al. 1979. Dead and down woody material. In Thomas, J.W. (technical editor), *Wildlife Habitats in Managed Forests: The Blue Mountains of Oregon and Washington*. USDA. For. Serv. Agric. Handb. No. 553.)

in the winter when other digestible foods are in short supply. Soft-mast production is greatest in most species where the fruit-producing plants are receiving full or nearly full sunlight (Wender et al. 2004). Plants in partial shade often allocate most energy to growth and not to fruit production. Consequently, providing patches of forest where sunlight can reach these plants may increase food availability and quality (Perry et al. 1999).

Dead and Damaged Trees

Trees provide a basis for food and cover resources for various species while they are alive and growing. The value of trees to some species of wildlife, however, extends well beyond this period, and for many species the value of a tree only begins after the tree has died. Dead trees in various stages of decay offer sites for nesting, resting, and foraging for many species of vertebrates and invertebrates (Figure 3.6). Species vary in their use of dead wood size and decay classes (McComb and Lindenmayer 1999). Those species that use standing dead trees, or *snags*, are often separated into two groups. *Primary cavity nesters* are those species that can excavate a cavity in dead wood or trees with heart rot decay (Figure 3.7). Woodpeckers are the best examples of this; indeed, most woodpecker species must excavate a cavity in a tree or snag before they will complete the nesting ritual (for instance, they will usually not nest in nest boxes unless the box is filled with sawdust so that they have something to excavate). *Secondary cavity users* use cavities that were either created by primary cavity nesters or in natural cavities (Figure 3.7). This group of species can be extremely diverse and includes parrots, tree frogs, tree hole mosquitoes, and black bears. All of these species rely on either primary cavity nesters or trees with natural cavities for survival. Snags go through a process of decay that allows primary cavity nesters to excavate cavities. Only a few species of woodpeckers can excavate cavities in snags that are not well decayed are still hard wood. As the snags decay and become soft snags, other species can then excavate cavities in the snags. If snags of both types are provided in a stand, then there are more potential nest sites for more species. Snags are particularly important in conifer forests. Hardwood forests, especially with large trees, often have large dead limbs that provide many of the same benefits as snags.

FIGURE 3.7   Primary cavity nesters have used this snag (left), and these cavities are now available to secondary cavity nesters. Secondary cavity nesters can also use cavities created by fungal decay, such as the one in this live tree used by porcupines (right).

Tree species vary in their propensity to decay following the death of a tree or a wound to a live tree. Some species tend to be more prone to forming *natural cavities* that result from tree injuries. Many of the conifers (e.g., pines) are poor producers of natural cavities. These tree cavities are important den sites for secondary cavity-using species (those species that cannot excavate their own cavity as woodpeckers do). Hollow trees are formed through top breakage and subsequent heart rot. Large hollow trees are especially important for species such as fisher, bears, and some species of bats and swifts. Species such as Oregon oak, Pacific madrone, and bigleaf maple in the western United States and red maples and blackgum in the southern United States seem to produce many natural cavities and dead limbs (effectively, elevated snags), which are used by cavity-nesting animals (Gumtow-Farrior 1991, McComb et al. 1986, Raphael 1987).

Fallen logs also provide cover and nesting sites for a wide range of species, including many amphibian and reptile species, small mammals, and a few species of ground-nesting birds (Butts and McComb 2000). Large logs provide more cover and nesting opportunities for a greater number of species than small logs, and so the production of large trees that can fall to the forest floor should be given consideration during silvicultural activities (McComb 2003). Hollow logs can only occur if a hollow tree falls to the ground (they do not decay into hollow logs after they have fallen); so retention of some decaying hollow trees in a forest is necessary to provide hollow logs. Trees that fall into streams and lakes also play a role in habitat quality for many aquatic and semiaquatic species (Naiman et al. 2002). Logs in a stream divert water and cause pools either from plunging over the log or scouring under the log. Dead wood also provides cover for fish and amphibians and is used as a substrate upon which some salamanders lay their eggs. Large logs also are often the basis for a beaver dam in a stream. More information on managing dead wood is provided in Chapter 8.

TREE SPECIES AND INVERTEBRATE ASSOCIATIONS

MacArthur and MacArthur (1961) also found that BSD is associated with plant species diversity within stands, probably because of the additional niche space provided in stands with more plant

FIGURE 3.8 Example of a rough-barked tree (white oak, left) that supports higher densities of overwintering invertebrates than a smooth-barked tree (red maple, right).

species. Vertebrates that feed on invertebrates associated with the leaves and needles of trees and shrubs select certain plant species over others for feeding (Holmes and Robinson 1981, Muir et al. 2002). Selection is probably dependent on the food resources available, competition among species, and the foraging adaptations of each species. Insect abundance and species richness tends to be higher in hardwood than conifer stands, but clearly there are species well adapted to gleaning insects from both types of trees (Muir et al. 2002). Hardwood composition in conifer stands is associated with the abundance and occurrence of several bird species in the northwest United States (Huff and Raley 1991).

Tree species also vary in their ability to support bark-dwelling insects, an important overwintering food supply for some bird species (Mariani 1987). Rough-barked trees provide more cover for these insects and hence support higher insect biomass than smooth-barked trees (Brunnell 1988; Figure 3.8).

TREE SIZE AND DENSITY

Trees of different sizes play various roles as vertebrate habitat elements in stands. Seedlings provide browse for deer (much to the chagrin of some foresters!), nest sites for shrub-nesting birds, shade for forest floor amphibians, and hiding cover for many species of birds and mammals. Saplings provide browse for larger herbivores such as moose, and pole-sized trees may provide cover for ungulates. Large trees, especially those that grow beyond marketable size, can significantly influence the quality of a stand as habitat for some species. Trees in some managed stands are designated as legacy trees and left to grow to maturity and die through natural processes (Carey and Curtis 1996). For instance, Douglas-firs more than 125 cm (50 in.) in dbh (diameter at breast height) are used by marbled murrelets (Singer et al. 1991), red tree voles, and northern spotted owls. Large trees also add to the vertical structure within forests.

Large trees that add large surfaces of deeply fissured or scaly bark are used by bark-foraging birds such as brown creepers (Mariani 1987), and they support lichens, an important food source for species such as northern flying squirrels (Martin 1994). Designating a variety of tree species as legacy trees in forests would provide a range of growth rates and bark surfaces and contribute to complexity in the stand. These legacy trees provide an ecological link to the previous stand structure and composition.

Tree density also influences the production of many habitat elements. Dense stands with many trees may exclude sunlight from the forest floor, producing an open subcanopy condition. Sparsely stocked savannah stands leave much sunlight and moisture available for grasses, forbs, and browse. Manipulation of stand density is probably the most significant influence that a forester can have on habitat availability for a wide variety of species (Carey 2003).

FOREST FLOOR LITTER AND SOIL

Forest floor and below-ground conditions influence habitat quality for ground-foraging and burrowing species. The type and depth of leaf litter has been shown to be associated with the community structure and abundance of invertebrates (Bultman and Uetz 1982). Consequently, leaf-litter characteristics are associated with species that find food or cover on the forest floor, such as ovenbirds in the northeastern United States (Burke and Nol 1998). Insectivorous mammals also are assumed to be associated with litter type and depth. Some terrestrial mammals and amphibians remain active below the ground during the summer. For instance, rough-skinned newts use logs and burrow systems of voles and shrews as summer daytime refugia (W.C. McComb and C.L. Chambers, Oregon State, University, unpublished data). Burrow systems of mountain beavers, gopher tortoises, and pocket gophers are used by many other species (Maser et al. 1981; Figure 3.9). These below-ground conditions are often not considered during forest management, and so the next time that you walk through a forest think of the unseen animal community that lives beneath your feet.

PROXIMITY TO WATER

Intermittent and permanent streams, seeps, springs, vernal pools, ponds, swamps, marshes, and lakes all provide water in a setting that can be critical to habitat quality for many species of aquatic and

FIGURE 3.9   Gopher tortoises are associated with certain soil conditions and their burrows are used by a wide variety of other species. (Photo by Mike Jones and used with his permission.)

semiaquatic organisms. Although we have little control over how close a stand or forest is to water, we do have control over the function of the water body as habitat for a variety of species. For aquatic and semiaquatic species, the temperature, sediment load, and chemical concentrations in water may be influenced by the surrounding forest. Trees and shrubs over the water influence the temperature of the water by providing shade, by the influx of nutrients through litter fall, and by the degree of erosion through root strength. Some species of amphibians require clear, cold water for survival and have greatest fitness in water bodies where there are no fish (predators) (Lowe and Bolger 2002). These nonfish-bearing streams often are overlooked as potential habitat for animals because they may be dry at some times of the year and may appear no different than the surrounding uplands. In many settings, especially on federal lands in the United States, buffer strips are provided to retain habitat for species associated with these sites (Vesely and McComb 2002). More information on managing riparian areas is provided in Chapter 9.

CASE STUDY: PLANT RESPONSE TO HERBIVORES, *OR* IT'S A (CHEMICAL) WAR OUT THERE!

Some plants produce chemicals in their leaves and twigs that reduce herbivory. Others produce spines, thorns, and physical barriers to herbivory. Consequently, the effect that herbivores have on plant communities can be altered depending on the ability of the plants to cope with or avoid being eaten. A well-known herbivore, beaver, cuts trees of a range of sizes to feed on the bark and to use in building dams. They are selective of certain sizes and species of plants that they cut, and so influence the riparian forest composition and structure considerably. A study by Martinesen et al. (1998) examined interactions between beaver, leaf beetles, and cottonwoods. Cottonwoods felled by beavers sprout vigorously, and these sprouts contain higher levels of defensive compounds than the original stem that was cut. This is an important chemical strategy for the cottonwood because it can repel generalist herbivores (those that eat a wide variety of plants), but not a specialist insect herbivore, the leaf beetle, which sequesters these chemicals for its own defense. Martinesen et al. (1998) found 15 times as many adult beetles on resprouts following beaver cuttings as on uncut cottonwoods. Resprout cottonwoods have twice the concentration of phenolic chemicals as uncut stems. Several indices of beetle fitness were also higher on resprout growth than on uncut cottonwoods. This is fascinating, but what does this have to do with managing habitat in forests? There are several implications from this work. Phenols are highly toxic to some mammals. As little as 1 g of phenols can kill a human (Budavari 1989). Phenolic concentrations are often higher in juvenile plants than in mature plants. For instance, hares prefer to feed on mature willows and poplars, avoiding juvenile trees. Through plant breeding or nursery practices if nursery stock can be developed that contain higher levels of phenolics, then seedling damage by herbivores could be reduced. But it may not be that simple. Deer saliva contains a glycoprotein containing large amounts of proline, glycine, and glutamate/glutamine that binds with tannins, and potentially other defensive compounds, to reduce the effectiveness of these compounds. So, although increasing defensive compound concentrations in seedlings may be a reasonable strategy to reduce some forms of herbivory, it may not work well for all herbivores.

In addition, this study points to the fact that browse is not browse is not browse. Species vary in their production of these defensive compounds and hence in the quality of browse for herbivores. And individual plants vary in production of these chemicals depending on if they have already been browsed, if they are growing rapidly or slowly, or growing in shade or sun (Martinsen et al. 1998). Consequently, we can create literally tons of browse per hectare following a clearcut, but if the species composition is such that the resulting browse is of low quality, then herbivores may be at a disadvantage eating this browse compared with another site with higher quality, but less quantity of browse.

SUMMARY

Managing habitat for vertebrates in forests often entails manipulating a set of habitat elements that are important to many species. The sizes, density, and distribution of plants; vertical structure; horizontal complexity; forage; dead wood; large trees; leaf litter; soil; and water contribute to habitat quality for many species. Habitat is not just vegetation, but also includes soils, water, and below-ground structure. But managers have control over the structure and composition of vegetation; so by manipulating the density, sizes, and distribution of trees and shrubs in a stand, foresters can have a tremendous influence on the availability of these habitat elements to vertebrates. Further, manipulation of vegetation can also influence the quality of these habitat elements. Browse resources that are high in lignin, cutin, phenols, and tannins reduce digestibility for many herbivores. Managing in a way that provides not only abundant browse resources but high-quality browse resources can have the biggest benefit to ungulates. Similarly, providing large pieces of dead wood or large decaying trees, and stands representing a range of vertical and horizontal complexities, can also benefit a wide variety of species.

REFERENCES

Austin, P.J., L.A. Suchar, C.T. Robbins, and A.E. Hagerman. 1989. Tannin-binding proteins in saliva of deer and their absence in saliva of sheep and cattle. *J. Chem. Ecol.* 15: 1335–1348.

Belsky, J. 1986. Does herbivory benefit plants? A review of the evidence. *Am. Naturalist* 127: 870–892.

Brunnell, A.M. 1988. Food selection and foraging behavior by white-breasted nuthatches on the Daniel Boone National Forest. M.S. Thesis, University of Kentucky, Lexington, KY.

Budavari, S. (ed.). 1989. *The Merck Index*. Merck, Rahway, NJ.

Bultman, T.L., and G.W. Uetz. 1982. Abundance and community structure of forest floor spiders following litter manipulation. *Oecologia* 55: 34–43.

Burke, D.M., and E. Nol. 1998. Influence of food abundance, nest-site habitat, and forest fragmentation on breeding ovenbirds. *Auk* 115: 96–104.

Butts, S.R., and W.C. McComb. 2000. Associations of forest-floor vertebrates with coarse woody debris in managed forests of western Oregon. *J. Wildl. Manage.* 64: 95–104.

Carey, A.B. 2003. Biocomplexity and restoration of biodiversity in temperate coniferous forest: inducing spatial heterogeneity with variable-density thinning. *Forestry* 76: 127–136.

Carey, A.B., and R.O. Curtis. 1996. Conservation of biodiversity: a useful paradigm for forest ecosystem management. *Wildl. Soc. Bull.* 24: 61–62.

Chapin, T.G., D.J. Harrison, and D.M. Phillips. 1997. Seasonal habitat selection by marten in an untrapped forest preserve. *J. Wildl. Manage.* 61: 707–717.

Cork, S.J., and G.J. Kenagy. 1989. Rates of gut passage and retention of hypogeous fungal spores in two forest-dwelling rodents. *J. Mammal.* 70: 512–519.

DeGraaf, R.D., and M. Yamasaki. 2003. Options for managing early-successional forest and shrubland bird habitats in the northeastern United States. *For. Ecol. Manage.* 185: 179–191.

du Toit, J.T., J.P. Bryant, and K. Frisby. 1990. Regrowth and palatability of acacia shoots following pruning by African savanna browsers. *Ecology* 71: 149–154.

Farentinos, R.C., P.J. Capretta, R.E. Kapner, and V.M. Littlefield. 1981. Selective herbivory in tassel-eared squirrels: role of monoterpenes in ponderosa pines chosen as feeding trees. *Science* 213: 1273–1275.

Flather, C.H., S.J. Brady, and D.B. Inneley. 1992. Regional habitat appraisals of wildlife communities: a landscape-level evaluation of a resource planning model using avian distribution data. *Landsc. Ecol.* 7: 137–147.

Friesen, C.A. 1991. The Effect of Broadcast Burning on the Quality of Winter Forage for Elk, Western Oregon. M.S. Thesis, Oregon State University, Corvallis, OR.

Germaine, S.S., S.H. Vessey, and D.E. Capen. 1997. Effects of small forest openings on the breeding bird community in a Vermont hardwood forest. *Condor* 99: 708–718.

Gumtow-Farrior, D. 1991. Cavity resources in Oregon white oak and Douglas-fir stands in the mid-Willamette valley, Oregon. M.S. Thesis, Oregon State University, Corvallis.

Hanley, T.A., C.T. Robbins, and D.E. Spalinger. 1989. Forest Habitats and the Nutritional Ecology of Sitka Black-Tailed Deer: A Research Synthesis with Implications for Forest Management. USDA For.n Serv. Gen. Tech. Rep. PNW-GTR-230.

Healy, W.M., A.M. Lewis, and E.F. Boose. 1999. Variation of red oak acorn production. *For. Ecol. Manage.* 116: 1–11.

Holmes, R.T., and S.K. Robinson. 1981. Tree species preferences of foraging insectivorous birds in a northern hardwoods forest. *Oecologia* 48: 31–35.

Horsley, S.B., S.L. Stout, and D.S. DeCalesta. 2003. White-tailed deer impact on the vegetation dynamics of a northern hardwood forest. *Ecol. Appl.* 13: 98–118.

Huff, M.H., and C.M. Raley. 1991. Regional patterns of diurnal breeding bird communities in Oregon and Washington. In Ruggiero, L.F., K.B. Aubrey, A.B. Carey, and M.H. Huff (eds.), *Wildlife and Vegetation of Unmanaged Douglas-Fir Forests.* USDA For. Serv. Gen. Tech. Rep. 285.

Lowe, W.H., and D.T. Bolger. 2002. Local and landscape scale predictors of salamander abundance in New Hampshire headwater streams. *Conserv. Biol.* 16: 183–193.

MacArthur, R.H., and J.W. MacArthur. 1961. On bird species diversity. *Ecology* 42: 594–598.

Mariani, J.M. 1987. Brown creeper (*Certhia americana*) abundance patterns and habitat use in the southern Washington Cascades. M.S. Thesis, University of Washington, Seattle.

Martin, K.J. 1994. Movements and habitat associations of northern flying squirrels in the central Oregon Cascades. M.S. Thesis, Oregon State University, Corvallis, OR.

Martinsen, G.D., E.M. Driebe, and T.G. Whitman. 1998. Indirect interactions mediated by changing plant chemistry: beaver browsing benefits beetles. *Ecology* 79: 192–200.

Maser, C., J.M. Trappe, and R.A. Nussbaum. 1978. Fungal-small mammal interrelationships with emphasis on Oregon coniferous forests. *Ecology* 59: 799–809.

Maser, C., R.G. Anderson, and K. Cromack Jr. 1979. Dead and down woody material. In Thomas, J.W. (technical editor), *Wildlife Habitats in Managed Forests: The Blue Mountains of Oregon and Washington.* USDA. For. Serv. Agric. Handb. No. 553.

Maser, C., B.R. Mate, J.F. Franklin, and C.T. Dyrness. 1981. *Natural History of Oregon Coast Mammals.* USDA For. Serv. Gen. Tech. Rep. PNW-133.

Mautz, W.H., H. Silver, J.B. Holter et al. 1976. Digestibility and related nutritional data for seven northern deer browse species. *J. Wildl. Manage.* 40: 630–638.

McCarter, J.M., J.S. Wilson, P.J. Baker, J.L. Moffett, and C.D. Oliver. 1998. Landscape management through integration of existing tools and emerging technologies. *J. For.* 96: 17–23.

McCarty, J.P., D.J. Levey, C.H. Greenberg, and S. Sargent. 2002. Spatial and temporal variation in fruit use by wildlife in a forested landscape. *For. Ecol. Manage.* 164: 277–291.

McComb, W.C. 2003. Ecology of coarse woody debris and its role as habitat for mammals. Pages 374–404 In Zabel, C.J., and R.G. Anthony (eds.), *Mammal Community Dynamics: Management and Conservation in the Coniferous Forests of Western North America.* Cambridge University Press, Cambridge, UK.

McComb, W.C., and D. Lindenmayer. 1999. Dying, dead, and down trees. Pages 335–372 In Hunter, Jr. M.L. (ed.), *Maintaining Biodiversity in Forest Ecosystems.* Cambridge University Press, Cambridge, UK.

McComb, W.C., S.A. Bonney, R.M. Sheffield, and N.D. Cost. 1986. Den tree characteristics and abundance in Florida and South Carolina. *J. Wildl. Manage.* 50: 584–591.

Muir, P.S., R.L. Mattingly, J.C. Tappeiner II, J.D. Bailey, W.E. Elliott, J.C. Hagar, J.C. Miller, E.B. Peterson, and E.E. Starkey. 2002. *Managing for Biodiversity in Young Douglas-Fir Forests in Western Oregon.* Biol. Sci. Rep. USGS/BRD/BSR-2002-0006. p. 76.

Naiman, R.J. 1988. Animal influences on ecosystem dynamics. *BioScience* 38: 750–752.

Naiman, R.J., E.V. Balian, K.K. Bartz, R.E. Bilby, and J.J. Latterel. 2002. Dead wood dynamics in stream ecosystems. Pages 23–48 In Shea, P.J., W.F. Laudenslayer Jr., B. Valentine, C.P. Weatherspoon, and T.E. Lisle (eds.), *Proceedings of the Symposium on the Ecology and Management of Dead Wood in Western Forests.* USDA For. Serv. Gen. Tech. Rep. PSW-GTR-181.

Pearson, D.L. 1975. The relation of foliage complexity to ecological diversity of three Amazonian bird communities. Condor 77: 453–466.

Perry, D.A., M.P. Amaranthus, J.G. Borchers et al. 1989. Bootstrapping in ecosystems. *BioScience* 39: 230–237.

Perry, R.W., R.E. Thill, D.G. Peitz, and P.A. Tappe. 1999. Effects of different silvicultural systems on initial soft mast production. *Wildl. Soc. Bull.* 27: 915–923.

Raphael, M.G. 1987. Use of Pacific madrone by cavity-nesting birds. Pages 198–202 In T.R. Plumb and N.H. Pillsbury, eds. *Symposium on Multiple-Use Management of California's Hardwood Resources.* USDA For. Serv. Gen. Tech. Rep. PSW-100.

Ripple, W.J., E.J. Larsen, R.A. Renkin, and D.W. Smith. 2001. Trophic cascades among wolves, elk and aspen on Yellowstone National Park's Northern Range. *Biol. Conserv.* 102: 227–234.

Robbins, C.T., S. Mole, A.E. Hagerman, and T.A. Hanley. 1987. Role of tannins in defending plants against ruminants: reduction in dry matter digestion? *Ecology* 68: 1606–1615.

Rooney, T.P., and D.M. Waller. 2003. Direct and indirect effects of white-tailed deer in forest ecosystems. *For. Ecol. Manage.* 181: 165–176.

Sinclair, A.R.E., and J.N.M. Smith. 1984. Do secondary compounds determine feeding preferences of snowshoe hares? *Oecologia* 61: 403–410.

Singer, S.W., N.L. Naslund, S.A. Singer, and C.J. Ralph. 1991. Discovery and observations of two tree nests of the marbled murrelet. *Condor* 93: 330–339.

Strole, T.A., and R.C. Anderson. 1992. White-tailed deer browsing: species preferences and implications for central Illinois forests. *Natural Areas J.* 12: 139–144.

Suzuki, N., and B.C. McComb. 2004. Associations of small mammals and amphibians with beaver-occupied streams in the Oregon Coast Range. *Northwest Sci.* 78: 286–293.

Taylor, R.J. 1992. Seasonal changes in the diet of the Tasmanian Bettong (*Bettongia gaimardi*), a mycophagous marsupial. *J. Mammal.* 73: 408–414.

Vesely, D.G., and W.C. McComb. 2002. Salamander abundance and amphibian species richness in riparian buffer strips in the Oregon Coast Range. *For. Sci.* 48: 291–297.

Wender, B.W., C.A. Harrington, and J.C. Tappeiner. 2004. Flower and fruit production of understory shrubs in western Washington and Oregon. *Northwest Sci.* 78: 124–140.

Zeller, N.S., and J.A. Collazo. 1995. Abundance and distribution of overwintering passerines in bottomland hardwood forests of North Carolina. *Wilson Bull.* 107: 698–708.

4 Physical and Cultural Influences on Habitat Patterns

Habitat elements, those pieces of the environment that in certain sizes, numbers, and distribution influence habitat quality for vertebrates, are not uniformly distributed across stands and forests. Some stands have high vertical diversity and others are rather simple in their vertical structure. Some support a deep litter layer and others virtually none. Browse quantity and quality varies tremendously depending on the plant species, growth rates, shade, and past browsing. So what is it that influences these patterns of habitat elements? In this chapter, we will explore the physical factors influencing patterns of habitat elements, as well as three dominant cultural influences, global warming, land use, and invasive species.

Although habitat is not simply vegetation, vegetation is shaped by natural disturbances and is the part of the environment that we can influence by silviculture and harvest planning. The patterns and dynamics of vegetation have a significant effect on the patterns and dynamics of habitat elements. Most of this chapter will focus on the physical and cultural processes influencing the pattern of vegetation and habitat elements associated with vegetation, and then in the next chapter we will address how disturbances change those patterns.

THE PHYSICAL ENVIRONMENT

Probably the greatest overriding effects of the physical environment on habitat quality for many species are those of climate, geology, soils, and topography on vegetation patterns. Associations between the physical environment and vegetation patterns are complex and they vary regionally. These associations have a significant influence on patterns of habitat elements such as vertical complexity, horizontal complexity, and forage resources. In the following subsections we will cover both direct and indirect effects of physical factors that influence habitat quality for vertebrates. Although major factors are presented independently, it should be clear that all of these factors interact to a greater or lesser degree to influence habitat quality for species.

GEOLOGY

Vegetation structure and composition are often highly associated with the underlying geology. For instance, in the Klamath-Siskiyou Province of northwestern California and southwestern Oregon, diverse geologic conditions and soils produce a rich array of plant communities (Coleman and Kruckeberg 1999). Soils developed from serpentinized rock may be relatively enriched in various toxic metals, including nickel, magnesium, barium, and chromium, and lacking in important nutrients, such as calcium (Smith and Diggles 1999). Consequently, serpentine geology provides the basis for a unique plant structure and composition that includes Jeffrey pine savannas, xeric shrub types, and serpentine barrens, as well as hygric Darlingtonia fens (Coleman and Kruckeberg 1999). Vegetation structure and composition, and hence habitat element characteristics, are also related to surficial geology and topography over larger areas. For instance, the complex surficial geology and topography of northeastern Ohio created a variety of conditions favoring presettlement forests composed of American beech and maples on some sites and oaks and hickories on others (Whitney 1982). The structure of these forests and their ability to produce mast, browse, and other resources varies in part because plant species composition differs between sites with different geologic histories, slope, and aspect.

In addition to the effects that geology has on topography and soil formation, the interaction of geology with water can lead to direct effects on habitat for some species. Wilkins and Peterson (2000) related amphibian occurrence and abundance in headwater streams draining second-growth Douglas-fir forests to two geologic substrates: sedimentary and basalt formations. Streams traversing basalt had almost twice abundance of the Pacific giant salamanders as streams flowing over marine sediments. They concluded that habitat quality for headwater amphibians in western Washington was strongly influenced by basin geology. In parts of North America, karst geology often occurs in association with limestone formations, or other carbonate rocks that dissolve more easily than surrounding rock formations (Figure 4.1). These karst geologies are prone to formation of caves, sink holes, and subterranean water flow. In Alaska, some species seem to seek karst features and the stable environment provided within caves (Baichtal 1993). Caves are used by a wide variety of species (Blackwell and Associates, Ltd. 1995). Deer are known to rest in the vicinity of caves during both the summer, when air from the caves is cooler, and in winter, when cave entrance air is generally warmer (Blackwell and Associates, Ltd. 1995). Cave systems also are used by many species of bats for roosting and hibernation. Cave environments provide specific air circulation patterns, temperature profiles, humidity, cave structure, and locations relative to feeding sites which some species of bats require (Hill and Smith 1992).

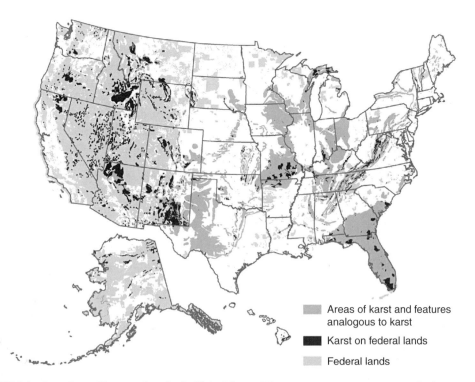

                  Areas of karst and features analogous to karst

                  Karst on federal lands

                  Federal lands

FIGURE 4.1 Locations of karst geology in the United States. These areas are prone to caves and subterranean streams and often require special consideration during forest management. (Reprinted with permission from the National Cave and Karst Institute, U.S. Dep. Interior National Park Service.)

Although manipulation of vegetation does not directly influence geological features such as caves and headwater streams, the microclimatic characteristics of these environments near the surface of the ground can be altered by forest manipulation. For instance, these associations provide managers an opportunity to consider headwater amphibian conservation strategies by prioritizing stream segments based on geology with respect to their likely amphibian fauna and providing either more or less shade to these systems depending on the needs of the species being managed.

TOPOGRAPHY: SLOPE, ASPECT, AND ELEVATION

Vertebrates use behavior to modify the ambient temperature around them and move themselves closer to their thermal neutral zone as they seek appropriate thermal cover. Quite often that thermal cover may not only entail vegetative cover but also be located on particular aspects and slope positions that provide warmth or cooling. Variation in slope and aspect affect the spatial variability in solar radiation incident on the ground, and also ground surface temperature, and wind speeds (Porter et al. 2002). These areas of slope and aspect can particularly influence habitat quality for ectotherms (Thomas et al. 1999). Endotherms also can use behavior to modify their ambient environment. For instance, Pearson and Turner (1995) studied ungulates (deer, elk, and bison) in Yellowstone National Park and found that grazing occurred most often in burned areas at low elevation, drier sites, and on steep southerly slopes. In addition to the direct effects of slope and aspect on ambient temperature, forest vegetation can moderate extreme temperatures on those sites receiving direct sunlight, or lack of vegetation can cause cooler sites to be warmer in the day and cooler at night. Slope and aspect in combination with vegetation can influence snow depths and winter habitat quality for some species of ungulates (D'eon 2001), and large mammals will use topography as a form of hiding cover (Ager et al. 2003). Hence, silvicultural activities that consider slopes and aspects as potential thermal environments or hiding cover for some species can significantly influence habitat quality for these species.

Vegetation patterns often are significantly influenced by changes that occur along elevational gradients. Changes in elevation of less than a meter influence vegetation patterns in bottomland hardwood forests (Wall and Darwin 1999). In mountainous terrain of the western United States and Canada, vegetation patterns are significantly structured by patterns of precipitation and temperature that change with elevation. As air moves over mountains, its elevation increases, it cools, and moisture precipitates from the air as rain or snow. This *orographic precipitation* produces marked patterns in vegetation, which in turn influences the occurrence and abundance of some vertebrates (e.g., Terborgh 1977). For instance, the distribution of several species of reptiles and amphibians, such as Cascades frogs, are highly associated with elevation and, hence, temperatures and snowfall in the Cascades Mountains. Some species of *Anolis* lizards also seem to separate along elevational gradients in tropical forests (Buckley and Roughgarden 2005). Elevation also can interact with other physical variables such as ultraviolet radiation due to ozone depletion (UV-B) to influence abundance of some species. Declines in California red-legged frog populations exhibit a strong positive association with elevation, as well as with percentage of upwind agricultural land use (i.e., chemicals), and local urbanization (Davidson et al. 2001). These declines in frog abundance along the elevational gradient are consistent with increased levels of UV-B radiation.

SOILS

Maintaining long-term soil productivity in managed forests is critical to producing wood products and habitat elements (Fox 2000). Given the short-term positive effects of forest fertilization (Brockway 1983) and negative long-term effects of soil compaction (on certain soils; Kozlowski 1999) on tree growth it is easy to see why soil structure and composition are critical to sustaining various ecosystem services (Knoepp et al. 2000). Bedrock and surfical geology interact with climatic variables and hydrology to produce soils varying in their ability to support growth of certain species and clearly alter growth rates of trees. On sites where trees grow faster, large trees and vertical complexity can develop more quickly. Soil structure, nutrient composition, moisture content, and temperature clearly influence the vegetation species composition and quality of habitat elements such as tree species composition and browse quality. For instance, the chemical constituency of flowering dogwood and red maple foliage changes over subtle gradients of soil moisture and nutrient availability such that the production of phenolic compounds was highest on sites of greatest plant stress (Muller et al. 1987).

Soils can also have direct effects on vertebrate habitat quality. Burrowing animals are clearly affected by soil structure. As an example of the extensive nature of burrow systems, Askram and Sipes (1991) found approximately 16,000 holes and 7.5 km (4.7 miles) of above-ground runways per 0.4 ha (1 acre) created by voles in an orchard in the northwestern United States. Gopher tortoises are adept at making burrows in sandy coastal plain soils (Ultsch and Anderson 1986; Figure 3.9), Gapper's red-backed voles in boreal forests develop elaborate networks of burrows, and mountain beavers develop extensive burrow systems used by many other species. Regosin et al. (2003) suggested that small mammal burrow densities could influence abundances of spotted salamanders in northeastern U.S. forests. Ground-based logging equipment and other forest management activities can disrupt soil structure and reduce burrow availability. In addition, the minerals available in some soils can meet direct dietary needs. Species such as moose, deer, caribou, and band-tailed pigeons seek sodium-rich soils and springs to meet demands for this nutrient (e.g., Kennedy et al. 1995, Sanders and Jarvis 2000). Hence, protection of soils from disruption, compaction, and erosion can be a key step in providing habitat for many species that spend their time both below and above ground.

CLIMATE

Climatic conditions are a predominant driving force influencing the distribution of organisms. Temperature and precipitation, separately and in combination, have significant effects on vegetation patterns. The physical environment of North America has been altered by climatic processes extending hundreds of thousands of years into the past. Most notably glaciation and subsequent climate change have influenced surficial geology, soils, and hydrology over large parts of North America. For instance, about 12,000 years ago there was a 2-km-thick chunk of ice sitting over much of the northern United States and Canada. At that time, the sea level was considerably lower and forests grew on parts of what is now the continental shelf. As climates changed, the receding glaciers left deeply scoured valleys, outwash plains, windblown deposits of sands, drumlins, and erosion-resistant ridges of granite and basalt. The distribution of tree species found in forests then was very different from what it is today, and patterns of forest species changed markedly as the glaciers receded (Figure 4.2; Jacobson et al. 1987). The soils that were left behind as glaciers receded represented a complex mosaic of sands, gravels, and clays that structures vegetative communities quite markedly.

In addition to those soil characteristics, volcanic extrusions in combination with glacial activities have left cliffs (nest sites for common ravens and peregrine falcons), caves (used by some bat species for hibernacula), and a diverse topography producing vegetation gradients over relatively small areas. Distributions of some animal species also follow the complex topographic features that resulted from glacial recession. For instance, the distribution of Bicknell's thrush is highly associated with the higher elevations in the Berkshires, Green and White Mountains, Adirondacks, and Catskills, which generally represent rock formations resistant to the forces of the past glaciers.

Climatic conditions not only influence the vegetation patterns of a region but also directly influence the ability of species to survive in an area. Consider the geographic distribution of the northern copperhead in New England (Figure 4.3a). The northern limit of the geographic range for this species, an ecotherm sensitive to prolonged cold, closely corresponds to the mean dates of the first and last frost (Figure 4.3b). Indeed many species of ectotherms are influenced by air, soil, or water temperatures in forests (e.g., Nussbaum et al. 1983, Welsh and Lind 1996). Some endotherms are also highly influenced by changes in ambient temperature. The temperature at roosts and maternity

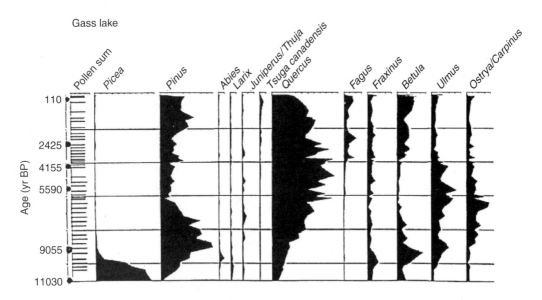

FIGURE 4.2 Patterns of tree pollen from the sediments in a Wisconsin lake. Note that as spruces decline about 10,000 years ago, pines increase in dominance, until about 6,000 years ago, and then oaks dominate. American beech did not begin to have a significant presence in the area until about 4,000 years ago. (From Webb, S.L. 1987. *Ecology* 68: 1993–2005. Reprinted with permission of the Ecological Society of America.)

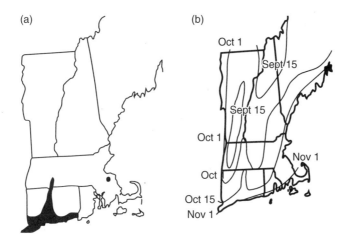

FIGURE 4.3 (a) The geographic range of northern copperheads and (b) the isotherms for date of first frosts. (From DeGraaf, R.M., and M. Yamasaki. 2001. *New England Wildlife: Habitat, Natural History, and Distribution*. University of Press of New England, Hanover, NH. Reprinted with permission.)

sites of some bat species seems to be particularly important for survival and growth of adults and young (Agosta 2002). In fact, there is even evidence that sex ratios of bats and some reptiles can be influenced by ambient temperatures (Ewert et al. 1994, Ford et al. 2002). Consequently, considering the effects of forest management on temperature regimes in forests can be quite important to providing adequate habitat for species such as these.

Moisture plays a critical role in structuring the patterns of vegetation and often is more important than temperature in structuring the regional environment (e.g., the Pacific Northwest). Clearly,

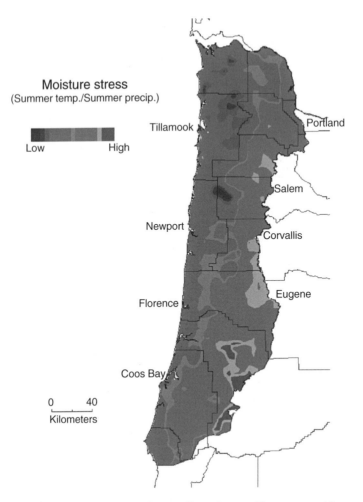

FIGURE 4.4 Map of moisture stress in the Oregon Coast Range. Plant communities seem to be structured in large part by this variable. (Reprinted from a figure developed by the USDA Forest Service Pacific Northwest Research Station Coastal Landscape Analysis and Modeling Study. With permission from Janet Ohmann.)

precipitation and temperature are interrelated, particularly as they influence moisture stress on plants. Consider the map of moisture stress in Figure 4.4 for western Oregon. Moving from west to east, patterns of tree dominance changes from Sitka spruce and western hemlock in areas with low moisture stress, to Douglas-fir and western hemlock, and to Douglas-fir and grand fir to Oregon oak and grand fir as you move east along this gradient. Moisture stress seems to be an important factor associated with vegetation patterns over much of the northwest (Ohmann and Spies 1998, Zobel et al. 1976) and possibly the continent (Goward et al. 1985).

Local patterns of vegetation are also driven by moisture stress gradients (Figure 4.5). In the eastern United States, oak species segregate along topographic features that are related to a moisture stress gradient. Sites with standing water are dominated by pin oak and swamp white oak, and as you move upslope, these are replaced with northern red oak and white oak, then black oak, and finally chestnut oak and scarlet oak at the ridgetops. Hence, the ability of a forest to produce mast is highly influenced by site conditions that reflect moisture stress. Further, it is important to recognize these patterns when regenerating forests during management. Matching plant species to sites is a key step in successfully regenerating a stand.

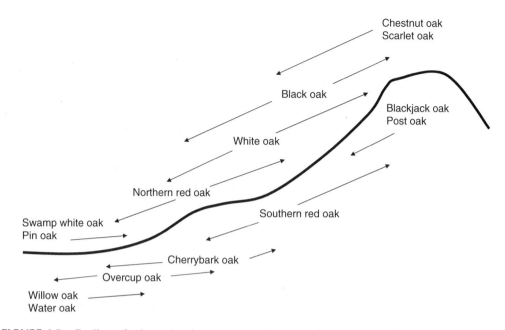

FIGURE 4.5 Gradient of oak species along a topographic and moisture stress gradient in the eastern United States.

Precipitation also influences soil moisture and hydrology for an area (see previous and following sections) and consequently plays a key role in structuring habitat quality for many species. Recently though, there also have been noticeable effects of acid precipitation on habitat quality for species such as amphibians and waterfowl (Stenson and Ericksson 1989). The effects of acid deposition on vertebrates is complex. There are direct effects of reduced pH such as reduced reproduction or survival of fish and probably aquatic amphibians. There also are indirect effects, such as a shift of top predators from fish to invertebrates and a reduced decomposition rate due to decreased abundance of detrivores (Stenson and Ericksson 1989). In addition, productivity and turnover rate of nutrients can be reduced, and there often is an increase in water transparency, which influences predation effectiveness. In addition, acid precipitation can have indirect effects on terrestrial systems through changes in vegetation structure and composition resulting from alteration of nutrient exchange capacity in the soils (Schreiber and Newman 1988).

There are also direct effects of rain (Waltman and Beissinger 1992) or snow (Kirchoff and Schoen 1987) on habitat quality for many species. Many species seek shelter from precipitation due to the evaporative cooling effect on their bodies. Snow in particular can influence movements and choice of foraging and resting areas by deer and other ungulates (Nelson 1998).

HYDROLOGY

Watersheds influence the physical structuring of the environment and consequently habitat elements and habitat quality for a number of species. We will address in greater depth the issues of riparian vegetation and management on habitat elements in a later chapter. Clearly though, the proximity of habitat elements to water and groundwater conditions can influence vegetation patterns markedly. Some tree species are very well adapted to growing in flood-prone or saturated soils (e.g., baldcypress, eastern cottonwood, and water oak). Considering management effects on the regeneration and growth of these species is the key to effective management of many bottomland species. In particular, waterfowl managers in the southern United States have used green tree reservoirs to provide timber and waterfowl habitat (Wigley and Filer 1989). Water levels are manipulated in green tree reservoirs to flood bottomlands in the winter to attract waterfowl feeding on acorns. Water is then drawn down during the growing season to allow rapid growth of oaks that are then harvested for timber.

The hydrologic features of an area can also have more direct effects on some species. Seeps and steep headwater streams that provide large boulder and rock substrates, cool water, and highly aerated water provide habitat for species such as spring salamanders in the east and torrent salamanders in the west (Sheridan and Olson 2003). As the watershed area increases, headwater streams that may be intermittent become permanently flowing. If there are no barriers to fish movement, we may find brook trout in the eastern streams and cutthroat trout in western streams, which in turn influences the distribution of salamanders in those streams. As the gradient (slope) of the stream declines and water volume increases, we see sediments deposited and more meandering stream courses. These midwatershed stream reaches provide the substrate for dens for muskrats and beavers. Farther along the gradient where wide valley floors and annual floods characterize the large river systems, backwater sloughs, swamps, side channels, and flooded wetlands become important nest sites for species such as wood ducks, great blue herons, and western pond turtles. In fact, the large bottomland hardwood forests of the Mississippi river floodplain and Gulf coast provide habitat for many species including the ivory-billed woodpecker, which may have been recently rediscovered (although that remains to be confirmed). Finally, as the river empties into an estuary, the interface between fresh and salt water provides a set of conditions that are ideal for many species such as American black ducks, muskrats, and many species of wading birds.

Kettle holes that resulted from the melting of large ice blocks embedded in glacial soils may allow the formation of vernal pools. Vernal pools hold water for a period in the spring, but dry out later in the summer. These sites are key breeding habitat for species such as marbled salamanders and wood frogs. Large mountain lakes that result from scouring from glaciers also provide habitat for species such as loons, cascade frogs, and common goldeneyes. Because each of these hydrologic features provides habitat for a different suite of species, each must be considered differently when managing adjacent forestlands.

Vegetation Patterns

Vegetation patterns are associated with these physical features of the environment. The potential vegetation of a region represents the dominant vegetation that would be present in a region in the absence of culturally induced changes (Figure 4.6). Clearly, these vegetation patterns influence the patterns of habitat elements over the region. Wildlife habitat types have been delineated in many states and provinces to facilitate the understanding of historic and current patterns of vegetation and topography. For instance, in Oregon, 33 habitat types representing vegetation, water, and geologic features are a way of understanding how species might respond to these features (Figure 4.7). It should be clear by now though that these broad categories driven by physical forces may only be crudely related to the distribution of some species because every species has its own habitat requirements.

Vegetation patterns that we see today were quite different historically. The changes in vegetation patterns resulting from glaciation and climate changes occurred relatively slowly in the past and encompassed thousands of generations of vertebrates. The slow rate of change usually allowed species to adapt to these changes. Habitat selection also certainly would have been influenced by these changes and led to genetic advantages for those individuals most adaptable to the new conditions. Habitat selection also would have facilitated the changes in vegetation pattern. Movement of heavy-seeded tree species (e.g., oaks) was facilitated by birds and mammals (e.g., jays and squirrels). The current distribution of vertebrates may be driven primarily by vegetation pattern (e.g., mink frogs are associated with boreal forest) or apparently by thermoclines (Figure 4.3) or by the distribution of vegetation in conjunction with temperature (e.g., Carolina wren; see DeGraaf and Yamasaki 2001). For many species it is the interplay among these factors that led to regional patterns of distribution in vegetation, habitat elements, and geographic ranges of species.

FIGURE 4.6 Generalized map of potential vegetation for North America based on climate and soils. (B.S. Arbogast and used with his permission.)

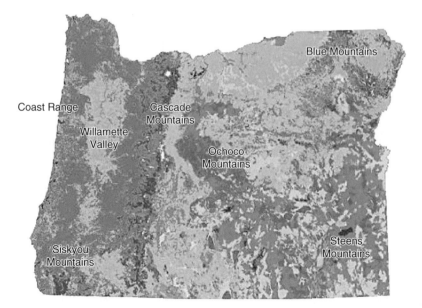

FIGURE 4.7 The mosaic of historic habitat types for the state of Oregon, United States. These patterns have been changed through recent cultural activities associated with land use. (Redrafted from a map produced by the Institute for Natural Resources, Oregon State University. With permission.)

The potential patterns of vegetation are particularly important for many species because they represent the potential of a system to provide energy to the trophic levels within the ecosystem. Systems vary considerably in their production of biomass. Biomass represents stored energy sources for consumers and detritivores. Mature deciduous forests can support over 475 ton/ha (190 ton/acre) of biomass (Whittaker et al. 1974), while deserts support only 5 to 50 ton/ha (2 to 20 ton/acre) of biomass (Noy-Meir 1973). Consequently, human activities that influence the direct physical factors of geology, soils, climate, or hydrology, or vegetation patterns through land use can have huge impacts on the distribution of vertebrates by changes in habitat quality.

CULTURAL EFFECTS ON HABITAT PATTERNS

In the following chapter we will spend more time discussing the role of disturbance in driving patterns of vegetation across large regions. But there is also a historical context associated with human activities that must be understood to explain changes in the vertebrate communities that we have observed over the past several hundred years and the patterns that we see today. Native Americans likely maintained more open landscape conditions through use of agriculture and fire than what may have been first described by European settlers (Boag 1992). Humans are a part of ecosystems and they have been for millennia. The influx of European humans into the North American environment led to changes in forest cover and distribution that were quite different from the historic conditions that occurred up to that point.

There are several factors that have occurred since the arrival of European humans into the North American continent, which set the context for management of habitat in North American forests. The distribution of vegetation is faced with three dominant current pressures that might change habitat quality for many species at a much more rapid rate than has occurred historically: (1) land use changes, (2) global climate change, and (3) invasive species. These forces represent a significant common ground between foresters and wildlife biologists. Discussions on how to manage forests for products and habitat for species fall silent when forests are replaced by other systems.

LAND USE

The effects of land use on habitat patterns have been apparent for centuries. But development, especially as reflected in urban sprawl, is occurring at a remarkable rate in many of our forests. In Massachusetts, 16 ha (40 acres) per day are converted from forest to housing (Foster et al. 2005). The rate is similarly as great in urbanizing areas across North America. One only needs fly over Mexico City, Phoenix, Seattle, or Vancouver to see the effects of development and sprawl on forest, grassland, and desert ecosystems. As human populations increase, the urban–rural interface expands and the effects of urbanization extend beyond that of the individual house footprints. The proliferation of roads, utility infrastructure, and human use of remaining fragments of forest land lead to marked changes in the function of these forests as habitat for many vertebrates (Theobald et al. 1997). Some species increase in abundance and expand their distribution in response to these changes. Two bird species, Carolina wrens and tufted titmice, have increased in abundance by 17 and 7% over the past 40 years in Massachusetts (Sauer et al. 2005). Many more species have declined significantly over that same time period, such as wood thrushes and black-and-white warblers (Sauer et al. 2005). Conversion of forest to subdivisions probably has at least some role in these changes, and consequently some native species face habitat loss from development. Certainly, we see loss of the potential production of wood products from these lands. The greatest threat to forest sustainability and biodiversity is conversion of forests to other land uses, which often results when markets value forest systems and the ecosystem services they provide less than the economic value of houses and industries (NCSSF 2005).

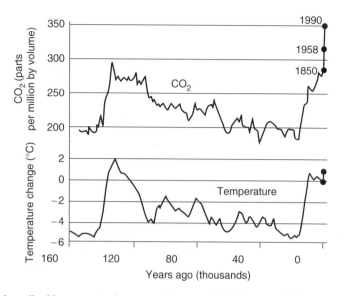

FIGURE 4.8 Carbon dioxide concentrations over the past 160,000 years. Circles represent changes due to human activities. (Reprinted from Schneider, S.H., and T.L. Root. 1998. Climate change. In Mac, M.J., P.A. Opler, E.P. Haecker, and P.D. Doran (eds.), *Status and Trends of the Nation's Biological Resources*. U.S. Dep. Interior U.S. Geol. Surv., Reston, VA, pp. 89–116.)

CLIMATE CHANGE

Climate change also has the potential to alter the distribution of vegetation. Consider the change in CO_2 concentration in the atmosphere over the past few thousand years (Figure 4.8). This change is quite likely to lead to dramatic changes in the ability of plants and animals to tolerate conditions as temperature and precipitation patterns change more rapidly than they have historically. The result will likely be a shift northwards for many southern species and a compression of the area that would be available to meet their needs.

These shifts in distribution are assumed to occur if the organism is mobile enough and adaptable enough to allow movement in response to these climatic changes. For example, changes in vegetation as a result of elevated temperatures have been linked to the current and likely future distribution of animal species (Figure 4.9). Clearly, plant species face issues of coping with movement rates that will keep pace with changing temperatures, but even some vertebrates (e.g., salamanders) likely will not respond quickly enough. Further, these organisms face obstacles as they are forced to change their geographic ranges (e.g., roads, farms, and cities). Of course, as urban areas expand and fossil fuel demands increase worldwide, the effects of climate change will likely worsen until alternative energy sources begin to dominate. Even with those changes, it will likely take decades or centuries before CO_2 comes back to levels seen before the 1900s.

INVASIVE SPECIES

Another current effect on the distribution of organisms is the increase in numbers of invasive species. Either intentionally or by chance, a species from a latitude in one location that finds itself at a similar latitude in another location where it historically did not occur may be able to dominate the site because of the lack of predators, competitors, or disease. These invasive species are often exotic, those that are brought into an area from other countries or continents. Chestnut blight fungus, gypsy moth, Scotch broom, European starlings, and house sparrows are examples of exotic species that have become invasive in North America. But species native to a continent can also be invasive

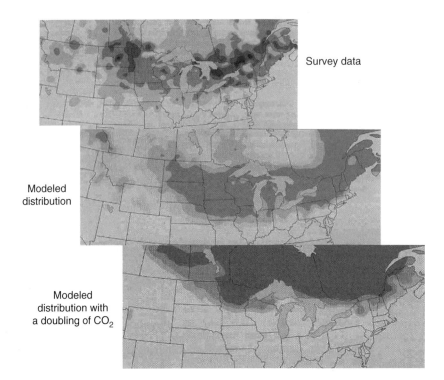

Survey data

Modeled
distribution

Modeled
distribution with
a doubling of CO_2

FIGURE 4.9   Current range of the northern bobolink (upper frame), predicted range based on current carbon dioxide levels (middle frame), and predicted range under doubled carbon dioxide concentrations (lower frame). (Reprinted from Schneider, S.H., and T.L. Root. 1998. Climate change. In Mac, M.J., P.A. Opler, E.P. Haecker, and P.D. Doran (eds.), *Status and Trends of the Nation's Biological Resources*. U.S. Dep. Interior U.S. Geol. Surv., Reston, VA, pp. 89–116.)

when they are placed in a new location. Bullfrogs moved to the west coast from eastern North America are likely responsible for declines in some western native amphibians (Figure 4.10), and not all exotics are invasive. In fact, most are not. Ring-necked pheasants are an example of an exotic species that colonized the Midwest, but probably does not displace any native species. The landscape planting industry annually uses thousands of species of plants that are not native, and only a few escape to become invasive. Increasing expansion of suburbia leads to increasing spread of plants that are planted around houses, such as Norway maple, and can become invasive. In addition, as climate changes, some invasive plants and animals that may not be a problem now may become a problem under new climatic conditions, or the current problem may be reduced under new climatic conditions.

How do these exotic plants and animals find their way into our forests? Most arrive on this continent at port cities (Figure 4.11). States that have ports rank high in the number of these species that have been introduced. Some were introduced for purposes of soil stabilization and wildlife habitat improvement: multiflora rose and autumn olive, for example. Now biologists are trying to eradicate the very species that were imported and planted for habitat purposes! The continuation of introductions of exotic species by the landscaping industry and some forest products industries is raising more and more concerns about homogenization of our globe, which could lead to a net loss in biodiversity (Richardson 1998). The direct effects of invasive species on habitat quality can be quite apparent. The competitive advantage that invasives have over other species can lead to homogenization of the site and the displacement of native species into more isolated patches. For instance, pines were once only found in the northern hemisphere, but over 19 species are now established in the southern hemisphere through use of exotic species in plantations and for erosion control (Richardson 1998). Australian paperbark tree was introduced to Florida in the early

FIGURE 4.10   Bullfrogs were introduced to the west coast from eastern North America and are likely responsible for declines in some western native amphibians. (Photo by Mike Jones and used with his permission.)

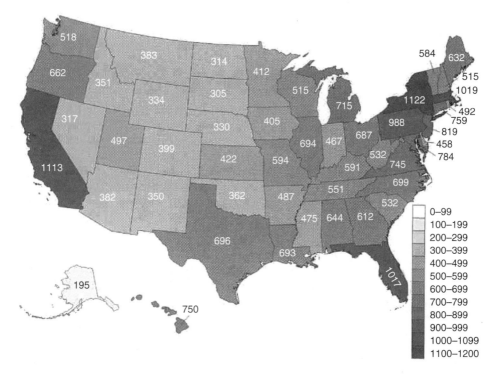

FIGURE 4.11   Numbers of non-native plant species that have been introduced into each state in the United States. (Reprinted from Williams, J.D., and G.K. Meffe. 1998. Nonindigenous species. In Mac, M.J., P.A. Opler, E.P. Haecker, and P.D. Doran (eds.), *Status and Trends of the Nation's Biological Resources.* U.S. Dep. Interior, U.S. Geol. Surv., Reston, VA, pp. 117–130.)

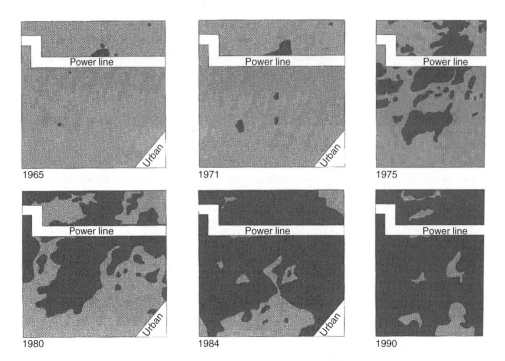

FIGURE 4.12   Changes in native plant species cover over time following invasion by Australian paperbark tree (dark gray). (Reprinted from Williams, J.D., and G.K. Meffe. 1998. Nonindigenous species. In Mac, M.J., P.A. Opler, E.P. Haecker, and P.D. Doran (eds.), *Status and Trends of the Nation's Biological Resources.* USDI, U.S. Geol. Surv., Reston, VA, pp. 117–130.)

1900s. Prolific seed production, flood tolerance, and rapid regrowth following fire enabled this species to invade wetlands and eliminate native plants and the animal species that rely on them (Figure 4.12; Hofstetter 1991). Once established, invasive species such as false brome can eliminate native vegetation and can be very difficult to control (Figure 4.13).

If society wishes to maintain habitat for various wildlife species, then biologists and foresters must first work together to address issues of development, climate change and invasive species. Or else the discussion of how to manage forests becomes moot.

CASE STUDY: PASSENGER PIGEONS, HUMANS, AND FORESTS

Contemporary decision makers could learn about the roles of physical and cultural influences on habitat quality for selected species from patterns and changes in habitat elements that have occurred during the recent history of the United States. The extinction of passenger pigeons represents a classic example of how a species was not able to persist when faced with a suite of pressures on the populations, especially changes in habitat conditions that were imposed by European humans.

Prior to European settlement, passenger pigeons were nomadic, occurring in flocks of millions of birds moving over vast areas of contiguous eastern deciduous forests (Ellsworth and McComb 2003; Figure 4.14). Acorn production in the forests varied considerably from year to year and place to place (Healy et al. 1999), and so the large flocks of pigeons provided many eyes to search for available food. Once a member of the flock found food, the remainder of the flock would follow, using a process known as social facilitation to locate patches of high acorn production near nesting and roosting areas. In the spring, pigeons followed the receding snowmelt northward to the nesting

FIGURE 4.13 Understory of a Douglas-fir stand dominated by the invasive exotic grass, false brome. Note the absence of native understory plants in this stand.

areas, relying largely on red oak acorns and beechnuts as food during these movements (Bucher 1992). Red oaks, unlike white oaks, undergo a winter stratification period and germinate in the spring. White oaks germinate in the fall, and so were less available to pigeons following snowmelt. Consequently, pigeons were well adapted to the extent and variability in patterns of acorn production in oak forests across the eastern United States and southern Canada.

European settlers cleared the forests initially for agriculture and eventually for cities and industries. Farmers often allowed domestic pigs to forage for mast (acorns, chestnuts. and beechnuts) in the forest, and pigs are a bit like 200 lb mammalian vacuum cleaners in the forest, eating mast and digging roots (Henry and Conley 1972). The combination of clearing of oak forests, foraging by domestic livestock for acorns, and increased levels of harvest of pigeons as food caused pigeons to be less abundant. Patches of food became more widely dispersed, and despite the pigeons' social facilitation behavior, food became more difficult to find. Hunting of the pigeons reduced their numbers, thereby making social facilitation as a mechanism for finding food patches even more ineffective because there were fewer eyes to find the more dispersed food. Nest sites also became less available, and because passenger pigeons only laid one egg per year, and both parents helped with incubation and rearing, there was a high energy investment in reproduction but a low reproduction rate. This low reproductive rate exacerbated the issues associated with reduced abilities to find food and nesting sites, and so populations began to decline. Declines accelerated as the population entered what population biologists call the *extinction vortex* (Westemeier et al. 1998). Before long the population was simply not able to persist. The last nesting birds were seen in the Great Lakes region in the 1890s. The last individuals were killed in the wild in 1900, but some individuals remained in captivity until 1914. Martha, the last passenger pigeon, died at the Cincinnati Zoo on September 1, 1914. The ultimate demise of the passenger pigeon was more a result of habitat loss than other factors, although overhunting contributed to the declines. Habitat loss occurred through cultural activities

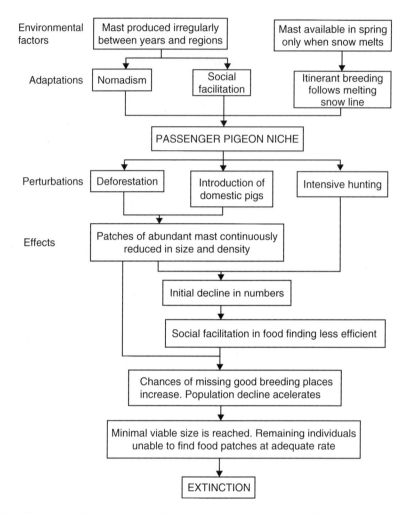

FIGURE 4.14   Factors leading to the extinction of the passenger pigeon in North America. Although many processes were at work, habitat loss was a primary driver. (Redrafted from Bucher, E.H. 1992. *Curr. Ornithol.* 9: 1–36. With permission from Springer Press.)

on a template of physical features and vegetation to which the species was well adapted. This classic example of species extinction should be one that we continue to learn from and consider how we should "save all the pieces" if we do not want additional species to have a similar fate.

SUMMARY

The physical factors of a region that include geology, soils, topography, climate, and hydrology interact to influence the potential for a region to support a plant community and the elements of habitat used by vertebrates. These physical factors also influence the quality of habitat for some species directly by their influence on providing suitable microclimates, burrowing substrates, cover, and stream features. The greatest effect of these physical factors is in structuring plant communities from continents to stands. It is the forest vegetation and the associated habitat elements that we can influence through forest management decisions. But the potential vegetation for a region is altered by disturbances (see the following chapter) and cultural changes to forests. Development pressures, proliferation of invasive species, and climate change all threaten the extent and function of forests in

North America, and hence will influence our ability to provide wood products and habitat elements. These pressures on our forests provide common ground for foresters and wildlife biologists to work together if society continues to demand both wood products and biodiversity conservation.

REFERENCES

Ager, A.A., B.K. Johnson, J.W. Kern, and J.G. Kie. 2003. Daily and seasonal movements and habitat use by female Rocky Mountain elk and mule deer. *J. Mammal.* 84: 1076–1088.

Agosta, S.J. 2002. Habitat use, diet and roost selection by the Big Brown Bat (*Eptesicus fuscus*) in North America: a case for conserving an abundant species. *Mammal Rev.* 32: 179–198.

Askram, L.R., and J. Sipes. 1991. Burrow-building strategies and habitat use of voles in Pacific northwest orchards. In Hygnstrom, S.E., R.M. Case, and R.J. Johnson (eds.), *Proceedings of the 10th Great Plains Wildlife Damage Conference*. University of Nebraska-Lincoln.

Baichtal, J.F. 1993. Evolution of karst management on the Ketchikan area of the Tongass National Forest: development of an ecologically sound approach. In *National Cave Management Symposium*. American Cave Conservation Association, Carlsbad, New Mexico.

Blackwell, B.A., and Associates, Ltd. 1995. *Literature Review of Cave/Karst Resources in Forest Environments*. Regional Recreation Office, Vancouver Forest Region, Nanaimo, B.C.

Boag, P.G. 1992. *Environment and Experience: Settlement Culture in Nineteenth Century Oregon*. University of California Press, Berkeley, CA.

Brockway, D.G. 1983. Forest floor, soil, and vegetation responses to sludge fertilization in red and white pine plantations. *Soil Sci. Soc. Amer. J.* 47: 776–784.

Bucher, E.H. 1992. The causes of extinction of the passenger pigeon. *Curr. Ornithol.* 9: 1–36.

Buckley, L.B., and J. Roughgarden. 2005. Effect of species interactions on landscape abundance patterns. *J. Anim. Ecol.* 74: 1182–1194.

Coleman, R.G., and A.R. Kruckeberg. 1999. Geology and plant life of the Klamath-Siskiyou Mountain Region. *Natural Areas J.* 19: 320–340.

Davidson, C., H.B. Shaffer, and M.R. Jennings. 2001. Declines of the California red-legged frog: climate, UV-B, habitat, and pesticides hypotheses. *Ecol. Applic.* 11: 464–479.

DeGraaf, R.M., and M. Yamasaki. 2001. *New England Wildlife: Habitat, Natural History, and Distribution*. University Press of New England, Hanover, NH.

D'eon, R.G. 2001. Using snow-track surveys to determine deer winter distribution and habitat. *Wildl. Soc. Bull.* 29: 879–887.

Ellsworth, J.W., and B.C. McComb. 2003. The potential effects of passenger pigeon flocks on the structure and composition of pre-settlement eastern forests. *Conserv. Biol.* 17: 1548–1558.

Ewert, M.A., D.R. Jackson, and C.E. Nelson. 1994. Patterns of temperature-dependent sex determination in turtles. *J. Exp. Zool.* 270: 3–15.

Ford, W.M., M.A. Menzel, J.M. Menzel, and D.J. Welch. 2002. Influence of summer temperature on sex ratios in eastern red bats (*Lasiurus borealis*). *Amer. Midl. Nat.* 147: 179–184.

Foster, D.R.; D.B. Kittredge, and B. Donahue. 2005. *Wildlands and Woodlands: A Vision for the Forests of Massachusetts*. Harvard Forest, Petersham, MA.

Fox, T.R. 2000. Sustained productivity in intensively managed forest plantations. *For. Ecol. Manage.* 138: 187–202.

Goward, S.N., C.J. Tucker, and D.G. Dye. 1985. North American vegetation patterns observed with the NOAA-7 advanced very high resolution radiometer. *Vegetatio* 64: 3–14.

Healy, W.M., A.M. Lewis, and E.E. Boose. 1999. Variation of red oak acorn production. *For. Ecol. Manage.* 116: 1–11.

Henry, V.G., and R.H. Conley. 1972. Fall foods of European wild hogs in the Southern Appalachians. *J. Wildl. Manage.* 36: 854–860.

Hill, J.E., and J.D. Smith. 1992. *Bats — A Natural History*. University of Texas Press, Austin, TX.

Hofstetter, R.L. 1991. The current status of *Melaleuca quinquenervia* in southern Florida. In Center, T.D., R.F. Doren, R.L. Hofstetter, R.L. Myers, and L.D. Whiteaker (eds.), *Proceedings of the Symposium on Exotic Pest Plants*, November 2–4, 1988, University of Miami. U.S.D.I. National Park Service, Washington, D.C.

Jacobson, Jr., G.L., T. Webb III, and E.C. Grimm. 1987. Patterns and rates of vegetation change during the deglaciation of eastern North America. Pages 277–288 In Ruddiman, W.F., and H.E. Wright Jr. (eds.), *North America During Deglaciation. The Geology of North America, DNAG vol. K3*, Geological Society of America, Boulder, Co.

Kirchhoff, M.D., and J.W. Schoen. 1987. Forest cover and snow: implications for deer in southeast Alaska. *J. Wildl. Manage.* 51: 28–33.

Kennedy, J.F., J.A. Jenks, R.L. Jones, and K.J. Jenkins. 1995. Characteristics of mineral licks used by white-tailed deer (*Odocoileus virginianus*). *Amer. Midl. Nat.* 134: 324–331.

Knoepp, J.D., D.A. Crossley Jr., D.C. Coleman, and J. Clark. 2000. Biological indices of soil quality: an ecosystem case study of their use. *For. Ecol. Manage.* 138: 357–368.

Kozlowski, T.T. 1999. Soil compaction and growth of woody plants. *Scand. J. For. Res.* 14: 596–619.

Muller, R.N., P.J. Kalisz, and T.W. Kimmerer. 1987. Intraspecific variation in production of astringent phenolics over a vegetation-resource availability gradient. *Oecologia* 72: 211–215.

NCSSF. 2005. *Science, Biodiversity, and Sustainable Forestry: A Findings Report of the National Commission on Science for Sustainable Forestry (NCSSF)*. National Commission on Science for Sustainable Forestry (NCSSF), Washington, D.C.

Nelson, M.E. 1998. Development of migration behavior in northern white-tailed deer. *Can. J. Zool.* 76: 426–432.

Noy-Meir, I. 1973. Desert ecosystems: environment and producers. *Ann. Rev. Ecol. System.* 4: 23–51.

Ohmann, J.L., and T.A. Spies. 1998. Regional gradient analysis and spatial pattern of woody plant communities. *Ecol. Monogr.* 68: 151–182.

Pearson, S.M., and M.G. Turner. 1995. Winter habitat use by large ungulates following fire in northern Yellowstone National Park. *Ecol. Applic.* 5: 744–755.

Porter, W.P., J.L. Sabo, C.R. Tracy, O.J. Reichman, and R. Navin. 2002. Physiology on a landscape scale: plant-animal interactions. *Integ. Comp. Biol.* 42: 431–453.

Regosin, J.V., B.S. Windmiller, and J.M. Reed. 2003. Influence of abundance of small-mammal burrows and conspecifics on the density and distribution of spotted salamanders (*Ambystoma maculatum*) in terrestrial habitats. *Can. J. Zool.* 81: 596–605.

Richardson, D.M. 1998. Forestry trees as invasive aliens. *Conserv. Biol.* 12: 18–26.

Sanders, T.A., and R.L. Jarvis. 2000. Do band-tailed pigeons seek a calcium supplement at mineral sites? *The Condor* 102: 855–863.

Sauer, J.R., J.E. Hines, and J. Fallon. 2005. *The North American Breeding Bird Survey, Results and Analysis 1966–2004. Version 2005.2*. U.S. Dep. Interior Geological Survey Patuxent Wildl. Res. Center, Laurel, MD.

Schneider, S.H., and T.L. Root. 1998. Climate change. In Mac, M.J., P.A. Opler, E.P. Haecker, and P.D. Doran (eds.), *Status and Trends of the Nation's Biological Resources*. U.S. Dep. Interior Geological Survey, Reston, VA, pp. 86–116.

Schreiber, R.K., and J.R. Newman. 1988. Acid precipitation effects on forest habitats: implications for wildlife. *Conserv. Biol.* 2: 249–259.

Sheridan, C.D., and D.H. Olson. 2003. Amphibian assemblages in zero-order basins in the Oregon Coast Range. *Can. J. For. Res.* 33: 1452–1477.

Smith, C., and M.F. Diggles. 1999. *Arabis macdonaldiana* in the Josephine Ophiolite, Six Rivers National Forest, Del Norte County, California. Final report to USDI USGS and USDA USFS. Menlo Park, CA.

Stenson, J.A.E., and M.O.G. Eriksson. 1989. Ecological mechanisms important for the biotic changes in acidified lakes in Scandinavia. *Arch. Environ. Contam. Toxicol.* 18: 201–206.

Terborgh, J. 1977. Bird species diversity on an Andean elevational gradient. *Ecology* 58:1007–1019.

Theobald, D.M., J.R. Miller, and N.T. Hobbs. 1997. Estimating the cumulative effects of development on wildlife habitat. *Landsc. Urban Planning.* 39: 25–36.

Thomas, J.A., R.J. Rose, R.T. Clarke, C.D. Thomas, and N.R. Webb. 1999. Intraspecific variation in habitat availability among ectotherm animals near their climatic limits and their centres of range. *Funct. Ecol.* 13: 55–64.

Ultsch, G.R., and J.F. Anderson. 1986. The respiratory microenvironment within the burrows of gopher tortoises (*Gopherus polyphemus*). *Copeia* 1986: 787–795.

Wall, D.P., and S.P. Darwin. 1999. Vegetation and elevational gradients within a bottomland hardwood forest of southeastern Louisiana. *Amer. Midl. Nat.* 142: 17–30.

Waltman, J.R., and S.R. Beissinger. 1992. Breeding behavior of the green-rumped parrotlet. *Wilson Bull.* 104: 65–84.

Webb, S.L. 1987. Beech range extension and vegetation history: pollen stratigraphy of two Wisconsin lakes. *Ecology* 68: 1993–2005.

Welsh, H.H., and A.J. Lind. 1996. Habitat correlates of southern torrent salamander, *Rhyacotriton variegatus* (Caudata: Rhyacotritonidae), in northwestern California. *J. Herpetol.* 30: 385–398.

Westemeier, R.L., J.D. Brawn, S.A. Simpson, T.L. Esker, R.W. Jansen, J.W. Walk, E.L. Kershner, J.L. Bouzat, and K.N. Paige. 1998. Tracking the long-term decline and recovery of an isolated population. *Science* 282: 1695–1698.

Whittaker, R.H., F.H. Bormann, G.E. Likens, and T.G. Siccama. 1974. The Hubbard Brook ecosystem study: forest biomass and production. *Ecol. Monogr.* 44: 233–254.

Whitney, G.G. 1982. Vegetation-site relationships in the presettlement forests of northeastern Ohio. *Bot. Gaz.* 143: 225–237.

Wigley, T.B., and T.H. Filer Jr. 1989. Characteristics of greentree reservoirs: a survey of managers. *Wildl. Soc. Bull.* 17: 136–142.

Wilkins, R.N., and N.P. Peterson. 2000. Factors related to amphibian occurrence and abundance in headwater streams draining second-growth Douglas-fir forests in southwestern Washington. *For. Ecol. Manage.* 139: 79–91.

Williams, J.D., and G.K. Meffe. 1998. Nonindigenous species. In Mac, M.J., P.A. Opler, E.P. Haecker, and P.D. Doran (eds.), *Status and Trends of the Nation's Biological Resources.* U.S. Dep. Interior Geol. Survey, Reston, VA, pp. 117–130.

Zobel, D.B., A. McKee, G.M. Hawk, and C.T. Dyrness. 1976. Relationships of environment to composition, structure, and diversity of forest communities of the central western Cascades of Oregon. *Ecol. Monogr.* 46: 135–156.

5 Disturbance Ecology and Habitat Dynamics

Stuff happens — fires, hurricanes, volcanoes, floods, earthquakes. On average, approximately 450,000 ha are burned in the United States annually, over 1 million ha are affected by hurricanes and over 20 million ha are affected by insects and pathogens (Dale et al. 2001). The economic cost to society is over $1 billion/year in the United States (Dale et al. 2001). To most people these events are catastrophes. They can kill people and destroy property. They can be catastrophic for other organisms too. Wind uproots trees, fires burn dead wood, and floods erode streambanks; they can also be events that renew habitat for other species. Indeed, biodiversity conservation depends on disturbance. Wind adds dead wood to a forest, fires open the tree canopy and initiate a new forest, and floods create a new seedbed for willows and cottonwoods. Disturbances to forests have occurred for as long as there have been forests. Animals living in forests have adapted to many of these disturbances, and some species rely on disturbances to provide food, cover, and water for survival. Understanding the characteristics of disturbances and how disturbances influence habitat elements in stands and over forests can provide information that forest managers can use to provide habitat for selected species or to aid in conserving biodiversity. Knowledge of natural disturbances can help when developing silvicultural systems that might meet the needs of forest-associated wildlife (Franklin et al. 2002).

There are three characteristics of disturbances that can be used to understand potential effects on forest development, forest function, and the sizes, numbers, and distribution of habitat elements: size, frequency, and severity. Disturbance type is also important, with changes in habitat elements being quite different depending on the cause of the disturbance (e.g., fire vs. wind). Estimating these characteristics of natural disturbances can facilitate prediction of forest recovery and the subsequent development of vegetation structure.

DISTURBANCE SIZE

Disturbances come in many shapes and sizes. The size of a disturbance can influence animal species that either remain in or recolonize after a disturbance (Rosenberg and Raphael 1986). Nearly 3 million ha of forest burned in the United States in 2002, a particularly bad "fire-year." Some, such as the Biscuit fire in southern Oregon, were 200,000 ha in size. Most however were much smaller. In fact, there are usually many more small fires than large fires (Figure 5.1). Similar negative exponential distributions of disturbance size have been reported for wind disturbances (Foster and Boose 1992) and tree-fall gaps (Foster and Reiners 1986). Generally, there are many more small nonhuman disturbances than large disturbances across most forested landscapes.

Species that benefit from a disturbance seem to be associated with disturbances of different sizes. Black-backed woodpeckers, elk, and bison colonize forests following large severe fires as millions of dead trees and millions of kilograms of forage become available (Figure 5.2). On the other extreme, white-footed mice are favorably affected by openings of 0.1 ha in size, and decrease in abundance in larger openings (Buckner and Shure 1985). A similar species, deer mice (Figure 2.1), were not found in these small openings, but increased in abundance in larger openings (Buckner and Shure 1985). Disturbances also can increase the probability that certain invasive species might become established (Hobbs and Huenneke 1992). For species adversely influenced by a disturbance, the likelihood that they would be displaced by severe disturbances increases as the disturbance approaches and surpasses

65

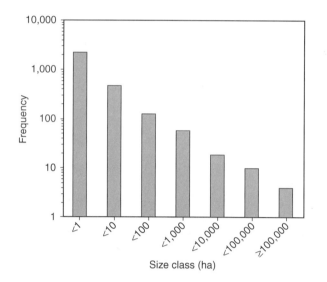

FIGURE 5.1   Histogram of the fire-size distribution for 2898 lightning fires during 1980–1993. Data are shown as counts on a logarithmic scale, in logarithmic bin widths. (From Cumming, S.G. 2001. *Can. J. For. Res.* 31: 1297–1303. With permission.)

FIGURE 5.2   Bison (shown here), elk, and other herbivores benefited from the large and severe fires in Yellowstone National Park.

the size of the its home range. But they may not be displaced if the disturbance is sufficiently small relative to the home range size (Hayward et al. 1999). Some species that inhabit forests have been able to persist by including fine-scale disturbances within their home ranges or by recolonizing stands of sufficient size that regrow to the point where necessary habitat elements become available.

Related to disturbance size is the shape of the area created by a disturbance. We will discuss edge effects in Chapter 11, but the function of an opening can be quite influenced by its departure from a circular shape because circles have the least edge per unit area of any regular shape. Consequently, two disturbances, both of 100 ha in size, may function quite differently for some species if one disturbance is circular and the other is shaped like an amoeba. More complex shapes can provide better opportunities for cover adjacent to food for some herbivores (Clark and Gilbert 1982) and can exacerbate the likelihood of colonization by invasive species (Cumming 2001).

Disturbance pattern is the spatial arrangement of the disturbance patches. Pattern is related to the size of a disturbance, but in addition to the areal extent of a stand or landscape affected by a disturbance, the pattern created by disturbance can influence the distribution of resources within and among potential home ranges of a species. Clumped distributions of fine-scale disturbances (e.g., clusters of root-rot pockets that kill trees) may result in a cumulative decrease in habitat availability within an individual's home range. A more random or uniform distribution of disturbances (e.g., lightening strikes) may allow that individual to tolerate the same disturbance density because only a small portion of any one individual's home range in the stand would be affected.

DISTURBANCE SEVERITY

Crown fires and volcanoes are severe disturbances. A tree-fall gap in a forest is not. The severity of a disturbance reflects the impact on the stand or forest. Is the stand completely replaced or are only a few trees removed providing more growing space to the remaining trees? The severity of a disturbance influences the amount of organic material destroyed and redistributed by the disturbance and hence the amount and form (living or dead) of material that remains after the disturbance. Mt. St. Helens was a severe disturbance, but even after its eruption, many pieces of the previous forest persisted. Trees were buried in ash, but so were seeds, fungi, and many species of animals that survived below the ground. These structures that remain following a disturbance may directly or indirectly provide the habitat elements needed for various species. Tree-fall gaps in later stages of forest development produce snags and logs, and sites where regeneration can become established in the openings allowing vertical complexity to increase. Similarly, following a hurricane or a fire, we see abundant dead wood and the establishment of a new cohort of plants and animals. The residual structures following severe disturbances such as these may persist into the next stand and provide the large trees and snags used by some species as the new stand grows (McComb and Lindenmayer 1999; Figure 5.3).

This legacy from the previous stand can be represented in the amount of dead wood, number of live trees remaining, depth of the leaf litter, and the vertical and horizontal complexity in the stand (Franklin et al. 2002, Spies 1998). The residual organic material that remains after a disturbance can influence the direction of succession and the rate of subsequent development following the disturbance (Franklin and Halpern 1989). Biological legacies such as these can also provide a seed source for the new stand and ensure sources of mycorrhizal fungi are available to reinnoculate the disturbed site (Dahlberg 2001). Legacies can also be particularly important to species later in the development of forests. Northern spotted owls, for instance, are associated with biological legacies such as remnant large old trees, which provide nest sites in many northwestern U.S. forests (North et al. 1998).

DISTURBANCE FREQUENCY

The frequency with which a disturbance occurs in a forest will influence the tree species composition and the amount of living and dead organic material present on the site over time. Hurricane frequency, for instance, can influence the proportion of large areas in early vs. late stages of forest development (Figure 5.4). Frequent, intense disturbances can delay the onset of forest development, or may

FIGURE 5.3 Legacy trees, snags, and logs retained following a timber harvest in the Blue River Ranger District, Willamette National Forest, Oregon. (Photo by Bruce McCune and used with his permission.)

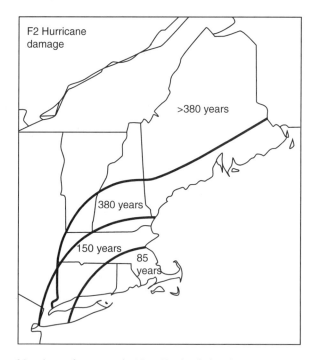

FIGURE 5.4 Zones of hurricane frequency in New England showing mean recurrence intervals between consecutive hurricanes of sufficient force to blow down forests. (From Boose, E.R. et al. 2001. *Ecol. Monogr.* 71: 27–48. With permission of the Ecological Society of America.)

preclude it. Infrequent, low-severity disturbances may lead to development of vertical structure in a stand. Frequency can be characterized in several ways: disturbance rate, percentage of a stand disturbed on an annual basis, or the return interval (time between disturbances). The return interval for disturbances of certain types and intensities varies considerably among forest types. Fire return intervals may be as frequent as once every 2 to 5 years in some savannah systems (Harrell et al. 2001), and as long as once every 300 to 400 years in northwestern coniferous forests (Wimberly et al. 2000). Return intervals between management events in managed forests also vary, and when the return interval for a managed forest departs significantly from the return interval that has occurred naturally for thousands of years, then the risk of losing habitat elements and associated species can increase (Hansen et al. 1991).

DISTURBANCE FREQUENCY, SIZE, AND SEVERITY RELATIONSHIPS

Imagine a forest 1000 ha in size and uniformly old and unmanaged. Then consider the frequency with which tree-fall gaps occur in that forest. Big trees die and fall, creating an opening in the forest probably many times each year. Disturbance sizes are small, severity is low (little biomass removed), and frequency is high. Now consider the frequency with which a stand-replacement fire or hurricane might occur in that forest. This may occur on average, perhaps, once every 100 years? 200 years? or longer? But when it does occur, it may affect the entire 1000 ha and be quite intense. Although a generalization, frequent small-scale disturbances are often of low severity. In general, large, severe disturbances are infrequent. In situations in which we may prevent a disturbance from occurring as frequently as it might ordinarily occur, the severity can be unusually high. For instance, by controlling and preventing fire in many western U.S. forests, trees become more susceptible to insect defoliation and the accumulation of large fuels. So when a fire does occur then it is unusually large and intense. Indeed, balancing disturbance frequency, severity and size through natural disturbances and silviculture is a key to providing habitat elements, water, timber, and other ecosystem services from these forests.

STAND DYNAMICS

A fire burns a forest, a hurricane blows over half of the trees in a stand, or an ice storm causes damage to a northern hardwood stand, reducing timber quality and value. How do stands respond to these disturbances? Early ecological perceptions of vegetation change following a disturbance provided the basis for the concept of ecological succession (Palmer et al. 1997), which presumed that there was a somewhat predictable change in the structure and composition of a stand following a stand-replacement disturbance. Although subsequent ecological research confirms that the recovery process in forests is not so deterministic (Palmer et al. 1997), simple concepts of ecological succession are a place to start to understand how forests change following a disturbance. Forest development is a continuum that we often break into arbitrary classes to help us simplify the complexity of forest change. After disturbances, forests develop through four general physiognomic stages: *stand initiation, stem exclusion, understory reinitiation*, and *old growth* (Oliver 1981, Oliver and Larson 1990, Smith et al. 1997) (Figure 5.5). Disturbance severity and frequency partially determine which species will dominate the forest during each of these stages. It is important to keep in mind, however, that these stages are merely convenient ways of understanding a sere or continually changing set of forest communities during stand regrowth following a disturbance.

STAND INITIATION

A disturbance usually kills or damages trees and consequently creates both growing space for the remaining trees or sites for germination and growth of new trees. Forests often contain large amounts of dead wood and live plants following most natural disturbances, and the disturbance often creates a suitable site for regeneration by seedlings (regeneration from seed) or sprouts (vegetative regeneration). This stage of stand development is called *stand initiation*. In many forests, regeneration

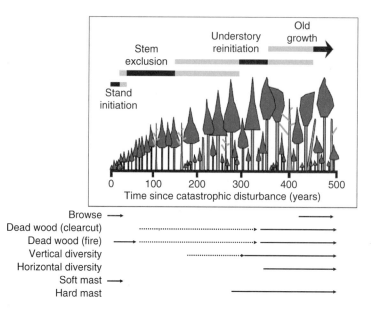

FIGURE 5.5 A conceptual timeline portraying developmental stages for temperate rain forests of southeast Alaska. Shaded bars represent temporal overlap among developmental stages. Relative importance of Hasitat elements indicated with arrows; dashed = present, solid = abundant. (From Nowacki, G.J., and M.G. Kramer. 1998. The effects of wind disturbance on temperate rain forest structure and dynamics of southeast Alaska. USDA For. Serv. Gen. Tech. Rep. PNW-GTR-421.)

occurs naturally through *advance regeneration* present in the stand prior to the disturbance or natural regeneration that occurs from seedling establishment or sprouting following the disturbance. Forest managers may choose to control the species and number of regenerating trees by planting seedlings at a particular spacing. This *artificial regeneration* also controls the spacing of seedlings. The early growth and survival of regeneration can be controlled ensuring that the future stand will be composed of the trees species and trees sizes desired by the land manager. Plantations of black spruce in Canada, Douglas-fir in the Pacific Northwest, and loblolly pine in the Gulf Coast are examples of artificial stand initiation. We will discuss this process in more detail in Chapter 6. In many forests, planting trees is unnecessary because it is expensive and there is abundant natural regeneration. Following an intense disturbance in central and northern hardwood forests of the eastern United States, it is common to expect 8,000 to 100,000 woody stems/ha (including trees and shrubs) within a few years following a disturbance (Annand and Thompson 1997, Schuler and Robison 2002). Unoccupied growing space for plants following a disturbance will become occupied if water, light, and nutrients are sufficient. From the standpoint of providing habitat elements, this early stage of stand development tends to be a grazing-based system, with much of the net primary production available to herbivores through forage. Foraging also occurs below the ground, with species such as pocket gophers feeding on root systems of the newly established plants. However, if the disturbance killed but did not remove trees from the previous stand as might occur in a low-severity fire or a windstorm, then much of the energy available in the dead wood is also available to other organisms through wood decomposition.

Trees remaining after a disturbance, or on the edges of disturbances, if not too severely damaged by the disturbance are suddenly free from competition by other trees and can respond by expanding their crowns or root systems and growing rapidly (Chen et al. 1992). In low-severity disturbances, which tend to be more frequent, this "thinning effect" in forests allows the remaining trees to continue to grow rather than succumb to competition. In so doing, large trees develop in these stands and, when they die, larger snags and logs are produced than would likely be produced at the same time in the absence of disturbance.

As plants begin to regenerate a site following a disturbance, eventually all the growing space in the site will be occupied. Foresters often use *basal area* as an index to the area of the stand occupied by trees. Basal area is the cross-sectional area of all trees on an acre or hectare. So, imagine a hectare of forest (an area of 10,000 m^2 or a square 100 m on each side; 2.47 acres). If you cut all of the woody stems off at 1.4 m (4.5 ft) above the ground and measured the area of all the cut surfaces and summed these areas, then you would have an estimate of basal area for the hectare. The maximum basal area that a site can support will depend on moisture, growing season length, tree species, and nutrients, among other things. Through stand development, basal area will increase rapidly at first and then slow and finally reach a point where it fluctuates around a certain upper level. Two very similar sites can support the same basal area but can have very different tree densities. One site can have high basal area in a few large trees and the other site can have the same basal area represented in many small trees. Once *stocking*, or the degree to which a site is occupied by trees of various sizes, reaches a certain combination of tree density and basal area, then competition between the trees begins to greatly influence the structure and dynamics of the stand.

STEM EXCLUSION

Once all of the growing space is occupied and plants begin to compete for light, soil moisture, or nutrients, then competition for those sparse resources begins. Because characteristics such as growth rates, shade tolerance, and moisture tolerance vary among species, some plants are better able to use the resources of the environment than others. The plants that are not as fit in this competitive environment begin to grow slower, and eventually die. The trees that remain during this process begin to stratify into crown classes that represent the various abilities of the trees in the stand to cope with competition (Figure 5.6). *Dominant* trees (D) are the most fit in this environment and they form the uppermost canopy. They typically receive sunlight from above and the sides, and have deep crowns and hence high leaf area (the area on the ground covered by leaves in the trees overhead), and so they can photosynthesize well and grow rapidly. *Codominant* trees (C) receive sunlight from above, and so grow well and form most of the primary canopy layer. Codominant trees may have somewhat smaller crowns and may not be growing as fast as the dominant trees, but still contribute significantly to stand structure. *Intermediate* trees (I) are those with smaller crowns and grow more slowly and may contribute little to the upper canopy. *Suppressed* trees (S) grow very slowly and have small crowns — they often die due to lack of light or moisture. *Wolf* trees (L) are those legacy trees from the previous stand that may have very deep crowns because they were open-grown during the stand-initiation phase.

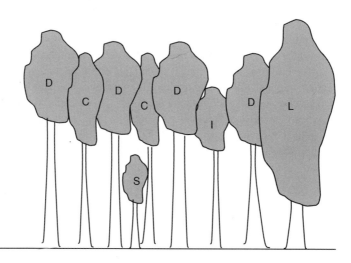

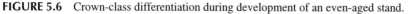

FIGURE 5.6   Crown-class differentiation during development of an even-aged stand.

The stand eventually develops a uniform canopy, further limiting the light available to plants beneath the canopy. During this stage, we see vertical structure become simplified into one dominant canopy layer and forage resources decline dramatically. In hardwood forests of the eastern United States, it is common to see 90 to 99% of woody stems die during these early stages of stand development. Although there can be instances where insect herbivory in these dense stands are high (e.g., spruce budworm irruptions), most of the energy available for animals is through decomposition and the forest is now a detrital-based system. However, because most of the trees that die are smaller than the dominant and codominant trees, pieces of dead wood produced during this phase are small and decay rapidly.

UNDERSTORY REINITIATION

As the trees in a stand begin to continue to differentiate into crown classes, and some of the larger trees begin to die, the openings created by the dead trees create gaps in the canopy, allowing sunlight to enter to the forest floor. As sunlight enters, the shade-tolerant understory plants can begin to establish and we begin to see a somewhat more complex forest structure. In many cases, seedlings will become established on the forest floor without a noticeable gap and survive until a gap is formed nearby, giving them a boost of growth. The gap may close and the shade-tolerant tree may have to wait for one or more additional gap events to reach the canopy. Less commonly, establishment may come from seeding into a tree-fall gap from the surrounding trees, or may be the result of seeds stored in the soil in a *seed bank* waiting for the correct light and moisture conditions to allow germination. Forage resources begin to return to the stand, although they are often low in quantity and quality. Any dead wood remaining from the disturbance that created this new stand has largely decayed by this time. Consequently, dead wood abundance is quite low at this stage, although a few large snags and logs start to form in the stand both through competition and from insects, disease, wind, and other small-scale disturbances.

OLD GROWTH

As regeneration in the tree gaps grows and eventually replaces the dominant and codominant trees that have died in the stand, a new stand structure begins to develop. Old growth or a *shifting-gap phase* allows a structurally more complex stand to develop. As trees age and die or are affected by diseases, insects, and other factors that lead to individual tree death in the stand, the regeneration periodically forms in gaps throughout the stand representing gap-phase regeneration. The stands begin to accumulate large pieces of dead wood, have trees of a variety of sizes and species, are vertically more complex, and increase in horizontal patchiness. Although the primary energy pathway in these stands is still through decomposition, small patches of forage develop. These forage patches in old forests can be quite variable in quality depending on the species and growth rates of the plants that fill the gaps. Frequent low-intensity fires, for instance, can create "gapiness" in forests and maintain a high level of forage availability (Figure 5.7). In some situations tree seedlings growing in partial shade have higher levels of defensive chemicals in their foliage than seedlings in direct sunlight (Tucker et al. 1976). However, Happe et al. (1990) found higher levels of defensive compounds in plants in clearcuts than in old-growth forests. Clearly these plant defensive strategies vary among forest types.

SUCCESSION AS A CONTINUUM OF HABITAT ELEMENTS

Stages are convenient ways of understanding stand dynamics. These stages are idealized and may occur at various spatial scales within and among stands. A continuous process of disturbance and

FIGURE 5.7   Repeated low-severity fires in some forests can increase forest gapiness, renewing forage availability for species such as this white-tailed deer.

regrowth occurs at various times and over various areas; so a continuum is the appropriate context in which to place these stages. Habitat elements and other aspects of forest structure and composition also change over this continuum (Figure 5.5). Following a stand-replacement disturbance such as a fire or hurricane, forage and soft-mast production are typically highest in the earliest stages of development and decline dramatically as crowns close. Once gaps begin to form as large trees die, a modest recovery of these elements can be expected. Large pieces of dead wood also are most abundant following a disturbance, but they decay and are absent in the middle stages of succession. Abundant dead wood is added once again as large trees begin to die late in stand development. Hard-mast production begins to occur once mast-bearing tree species reach maturity and develop large crowns during middle to late stages of development. Spies (1998) provided two generalized curves to understand some of these changes in forest structure and composition over time following a stand-replacement disturbance in Douglas-fir forests (Figure 5.8). Similar trends can be expected in other forest types, although the time scale would likely be different for other forests. Elements of forest structure that follow curve 1 (U-shaped) include the amount of dead wood, horizontal complexity, plant and animal species diversity, and susceptibility to fire. Factors that follow curve 2 (sigmoidal) include the diversity of tree sizes, vertical complexity, average tree size, incidence of tree damage and hollow trees, leaf litter depth, surface area of bark per tree, and live plant biomass. In some forest types, curve 2 can also represent changes in dominance by shade-tolerant species.

SUCCESSIONAL PATHWAYS

The discussion of stand dynamics so far has assumed that there is a reasonably predictable change in structure and composition over time following a disturbance. This is consistent with the view of succession proposed by Clements (1916) nearly 100 years ago. But succession is not deterministic. Consider the theoretical changes in biomass following a disturbance that leaves soil bare (Figure 5.9). As plants occupy the site and grow, the plant community increases in biomass. Simultaneously, the

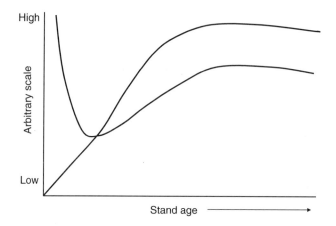

FIGURE 5.8   Generalized patterns of habitat element change over time in Douglas-fir forests. The time scale is different in other forest types. (From Spies, T.A. 1998. *Northwest Sci.* 72: 34–39. With permission.)

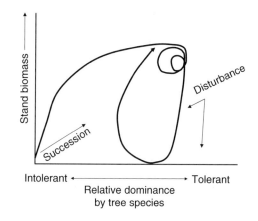

FIGURE 5.9   Theoretical changes in forest states during succession and disturbances of varying frequencies and intensities. (From Spies, T.A. 1997. Forest stand structure, composition, and function. In: Kohm, K.A., and J.F. Franklin (eds.), *Creating a Forestry for the 21st Century: The Science of Ecosystem Management.* Island Press, Washington, DC. With permission.)

plant community composition changes over time from one dominated by shade-intolerant species to more shade-tolerant species. An intense disturbance may set succession back to a point where there is no living biomass, but roots and seeds persist in the soil (e.g., an intense fire in a hardwood forest). If there are existing seedlings, a seed bank or a source for sprouts, then sprouts and seedlings may allow the site to recover with dominance by more shade-tolerant species than would have been there if these features were removed in a disturbance (a volcano or landslide). Disturbances that remove less biomass also recover but tend to be dominated by more shade-tolerant species. The availability of seed species in the soil seed bank, changing climate, soil pH, nutrient availability, and other factors continue to shift the final condition represented following full recovery after a disturbance (Figure 5.9). Variability in vegetation structure and composition following disturbances is to be expected because of the inherent variability in the severity, frequency, and size of disturbances affecting a forest as well as continual changes in climate, soils, and hydrology. Consequently, we often see a set of successional pathways that vegetation on a site could follow over time as disturbances occur and vegetation responds (Cattelino et al. 1979).

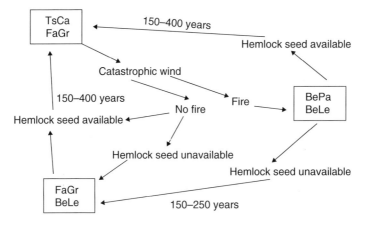

FIGURE 5.10 Successional pathways for a southern New England Forest. TsCa = eastern hemlock, FaGr = American beech, BeLe = black birch, BePa = paper birch. (Diagram courtesy of Anthony Damato and used with his permission.)

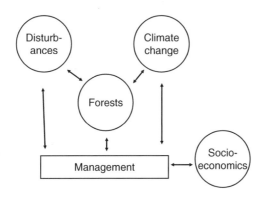

FIGURE 5.11 Interactions among forest disturbance, climate change, and management. (Redrafted from Dale, V.H., et al. 2001. *BioScience* 51: 723–734. With permission.)

In Figure 5.10, a site in which the potential vegetation is eastern hemlock–American beech may see a number of different vegetation states resulting from disturbance and regrowth. Following an intense hurricane and a subsequent fire or other disturbance that leaves bare soil, paper and black birch usually are the first tree species to dominate the site. Over time in the absence of a hemlock seed source, the site may move to a beech–birch stand. Should hemlock seed be available, then a beech-hemlock stand may develop. From this diagram, it should be clear that the same site can see several relatively long-lasting forest conditions that develop following disturbances and that these conditions each provide a different suite of habitat elements.

MANAGEMENT IMPLICATIONS FROM DISTURBANCES

Management activities do not replace natural disturbances but can be complementary to them. Knowledge of the frequency, severity, and size of natural disturbances, and the various successional pathways that emerge from them, can offer clues to management strategies that might be effective in achieving a variety of goals. Disturbances influence forest structure and composition, as do the myriad of physical factors such as climate (Figure 5.11). Management actions driven by the goals of society, be they economic, cultural, or spiritual, interface with the dynamic processes of disturbance

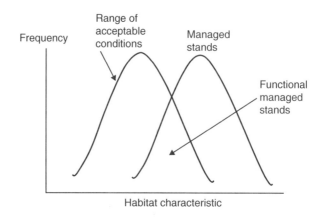

FIGURE 5.12 Hypothetical range of conditions in a habitat element that might be seen during stand development following a natural disturbance and during stand development using silvicultural practices. (From McComb, W.C. 2001. Management of within-stand features in forested habitats. In Johnson, D.H., and T.A. O'Neill (managing editors), *Wildlife Habitat Relationships in Oregon and Washington*. Chapter 4. OSU Press, Corvallis, OR.)

and climate change. When considering the diverse suite of habitat elements that might be needed to maintain biodiversity within a forested region, it is often useful to use variability in forest structure and composition to our advantage whenever possible. Forest managers producing wood products to meet industry and societal needs want to minimize uncertainty in the production process. But from the standpoint of providing a suite of habitat elements, uncertainty and the variability that it produces is something that should be embraced and worked with and not avoided. No single stand-management system will precisely match the variability inherent in natural forests that result from a variety of disturbance regimes. The compatibility of management and habitat goals is scale dependent both in space and time, and management often occurs over a much narrower range of space and time than scales associated with natural disturbances. But some of the variation can be incorporated into the managed forest landscapes of the region by using a variety of silvicultural systems and forest-management strategies. The choice of these systems will depend on the biological, social, and economic objectives for the stand and the landscape, and they will imitate natural disturbances to varying degrees (Figure 5.12). Indeed, the basis for development of existing silvicultural systems for timber objectives was that these systems reflect the regeneration and growth strategies of the commercially important tree species in a region. Intensive timber management as currently practiced leaves less dead wood and noncommercial plant species than natural disturbances (Hansen et al. 1991), and so it may not imitate natural disturbances for other forest resources as well as it does for timber. Hence, the management strategies that include goals for habitat elements for certain species or for biodiversity conservation goals require consideration of more factors than are necessary for production of commodities. But it is commodity production that can pay for the management activities needed to achieve certain habitat goals. Hence, the two goals should be complementary. In the following chapters we will explore methods of stand and forest management that can achieve both commodity and habitat goals.

SUMMARY

On the template of the physical and cultural landscape from which vegetation arises, vegetation is further altered by disturbances. Disturbance severity, size, and frequency interact to influence the sizes, amounts, and distribution of habitat elements such as vertical complexity, forage, dead wood, horizontal complexity, and plant species composition. Vegetation regrowth following disturbance in the very simplest sense follows several stages of stand development following a stand-replacement disturbance: stand initiation, stem exclusion, understory reinitiation, and shifting-gap phase. These conditions occur in various scales of time and space and represent somewhat arbitrary points on the continuum of successional change. Indeed, a variety of successional pathways can be seen following a disturbance depending on the severity and frequency of disturbance and the regrowth potential of the vegetation. Seed and sprout availability, shade or moisture tolerance, and time interact to determine what plant community is likely to develop and be maintained on a certain site following disturbances of varying intensities and frequencies. This knowledge of disturbance and succession can be used to craft management strategies to achieve multiple goals.

REFERENCES

Annand, E.M., and F.R. Thompson. 1997. Forest bird response to regeneration practices in central hardwood forests. *J. Wildl. Manage.* 61: 159–171.

Boose, E.R., K.E. Chamberlin, and D.R. Foster. 2001. Landscape and regional impacts of hurricanes in New England. *Ecol. Monogr.* 71: 27–48.

Buckner, C.A., and D.J. Shure. 1985. The response of *Peromyscus* to forest opening size in the southern Appalachian Mountains. *J. Mammal.* 66: 299–307.

Cattelino, P.J., I.R. Noble, R.O. Slatyer, and S.R. Kessell. 1979. Predicting the multiple pathways of plant succession. *Environ. Manage.* 3: 41–50.

Chen, J., J.F. Franklin, and T.A. Spies. 1992. Vegetation responses to edge environments in old-growth Douglas-fir forests. *Ecol. Applic.* 2: 387–396.

Clark, T.P., and F.F. Gilbert. 1982. Ecotones as a measure of deer habitat quality in central Ontario. *J. Appl. Ecol.* 19: 751–758.

Clements, F.E. 1916. Plant succession: an analysis of the development of vegetation. Carnegie Inst., Publication 242. Washington, DC.

Cumming G.S. 2001. A parametric model of the fire size distribution. *Can J. For. Res.* 31:1297–1303.

Dahlberg, A. 2001. Community ecology of ectomycorrhizal fungi: an advancing interdisciplinary field. *New Phytol.* 150: 555–562.

Dale, V.H., L.A. Joyce, S. McNulty et al. 2001. Climate change and forest disturbances. *BioScience* 51: 723–734.

Foster, D.R., and E.R. Boose. 1992. Patterns of forest damage resulting from catastrophic wind in central New England, USA. *J. Ecol.* 80: 79–98.

Foster, J.R., and W.A. Reiners. 1986. Size distribution and expansion of canopy gaps in a northern Appalachian spruce–fir forest. *Vegetation* 68: 109–114.

Franklin, J.F., and C.B. Halpern. 1989. Influence of biological legacies on succession. Pages 54–55 In D.E. Ferguson, P. Morgan, and F.D. Johnson (eds.), *Proceedings: Land Classifications Based on Vegetation — Applications for Resource Management.* USDA For. Serv. Gen. Tech. Rep. INT 257.

Franklin, J.F., T.A. Spies, R. Van Pelt, A.B. Carey, D.A. Thornburgh, D.R. Berg, et al. 2002. Disturbances and structural development of natural forest ecosystems with silvicultural implications, using Douglas-fir as an example. *For. Ecol. Manage.* 155: 399–423.

Hansen, A.J., T.A. Spies, F.J. Swanson, and J.L. Ohmann. 1991. Lessons from natural forests. *BioScience* 41: 382–392.

Happe, P.J., K.J. Jenkins, E.E. Starkey, and S.H. Sharrow. 1990. Nutritional quality and tannin astringency of browse in clear-cuts and old-growth forests. *J. Wildl. Manage.* 54: 557–566.

Harrell, W.C., S.D. Fuhlendorf, and T.G. Bidwell. 2001. Effects of prescribed fire on sand shinnery oak communities. *J. Range Manage.* 54: 685–690.

Hayward, G.D., S.H. Henry, and L.F. Ruggiero. 1999. Response of red-backed voles to recent patch cutting in sub-alpine forest. *Conserv. Biol.* 13: 168–176.

Hobbs, R.J., and L.F. Huenneke. 1992. Disturbance, diversity, and invasion: implications for conservation. *Conserv. Biol.* 6: 324–337.

McComb, W.C. 2001. Management of within-stand features in forested habitats. In Johnson, D.H., and T.A. O'Neill (managing editors), *Wildlife Habitat Relationships in Oregon and Washington.* OSU Press, Corvallis, OR.

McComb, W.C., and D. Lindenmayer. 1999. Dying, dead, and down trees. In Hunter Jr., M.L. (ed.), *Maintaining Biodiversity in Forest Ecosystems*. Cambridge University Press, Cambridge.

North, M., J. Chen, G. Smith, L. Krakowiak, and J. Franklin. 1996. Initial response of understory plant diversity and overstory tree diameter growth to a green tree retention harvest. *Northwest Sci.* 70: 24–35.

Nowacki, G.J., and M.G. Kramer. 1998. The effects of wind disturbance on temperate rain forest structure and dynamics of southeast Alaska. USDA For. Serv. Gen. Tech. Rep. PNW-GTR-421.

Oliver, C.D. 1981. Forest development in North America following major disturbances. *For. Ecol. Manage.* 3: 153–168.

Oliver, C.D., and B.C. Larson. 1990. *Forest Stand Dynamics*. McGraw-Hill, New York.

Palmer, M.A., R.F. Ambrose, and N.L. Poff. 1997. Ecological theory and community restoration ecology. *Restor. Ecol.* 5: 291–300.

Rosenberg, K.V., and M.G. Raphael. 1986. Effects of forest fragmentation on vertebrates in Douglas-fir forests. In Verner, J., M.L. Morrison, and C.J. Ralph (eds.), *Wildlife 2000: Modeling Habitat Relationships of Terrestrial Vertebrates*. The University of Wisconsin Press, Madison, WI.

Schuler, J.L., and D.J. Robison. 2002. Response of 1- to 4-year-old upland hardwood stands to stocking and site manipulations. In Outcalt, K.W. (ed.), *Proceedings of the Eleventh Biennial Southern Silvicultural Research Conference*. USDA For. Serv. Gen. Tech. Rep. SRS-48.

Smith, D.M., B.C. Larson, M.J. Kelty, and P.M. Ashton. 1997. *The Practice of Silviculture: Applied Forest Ecology*, 9th ed. John Wiley & Sons, New York.

Spies, T.A. 1997. Forest stand structure, composition, and function. In Kohm, K.A., and J.F. Franklin (eds.), *Creating a Forestry for the 21st Century: The Science of Ecosystem Management*. Island Press, Washington, DC.

Spies, T.A. 1998. Forest structure: a key to the ecosystem. *Northwest Sci.* 72: 34–39.

Tucker, R.E., W. Majak, P.D. Parkinson, and A. McLean. 1976. Palatability of Douglas-fir foliage to mule deer in relation to chemical and spatial factors. *J. Range Manage.* 29: 486–489.

Wimberly, M.C., T.A. Spies, C.J. Long, and C. Whitlock. 2000. Simulating historical variability in the amount of old forests in the Oregon Coast Range. *Conserv. Biol.* 14: 167–180.

6 Silviculture and Habitat Management: Even-Aged Systems

The forest disturbances described in the previous chapter continue to influence the structure and composition of forests. Shift happens. But, how contemporary human societies chose to use the forest and manage them for various species has an ever-increasing influence on the structure, function, and composition of forests. Although culturally induced pressures of development, invasive species proliferation, and climate change will continue to determine if we have forests to use for other purposes, two forest management practices currently have a significant impact on habitat for many species: silviculture and fire management (both fire protection and prescribed burning). *Silviculture* is the art and practice of managing forest stands to achieve specific objectives for the landowner or land manager. These objectives could include timber production, recreation, habitat for various wildlife and fish species, biodiversity conservation, aesthetics, and nontimber forest products. Indeed many nonindustrial forest landowners in the United States own and manage land for reasons other than timber production. But it is the economic value of the forest that allows many landowners to manage their forests to achieve other objectives. Consequently, managing forest lands for habitat requires a basic understanding of silvicultural principles to be effective in achieving habitat goals. Similarly, foresters charged with recruiting, maintaining, or removing various habitat elements must understand how their silvicultural prescriptions are likely to influence habitat elements.

SILVICULTURE AS A FOREST DISTURBANCE

Silvicultural activities are forest disturbances and as such have particular sizes, severities, frequencies, and patterns across a landscape. These human-caused activities interface with the suite of natural disturbances, which also have various sizes, severities, and frequencies. Forest managers can select the types and rates of disturbances that will meet specific resource objectives. Some habitat management issues that foresters and wildlife biologists face may result from insufficient consideration of the size, frequency, severity, and patterning of silvicultural disturbances on a landscape. A range of management decisions can be made on any given site that will result in stand conditions and plant communities that support only certain species (Figure 6.1).

Consider, for example, an Oregon Coast Range site managed with the following combination of decisions: clearcut; no retention of logs, snags, or green trees; no site preparation; rely only on natural regeneration; no vegetation management; and no precommercial thinning. The result would probably be a red alder or salmonberry plant community with few Douglas-firs or other conifers. A similar approach in the Gulf Coastal Plain would likely result in an area dominated by oaks and sweetgum but with few pines. In northern hardwoods, however, birch, beech, and maple would probably dominate the site, and in spruce-fir-dominated boreal forests where balsam fir advance regeneration occurs, the next stand would be dominated by balsam fir. Now consider the same sites managed with the following decisions: clearcut, snag legacy, plant the commercially important conifer of the region, seedling release with herbicides, and thin to 750 trees/ha (300 trees/acre). The result would probably be a conifer-dominated stand condition with a grass–forb understory (sparse shrubs) during the early stages of stand development. Each condition is habitat for a different suite

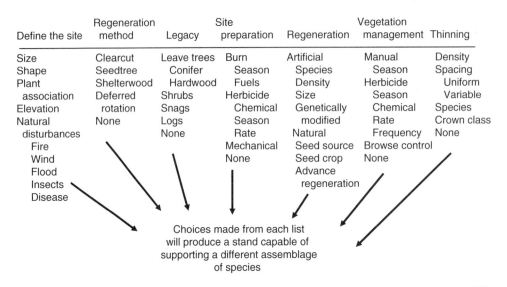

FIGURE 6.1 Decisions made by a land manager from each list will lead to development of very different forest structure and composition on a site. These choices are reflected in the availability, number, and sizes of habitat elements. (Redrafted from McComb, W.C. 2001. Management of within-stand features in forested habitats. In Johnson, D.H., and T.A. O'Neill (managing editors), *Wildlife Habitat Relationships in Oregon and Washington*, Chapter 4. OSU Press, Corvallis, OR. With permission from the Oregon State University Press.)

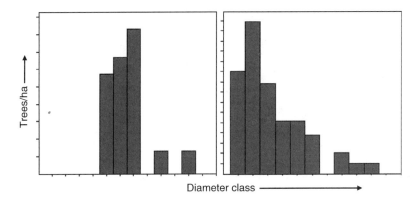

FIGURE 6.2 Distribution of live tree diameters in an even-aged stand with one species, white pine (left) and with multiple species (right) at 60 years of age as predicted using the Landscape Management System (McCarter et al. 1998).

of species on the same site managed in one of two ways. Many possible decisions could be made early in stand development that could produce a wide range of stand conditions (Figure 6.1).

Once the site characteristics and stand objectives have been identified, silviculturists usually first decide whether to use an even-aged or uneven-aged silvicultural system to regenerate and maintain the stand. These two systems ensure that the desired species of new young trees will replace the trees that are cut and that they grow to achieve the landowner's objectives. Even-aged stands are those in which all or most of the trees in the stand are approximately the same age. If the trees in an even-aged stand are also all the same species, then at least during the early stages of development most of the trees will be of similar diameter and height (Figure 6.2). Hence, the ideal distribution of tree diameters in an even-aged stand is a bell-shaped curve, but often there is some departure from a normal curve. If the new stand is composed of a variety of tree species, all with somewhat different growth rates, then a different pattern emerges. Within the first few decades, the distribution of tree diameters will begin to depart from a bell-shaped curve as some trees grow rapidly in diameter and height and others grow more slowly (Figure 6.2). From the standpoint of considering vertical structure, horizontal complexity, forage resources, and other habitat elements, it is often more useful to think about the stand size-class distribution than the age distribution in mixed-species stands.

Uneven-aged stands are those in which there are typically three or more cohorts or age groups represented simultaneously in the same stand. Typically, there are many more small trees than medium-sized trees and many more medium-sized trees than large trees, and so the distribution of tree sizes is similar to an even-aged, mixed-species stand (Figure 6.2). We will cover uneven-aged management in the next chapter.

CHARACTERISTICS OF EVEN-AGED STANDS

Even-aged stands typically develop through the stages of stand development described in the previous chapter: stand initiation, stem exclusion, understory reinitiation, and shifting-gap phase. Depending on the goals for a stand, such as pulpwood, sawtimber, aesthetics, or habitat for certain species, this sequence of stand developmental stages may or may not be truncated. For instance, a manager growing pulpwood and providing habitat for species typically found in the stem exclusion stage may never allow the stand to enter the understory reinitiation or shifting-gap-phase stages.

There are three classic methods to establish even-aged stands that will be described in more detail in the following section: clearcut, seed tree, and shelterwood. These methods can be used to initiate the stand development from stand initiation through shifting-gap stages. Depending on the goals for the stand, natural disturbances can be allowed to further modify the stand structure over time if the manager decides not to salvage the dead and damaged trees in the stand.

In this chapter, we will also cover two-aged stands as a modification of even-aged systems. Two-aged stands are not explicitly planned for in traditional silviculture methods, but can be quite useful when trying to retain a legacy from the previous stand or to recreate some aspects of natural disturbances such as fire and wind. Because these structures apparently occur quite commonly following natural disturbances and because they have many favorable aspects from the viewpoint of habitat element complexity, they should be standard options to be considered when planning silvicultural treatments.

CONSIDERING THE CAPABILITIES OF THE SITE

Before deciding which regeneration method to use to achieve silvicultural goals, the site must be evaluated. The size of the stand, its shape, access, slope, aspect, soils, and all of the other physical variables discussed in Chapter 4 must be taken into consideration. These factors will not only influence the type of vegetation the site can support but also the function of the vegetation in providing habitat elements on that site. For instance, an opening placed on a dry, south-facing slope with rocky ledges may be ideal for increasing habitat quality for some reptile species, but may not be nearly so effective on a north-facing slope. Forest management along streams may influence the water temperature and sedimentation rates, but future stand development may be affected by beavers. So, before planning any silvicultural activities, the potential for the site to support certain plant and animal species should be assessed. Ask yourself if the goal that you set for the area can be achieved given the physical environment in which it occurs. If your goal could be achieved, then you should ask if it can be achieved at an appropriate cost of time and money. For instance, converting a forest in Washington to a salmonberry field, one in New England or a mountain laurel field, may be easy and cost-effective (at least in the short run) when harvesting all trees and releasing these shrubs. On the other hand, converting these shrub fields to forests may simply not be possible without a significant initial investment of time and money.

The dominant disturbances in the area must also be considered. If your goals for the site are compromised by a high probability of incidence of root disease, insect defoliation, browse, fire, ice, or wind, then alternative goals might be more appropriate. An excellent example is the inability to establish adequate regeneration of many commercial tree species in hardwood stands in many parts of Pennsylvania simply because of deer browsing (Marquis 1974). Attempting to grow eastern hemlock where hemlock wooly adelgid is likely to occur is another example of an impending disturbance that would compromise your ability to achieve a habitat goal. Similarly, growing dense conifer stands in wind-prone or ice-prone environments may cause much of the stand to break or be blown over (Bragg et al. 2003). So, be sure that your goals for an area are consistent with the ability of the site to allow you to achieve those goals effectively and efficiently.

CHOOSING A REGENERATION METHOD

Clearcutting is probably the most commonly used silvicultural regeneration method designed to initiate an even-aged stand. This term should not be confused with a harvesting system. Harvesting systems are the means of removing the trees from the site to a landing where they can be taken to a mill. Harvesting systems include horses, skidders, cable systems, and helicopters. *Clearcutting* is a silvicultural decision designed to ensure adequate growing space for regeneration of a new stand. It is usually done to allow establishment of a plantation, though occasionally direct seeding is used to establish a new stand.

The boundary of the area to be clearcut is typically marked and then all commercial (and often noncommercial) trees are cut. If noncommercial trees and shrubs are not cut, then they are usually killed after harvest to allow rapid growth of the desired species. The size of the clearcut is usually determined by a combination of ownership boundaries, economics, and law. From simply an economic standpoint, larger clearcuts are more efficiently established than smaller ones. There are fixed costs associated with harvesting, planting, vegetation management, and thinning, which are all reduced when work is in one large area rather than many small ones. State policies and federal land management plans often dictate the upper end of clearcut sizes that are allowed by law. Habitat concerns, especially minimum opening sizes for some species or adequate edge conditions for others, will also influence clearcut size and shape.

If natural regeneration can be assured by leaving some trees after the harvest to provide a seed source to the newly created growing space, then the *seed-tree regeneration method* can be used. This approach is particularly effective for species that produce abundant seeds on a regular basis. Loblolly and shortleaf pines may be regenerated using this approach because they have winged seeds that disperse from the seed trees, they produce seeds regularly, and, if the site is appropriate, sufficient seeds germinate and grow. Using this approach, dominant trees in the stand that have deep crowns and seem to have regular cone or seed production are identified prior to harvest and marked for retention. The number of seed trees retained depends on the dispersal capabilities of the seeds. For instance, in shortleaf pine, approximately 85% of the seeds fall within 50 m (155 ft) of seed-producing trees (Burns and Honkala 1990). Seed-tree spacing for this species can probably be as much as 50 m among trees, but closer if there is concern regarding seed production following harvest. Once regeneration is established, the seed trees can be removed or retained depending on the goals and objectives for the stand. Retaining seed trees can have beneficial effects for species that use widely scattered mature trees in a savannah-like structure, such as red-cockaded woodpeckers (Conner et al. 1991). If seed trees are retained, then there can be a reduction in growth of new seedlings for some tree species (e.g., longleaf pine; Boyer 1993).

Where newly germinated seedlings need some protection from direct sunlight, dessication, or frost, a *shelterwood* regeneration method may be needed. A shelterwood system is designed to provide shelter or protection of the regeneration (newly established trees) by leaving a sparse canopy cover — enough to provide protection but not so much canopy cover that seedlings do not receive enough sunlight to survive and grow. Hence, this technique is most often used with tree species that are not too shade intolerant. This system is used in central hardwood oak forests (Annand and Thompson 1997), ponderosa pine (Anderson and Crompton 2002), upland mixed pine hardwood stands, and mixed-species jack pine stands, among others. By retaining a somewhat denser overstory to provide protection, some species that cannot survive in seed-tree stands can tolerate shelterwood stands (Taulman and Smith 2004). Establishing regeneration using the shelterwood system usually takes two to three steps. First, a *preparatory harvest* is made to thin the stand and allow sunlight to strike the crowns of the dominant trees. This allows trees to increase in diameter and crown size, which increases the probability of having adequate seed production for many tree species (Dey 1995). Once seed crops seem likely or sufficient, an *establishment or seed cut* is made to further release the trees with abundant seed production and to provide growing space for the new even-aged stand of seedlings. Finally, once the seedlings are well established and growing, and can tolerate full sunlight or frost, the overwood, or overstory trees, are removed. At least some *overwood removal* is usually necessary to allow the newly established regeneration to grow rapidly, but the level of removal is dependent on the goals for the stand.

Another type of even-aged method that has been proposed as a way of retaining some structure in these even-aged stands is the *deferred rotation* method described by Smith et al. (1986). Silviculturists also will refer to this as *clearcut with reserves*. This system retains some trees through two complete growing cycles, or rotations, in the stand, which benefits certain species of animals by allowing some open-grown trees to grow large and old (Chambers et al. 1999a, Thompson and Dessecker 1997). Whichever system is chosen, the stand usually proceeds through site preparation, stand reestablishment, vegetation management, and stand-density management before it is ready for harvest at the end of the *rotation*, or growing cycle. Decisions at each stage influence stand structure and composition and, in turn, habitat quality for the wildlife species present at various stages of stand development (Figure 6.1).

The selection of a regeneration method will have an effect on the stand structure during the early stages of stand development. Seed-tree and shelterwood systems at least initially leave vertical structure until the seed trees and overwood are removed wholly or in part. Regardless of whether natural or artificial regeneration is used to establish the new stand, the newly cleared area produces a flush of grasses, forbs, shrubs, as well as small trees. The duration of this stand-initiation phase, where herbaceous plants occur and can dominate also, can be influenced by the choice of regeneration method. Shade provided by overwood can limit production of herbs and forbs (Graham and Jain 1998). Because of the difficulties associated with vegetation management when residual trees are present, early vegetation control may not be possible in shelterwood stands. Consequently, the duration of the shrub condition in shelterwood stands may be longer than in clearcuts where shrubs are often controlled using herbicides.

Woody and herbaceous plant species associated with these even-aged regeneration methods are often very to moderately shade intolerant, although this can be adjusted with overwood retention in shelterwood systems and with artificial regeneration. Consequently, both the structure and composition of stands can be quite variable and diverse in the early stages of stand development depending on the degree to which trees are retained from the previous stand. Even-aged systems also produce stands that often create abrupt edges depending on the stature of adjacent stands. These high-contrast edges are beneficial to some species (McGarigal and McComb 1995) but not to others, especially some species of amphibians (Martin and McComb 2003).

IDENTIFYING LEGACY ELEMENTS TO RETAIN

During a harvest, the land manager may decide to leave certain structural components of the previous stand on the site and into the next rotation. This legacy from the previous stand is a means of adding more complexity to the new stand and allowing some of the structure, composition, and processes from the previous stand to be carried forward into the new stand. Snags are the most visible type of legacy left or created on many sites and they can significantly influence occurrence of a number of cavity-nesting species (Bull and Partridge 1986, Chambers et al. 1997, McComb and Lindenmayer 1999, Schreiber and DeCalesta 1992). Logs, living conifers, and hardwoods (and the lichens, bryophytes, fungi, and other species associated with them) left on the site can provide structural and compositional features that create conditions in the new stand more typical of those found after natural disturbances. Recall that, after most stand replacement natural disturbances, there is a considerable carryover of dead wood, and also some live trees and shrubs into the new stand. Animal communities associated with stands that include these features probably would be more complex than communities in stands that lack similar components (Chambers et al. 1999a, Thompson and Dessecker 1997). These features, however, may interfere with site preparation, vegetation management, and growth rates of the trees in the new stand, and so they may only be desirable where land-management goals include resources besides timber.

Legacy trees are retained during overstory removal cuts from one rotation through most or all of the next rotation, with the new stand growing up around the reserve trees. These trees are not the same as seed trees. Seed trees are specifically left to provide a seed source for the next stand, and then they may or may not be removed. When seed trees are retained, they function in much the same way that a green-tree retention stand might (Sullivan and Sullivan 2001). But retention trees are specifically left as legacy structure in a stand, though they may also provide seeds to the site. Green-tree retention and clearcut stands in Oregon seem to provide similar habitat for many bird species when <25 trees/ha (<10 trees/acre) are retained in retention stands (Chambers et al. 1999b). However, more species of birds typically use retention stands than clearcuts, at least in the breeding season (Chambers et al. 1999b, Vega 1993). However, more species were detected in uncut stands than in clearcut or retention stands during the winter (Chambers et al. 1999b). Other investigators reported similar findings in that animal response to the retained structures were variable and dependent on the number of trees retained and their spatial pattern. The degree to which retention trees function in older stands that are nearing rotation age have not been so well studied.

SITE PREPARATION EFFECTS ON HABITAT ELEMENTS

Once the site is harvested and legacy structures have been left, often the site must be prepared as a suitable seedbed or planting spot for the new regeneration. Site preparation may not be necessary on some sites, especially if advance regeneration is adequate. On the other hand, to ensure successful survival and growth of new seedlings in plantations, site preparation can be quite intense. When managing southern pines or Douglas-fir, it is not uncommon to see a site burned and then mechanically manipulated to prepare the site. In the southern United States, mechanical site preparation may include bedding (plowing soil into raised beds where the trees are planted), chopping with a large rolling drum behind a bulldozer, or scraping the unwanted vegetation into piles or windrows with a bulldozer. This type of scarification may significantly affect the below-ground structure of the stand by temporarily removing burrows and compacting soils. Burrowing species such as gopher tortoises in the southern Gulf Coast (Figure 3.9) may be particularly affected by such activities. The intensity of site preparation also may affect plant communities that develop after the disturbance. Intense scarification, burning, or some herbicides may reduce shrub development in subsequent stages of stand development. Alternatively, use of light fires without mechanical site preparation may proliferate the sprouting of shrubs and affect the presence or abundance of grasses and forbs in a young stand. The choice of site preparation influences not only the trajectory of the plant community that develops on the site but also the level of residual "legacy" that remains after the treatment. Intense burns or mechanical scarification, for example, will usually either reduce levels of dead wood on the site or concentrate it into piles and windrows. These intense treatments may be a desirable method for manipulating the habitat of some species that may influence regeneration success (such as mountain beaver), but it also may have adverse consequences for other species. In Oregon, species such as creeping voles and vagrant shrews increase after intense site preparation, but others such as Pacific and Trowbridge's shrews, ensatina salamanders, and Pacific giant salamanders decrease after site preparation (Cole et al. 1997, 1998).

NATURAL REGENERATION AND PLANTING OPTIONS

Land managers can determine the composition of developing stands by deciding which plant species will be reestablished after site preparation. In many northern hardwood and boreal forests, advance regeneration may already be present. Clearcutting simply releases the existing regeneration (actually an overstory removal) and provides adequate light, water, or nutrients for the small trees to begin growing. In addition, the added light and heat striking the newly exposed forest floor can lead to abundant natural regeneration through germination of seed in the soil (i.e., seed bank) and through sprouting of roots and stumps. The species of plants that regenerate following a clearcut can be quite diverse because of the availability of light, moisture and nutrients, and the relative lack of competition. Within a short period of time, however, all growing space is occupied and there are simply too many plants competing for resources for all to grow quickly. In some instances, these plants may not be the species desired to meet habitat or timber objectives. Consequently, artificial regeneration is often used in conjunction with even-aged management approaches.

Plantations help to ensure that the appropriate species of trees are regenerated on a site to meet future objectives. Plantations can be, and often are, of a single species. From the standpoint of growing a certain product for a market, growing one species to a uniform size during one time period is economically efficient, and for some animal species the conditions provided in these uniform stands is desirable. But for many species the simple structure and composition found in monoculture plantations does not provide the habitat elements needed for survival and reproduction (Kerr 1999). For a variety of reasons, mixed-tree-species plantations are becoming more and more common. Planting one species over large areas can lead to increased risk of crop loss due to insects or disease. Swiss needle cast causes Douglas-fir trees to become defoliated, but does not seriously affect western hemlock or Sitka spruce (Hansen et al. 2000). A species of invasive root rot is fatal to Port-Orford cedar but not Douglas-fir (Jules et al. 2002). Southern pine beetle may be a serious threat to loblolly or slash pines, but not longleaf pines (Burns and Honkala 1990). So, depending on the goal for the stand and the risks associated with losing certain species of trees or shrubs during stand development, mixed-species stands may carry less risk and also provide certain benefits in terms of plant diversity and vertical profile diversity as the stand develops.

Plantation establishment must not only consider the species to be planted but also the nursery stock to be used and the spacing of seedlings in the plantation. Typically tree seedlings that are lifted from a nursery are out-planted after 1 to 2 years in nursery. Seedlings with larger root collar diameters (stem size at the ground line) survive and grow faster than smaller seedlings (Rose and Ketchum 2003, South et al. 1995). Genetic variability and genetic resistance to insects and disease can also be considered at this period of plantation establishment. By selecting planting stock from local seed sources adapted to the local conditions, trees are more likely to survive and grow well. Genetically modified (GM) tree seedlings may also be used, and it is unclear what the long-term implications of using GM plants might be to development of habitat elements in plantations. Current efforts are focused on developing herbicide- and disease-resistant strains of trees so that competing vegetation can be more effectively controlled and trees can survive common diseases. If control of competing vegetation can be made more effective, then availability of grasses, forbs, and shrubs would be reduced significantly during the earliest stages of stand reestablishment. There is also the potential to use GM trees to influence the concentrations of secondary metabolic compounds such as phenols and tannins in seedlings and make them less palatable to vertebrate and invertebrate herbivores. The risks associated with using GM materials include the potential for gene escapement to native plants that may have unwanted and unexpected effects on their growth, survival, and value as food to herbivores (Johnson and Kirby 2001).

Seedlings grown in a nursery are often rich in nutrients to give them the best opportunity to grow and survive when they are out-planted. In addition, newly planted seedlings may be fertilized or mulched to ensure survival and growth. These treatments tend to increase their value as browse for some species because they represent a nutrient-rich source of food for some herbivores (Crouch and Radwan 1981). In many circumstances, seedlings must be protected from browse by deer, elk, moose, voles, gophers, mountain beavers, and American beavers. But even intensive protection treatments such as placing vexar tubes around seedlings may not be effective in all circumstances (Brandeis et al. 2002).

One factor that can have a significant effect on the production of habitat elements later in plantation development is planting density. Planting density is easily adjusted during plantation establishment. Typically 750 to 1250 trees/ha are planted to establish a plantation. If trees are planted on a wide spacing then it takes longer for the crowns of the growing trees to begin to overlap one another. When crowns begin to overlap, sunlight is no longer available to other plants beneath them. Wide spacing can lead to a longer-lasting grass–forb–shrub phase during the initiation stage of stand development. It can also lead to deep crowns and abundant limbs low on the trees, thereby increasing the number of knots in the wood when it is harvested and reducing its value unless the limbs are pruned. Variable-density planting represents an even more creative way of building horizontal complexity into a stand and can lead to quite variable vertical complexity in the stand. Mixed-species planting or variable-density planting or both can provide heterogeneity in an otherwise homogenous system.

VEGETATION MANAGEMENT EFFECTS ON HABITAT ELEMENTS

Management of competing vegetation can significantly affect the availability of certain plant species as food and cover. Herbicide applications that release conifers can temporarily decrease the availability of shrubs for shrub-nesting birds (Easton and Martin 1998, Morrison and Meslow 1984). For small mammals and amphibians, site preparation seems to have a greater effect on animal abundance than spraying of glyphosate herbicide following plantation establishment (Cole et al. 1997, 1998). Herbicide applications that release conifers can temporarily decrease the availability of browse during the early stages of stand development, but they can also increase the availability of browse when applied later in stand development. Either chemical or manual vegetation control can influence the heterogeneity of developing stands. Spot control of competing vegetation could lead to a more heterogeneous stand than what is possible with broadcast application of a treatment. Manual control of many shrub species can lead to a proliferation of sprouts that increase amounts of available browse, but this may lead to an increase in the concentration of phenolic compounds in leaves and twigs.

How vegetation is controlled has different effects on different species. Santillo et al. (1989) found fewer small mammals on glyphosate-herbicide-treated clearcuts for 1 to 3 years after spraying compared with untreated clearcuts in Maine. Insectivores seemed to be most significantly affected by herbicide spraying in this study, probably because of the effect that herbicides had on the vegetation and habitat for forest floor invertebrates. Herbivores also seemed to be adversely affected until vegetation recovered from spraying (Santillo et al. 1989). Santillo et al. (1989) suggested that changes in mammal abundance seemed to be associated with herbicide-induced reductions in invertebrates and plant food and cover. Patches of untreated vegetation within herbicide-treated clearcuts may provide a source of invertebrates and plants for those species adversely affected by herbicide spraying.

Sullivan and Sullivan (2003) found that plant species diversity was not adversely affected by glyphosate spraying in young conifer plantations in British Columbia. Responses of plant species to herbicides in forest ecosystems differ from those in agroecosystems where glyphosate is used to repeatedly reduce noncrop vegetation (weeds) in most situations. In forests, some species of birds and small mammals decline temporarily following spraying, while other species increase in abundance. Management for a mosaic of vegetation conditions within forested landscapes should help ameliorate the short-term changes in species composition accompanying vegetation management using herbicides.

Perhaps the most profound effect of vegetation management is the influence that such activities have on the future composition and structure of developing stands. Lack of any vegetation management in many northwestern or Gulf coastal sites may lead to stands with a large component of shrubs or hardwoods that will benefit some species of wildlife (Huff and Raley 1991), but interferes with tree growth and stand development. Intensive vegetation management may lead to a conifer-dominated stand early in stand development with little shrub development unless stand density is manipulated as the stand develops. Hence, the balance between competition control and maintenance of nontimber plants requires careful thought and planning on the part of the silviculturist.

Prescribed burning is often used as a management tool to prepare sites for planting, prepare seedbeds, reduce competing vegetation, or reduce fuels and hence future fire risks. Fire is a natural and frequent event in many forest systems, and so prescribed fire would seem to be a natural surrogate for uncontrolled wildfires. The choice of using fire, herbicides, or mechanical tools in forest management, either individually or collectively, is highly dependent on the types of habitat elements that a manager wishes to retain, remove, or develop in a stand. The effects that fire has on forest floor litter and dead wood can lead to adverse effects on forest floor amphibians and mammals (Randall-Parker and Miller 2002, Schurbon and Fauth 2003). On the other hand, forage quality and availability is often enhanced for herbivores following prescribed burning (e.g., Canon et al. 1987, Sparks et al. 1998). Again, as with any other forest management approach, there are species that will likely benefit from the treatment and others that will not. The winners and losers in the proposed management must be carefully considered prior to initiating a prescription, or plan, for a stand.

PRECOMMERCIAL THINNING

Precommercial thinning involves cutting or killing trees to achieve the desired density that promotes the growth of the residual trees and the development of desired stand structure and composition. This type of treatment is done early in stand development before the trees are large enough to be sold for a profit or at least offset the cost of the thinning. Hence, this activity is a net cost to the landowner. The value of precommercial thinning is to concentrate growth on the remaining trees and allow them to grow faster. Further, it increases the space between the crowns of the trees in the stand and may prolong the stand initiation (grass–forb–shrub) stage of forest development. This type of management can lead to extending the period of stand development that provides habitat for many early successional vertebrate species (DellaSala et al. 1996). Openings of less than 0.1 ha (0.25 acres) in second-growth stands of spruce–hemlock in southeast Alaska allowed access to food close to cover from snow for Sitka black-tailed deer and allowed certain bird species to occur in otherwise unsuitable habitat (DellaSala et al. 1996).

But timing of the thinning is critical. If thinning occurs just as crowns are closing but before early successional plants have been shaded out of the stand, then the early successional conditions will be maintained later into stand development. If thinning is done after crowns have closed and shaded out the grasses, forbs, and shrubs, then the newly created openings in the stand will be filled by those species found in the seed bank in the soil. Some of these species may be different from those found early in stand development. Precommercial thinning influences the structure and composition of the understory and may consequently influence the vertical and horizontal complexity in a stand. Precommercial thinning can also provide the opportunity to shift the dominant plant species in a stand. Preferential cutting of hardwoods in a mixed-species stand can shift stand development more to conifers, or vice versa.

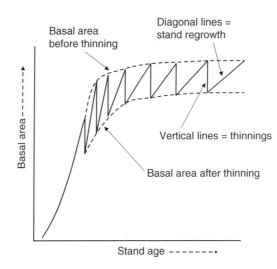

FIGURE 6.3 Stands increase in basal area over time; thinning reduces basal area and reduces the number of trees in a stand. The residual trees grow more rapidly and recover basal area, thereby focusing increased growth on fewer, larger trees. (Redrafted from Smith, D.M. et al. 1997. *The Practice of Silviculture: Applied Forest Ecology*, 9th edn. John Wiley & Sons, New York. With permission.)

COMMERCIAL THINNING

Commercial thinning is a more common intermediate treatment used during stand management. Commercial thinning manipulates stand density by harvesting trees that can be sold and thereby provide some income to at least offset the cost of the stand-management treatments. As illustrated in Figure 6.3, basal area declines abruptly as trees are removed during thinning, and then the stand regrows to recover the growing space that was lost by the trees that were removed. It is also important to realize that stands that may begin with 10,000 trees/ha (4,000 trees/acre) at year 10 will likely have only 500 trees/ha (200 trees/acre) at age 100. So, thinning can occur naturally (self-thinning caused by competition or fire or wind) or by humans. In most forests, however, most plants that establish after a stand-replacement disturbance die in the earliest stages of stand development through competition for resources, regardless of whether humans cut or spray them. However, the plants that do survive this period of plant competition may not always be the plants that humans want. Hence, management action may be necessary. Management is not an ecological necessity; it is done to meet human needs and desires.

How many trees should be cut? How many should be left? One way of estimating the effects of thinning on stand growth and structure is to understand the stocking in the stand, or the amount of the stand that is covered by trees in relation to the density of the stand. Stocking charts have been developed for even-aged stands in many forests in the world. Figure 6.4 provides an example of a stocking chart for an eastern white pine stand in the United States. To understand this diagram, work through an example. On the x-axis find stand density 200 — that is 200 trees/acre (500 trees/ha). Draw a line vertically (parallel to the y-axis) from this point. On the y-axis find 200 — that is 200 ft^2/acre (46 m^2/ha) of basal area. This is the amount of the stand covered by the boles of growing trees. Draw a horizontal line from 200 ft^2/acre parallel to the x-axis. Where these two lines intersect is the stocking level for a stand with 200 trees/acre that occupy 200 ft^2 of basal area per acre. Because most trees in a single-species even-aged stand are approximately the same size, we can also estimate the average diameter of trees where these two lines cross (about 14 in. [36 cm] dbh [diameter at breast height] in this case).

There are several things to notice. First, if you follow your vertical line up (increase basal area at the same number of trees per acre), tree diameters increase, up to a point. As trees get larger,

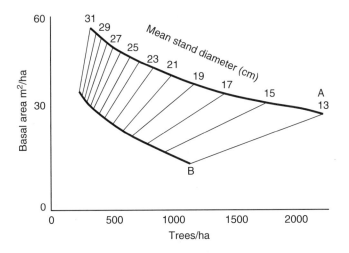

FIGURE 6.4 Eastern white pine stocking chart. Stands above the A line are considered to be overstocked (growing slowly, trees likely to die from competition). Stands below the B line are understocked (not all growing space is being used). Stands between the A and B line are considered to be fully stocked (growing well, all growing space is occupied). (Redrafted from Wendel, G.W., and H.C. Smith. 1990. Eastern white pine. In Bums, R.S., and B.H. Honkala (eds.), *Silvics of North America*, Vol. 1, Conifers. USDA For. Serv. Agric. Handbook 654. pp. 476–488.)

eventually all growing space in the stand is occupied by trees and they begin to compete with one another. You can pack only so many trees into a hectare before some start to die. Hence, the relationship between basal area and density can also be portrayed as the relative density, that is the density of the stand at a given basal area. For example, in Figure 6.5, draw a vertical line up from 1000 trees/ha on the *x*-axis. Eventually, that line will cross the upper diagonal line on this chart. This upper line indicates the maximum number of trees at a given diameter that can occur in the stand. Adding any more basal area (growing trees) at that density results in tree death from competition. So, as trees in the stand grow, some have to die and stand density will decrease. If natural disturbances or humans cutting trees do not thin the stands, then intertree competition leads to tree death within the stand. Trees that die from suppression mortality usually are about half the diameter of the dominant and codominant trees in the stand.

Note that neither of these charts includes anything about tree age. The rate at which trees grow to certain sizes at a give stand density and basal area is dependent on the quality of the site to grow trees. Some sites grow trees very rapidly and others grow the same species of tree much more slowly. Silviculturists use the term *site index* to indicate the potential of a site to grow trees in height. Why height and not diameter? Tree diameter growth is highly influenced by stand density. A tree growing among many other trees is competing for resources, and so allocates the resources it is able to capture first into height growth and then into diameter growth. A tree growing in the open may not grow in height much differently than one growing in a dense stand, but the open-grown tree will have the resources that enable it to grow rapidly in diameter. Consequently, the better indicator of growth potential of a site is height because it is more independent of stand density and is dependent on site quality (Figure 6.6). Foresters use the height of the dominant trees in an even-aged stand at a specified age (usually 50 years) as an index to site quality. Figure 6.6 is an example of site index curves for red oak in the eastern United States. Stand development, such as diameter growth, volume growth, and other features, will occur more slowly on sites with a lower site index. On high-quality sites, trees grow in diameter faster and stands increase in basal area (i.e., move through a stocking chart) more rapidly than on lower-quality sites. The rate at which a stand gains volume and some habitat elements is dependent on the quality of the site.

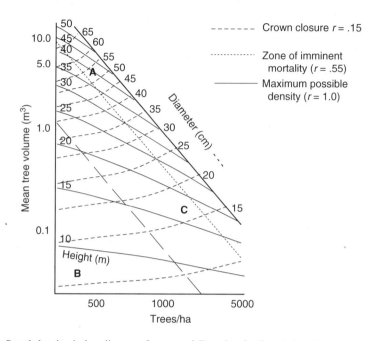

FIGURE 6.5 Stand density index diagram for coastal Douglas-fir. Stand density cannot exceed the upper diagonal line as trees compete for growing space. (Redrafted from Cameron, I.R. 1988. *An Evaluation of the Density Management Diagram for Coastal Douglas-Fir.* B.C. Ministry of Forests and Lands, Victoria, B.C. With permission from the BC Ministry of Forests.)

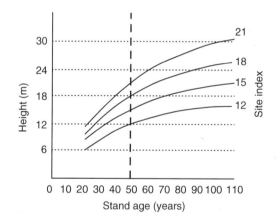

FIGURE 6.6 Site index curves for northern red oak in the Lake States. (Redrafted from Gevorkiantz, S.R. 1957. *Site Index Curves for Red Oak in the Lake States.* USDA For. Serv. Tech. Note 495.)

A stocking chart (Figure 6.4) or density management diagram (Figure 6.5) can be a useful mechanism for biologists and silviculturists to work together to design stands to meet certain needs. If managers wish to produce large snags or logs, it is best to allow the trees in the stand to grow as quickly as possible until the codominant trees reach a size that is desired, and then let competition mortality kill some of those trees. In Figure 6.5, location A in the diagram would be a stand condition where 55 cm trees are competing with one another and some will have to die to allow the rest to grow. Hence, snags (dead trees), which would be about half the diameter of the live trees or 25 to

30 cm in diameter, would begin to die and add dead wood to the stand. Opening the crowns by thinning also influences habitat quality for those species that find cover and food in tree crowns and in the flush of understory vegetation that might occur following thinning (location B in Figure 6.5). For instance, species that feed among tree crowns, such as Hammond's flycatchers, which perch on a branch and sally out to a space between crowns to catch flying insects, may benefit from thinning (Hagar et al. 1996). Thinning densely stocked conifer stands in landscapes dominated by younger stands enhances habitat suitability for several species of mammals and birds, but some unthinned patches and stands may be retained to provide refugia for bird species that are impacted by thinning (location C in Figure 6.5; Hayes et al. 2003, Suzuki and Hayes 2003). Variable-intensity thinning can produce a wide range of tree diameters and greatly influence the production of small snags early in stand development (Carey and Wilson 2001).

FERTILIZATION

Fertilization of forest stands has two dominant potential effects on habitat elements: increased diameter growth on the dominant trees and increased nutrient content in browse and forage. Added tree growth can be particularly important from a habitat complexity standpoint if some trees with added growth are retained into the next rotation. The returns in tree growth compared to the investments of fertilizer application are often most apparent on nutrient-poor sites.

On nutrient-deficient sites, fertilizer applications can increase tree growth and forage production. Nutrient-poor sites in the pine flatwoods of the Gulf coast are sites where fertilization seems to provide benefit to growing trees for forest products and should benefit forage for herbivores as well (Tiarks and Haywood 1996). The added nutrients that find their way into the leaves and twigs of browse plants and herbaceous forage seem to provide benefit to herbivores in a wide range of forest types. In Scandanavian conifer stands during the winter and summer following fertilization, moose strongly selected fertilized plots over unfertilized sites and hares left more fecal pellets in the fertilized plots than in untreated sites, indicating that they probably were using these fertilized patches more than unfertilized ones (Ball et al. 2000). Gibbens and Pieper (1962) found that fertilization increased growth and palatability of the shrubs in deer winter range in California resulting in selective thinning and browsing by herbivores. Increased browse production and use by elk has been reported following fertilizer application in Washington (Pierson et al. 1967). Selective browsing by deer on fertilized plants can reduce deer damage to fir seedlings on fertilized areas (Rieck and Jeffrey 1964 in Scotter 1980). The use of fertilizer and herbicides in combination can also benefit ungulate food resources by stimulating plant growth and by making plants more palatable (Carpenter and Williams 1973).

ROTATION LENGTH: ECOLOGICAL AND ECONOMIC TRADE-OFFS

Even-aged stands typically gain volume over time following a logistic or S-shaped curve. The average volume growth per hectare per year (mean annual increment, MAI) will peak at a point where it is economically most efficient to harvest the stand. This is called the culmination of MAI (Figure 6.7). The stand will have peaked in volume growth, and delaying harvest beyond this point means that you are investing money in maintaining a stand that is no longer maximizing profit. The point in stand development when the manager decides to harvest the stand and regenerate a new stand is the *rotation age*. The *economic rotation* age is often determined by the culmination of MAI. This peak will vary depending on the site quality for the stand and the products being managed (because volumes are estimated based on product values). The culmination of MAI will occur earlier on high-quality sites (high site index) and for products requiring small tree diameters (e.g., pulp) than on poor quality sites or for large sawtimber or veneer products. Most industrial forests will harvest stands nearing

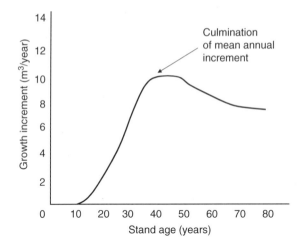

FIGURE 6.7 Change in MAI with stand age. As MAI begins to decline stand volume is increasing at a decreasing rate and will reduce net income to the landowner over time. (Redrafted from Smith, D.M. et al. 1997. *The Practice of Silviculture: Applied Forest Ecology*, 9th edn. John Wiley & Sons, New York. With permission.)

the culmination of MAI, and try not to carry too many stands longer than the culmination of MAI unless there is a need to provide larger products to a mill.

The *ecological rotation* age is the average interval between stand-replacement disturbances that are likely to regenerate a stand. There often is a very large difference between an economic rotation and an ecological rotation. Species of plants and animals that typically occur in stands older than the economic rotation age will lose habitat in forests managed to maximize profit. However, the culmination of MAI can be extended by thinning and by changing product goals. Further, the use of variable-density plantings and thinnings, retention of legacy, and use of mixed-species stands can provide many of the elements in young stands that normally might be found in older or unmanaged stands.

CASE STUDY: DOUGLAS-FIR PLANTATION

This case study provides an example of the difference in the potential structure of even-aged stands managed using approaches that consider habitat complexity. These two stands are from the mid-elevations of the western Cascades in Washington (data and projections based on Landscape Management Systems; McCarter et al. 1998). Traditional management of Douglas-fir plantations involves clearcutting followed by site preparation and vegetation management, which prepares a planting site and removes as much competing vegetation as possible. In this example, 750 seedlings/ha (300 trees/acre) were planted and allowed to grow to stand age 30 as an even-aged single-species stand (Figure 6.8). At age 30, there was approximately 28 m²/ha (120 ft²/acre) of basal area, and 675 trees/ha with an average dbh of almost 23 cm (9 in.) (Figure 6.9). At this point, a commercial thin could be implemented removing about 2 MBF (thousand board feet) per acre (5 MBF/ha) by removing those trees that would likely die through competition mortality and concentrating the remaining growth on the residual trees. The simulated thinning represents a reduction in basal area to 16 m²/ha (70 ft²/acre), and by stand age 50, there would be about 69 MBF/ha (28 MBF/acre) of volume to harvest in 160 trees/ha. The trees would average about 50 cm (20 in.) in diameter. If sawtimber sold for $400/MBF, then the landowner would gross approximately $27,664/ha ($11,200/acre) from which harvesting, planting, and site preparation costs must be deducted.

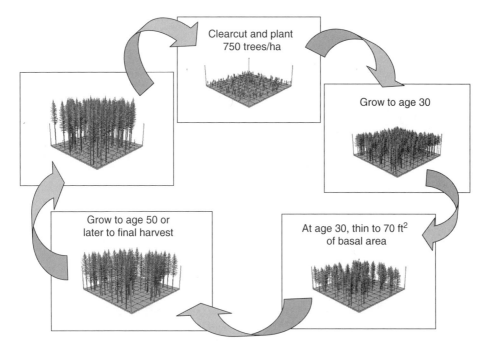

FIGURE 6.8 Example of plantation establishment and management in a Douglas-fir stand in the western Cascades of Washington. Data from Landscape Management Systems, University of Washington, Seattle. (McCarter, J.M. et al. 1998. *J. For.* 96: 17–23.)

FIGURE 6.9 Plantation of Douglas-fir on forest industry land in western Oregon. Note the uniformity in tree size and the lack of vertical structural complexity in this heavily stocked, even-aged stand.

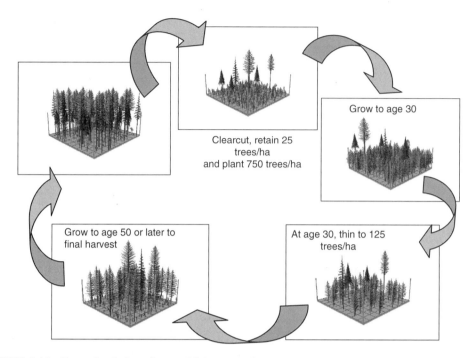

FIGURE 6.10 Example of plantation establishment and management in a Douglas-fir stand in the western Cascades of Washington that includes legacy trees, a multispecies plantation, and thinning to maintain structural complexity and tree species diversity. Data from Landscape Management Systems, University of Washington, Seattle. (McCarter, J.M. et al. 1998. *J. For.* 96: 17–23.)

FIGURE 6.11 Seed-tree regeneration method implemented in the north central Cascades of Washington. The harvest was designed to retain seed trees of ponderosa pine and western larch, with a legacy of dead trees and other species, leading to increased structural and compositional complexity early in stand development.

FIGURE 6.12 Green-tree retention stand in western Oregon 15 years after harvesting and planting. Note the increased level of structural complexity provided by regeneration, retained green trees that are still growing in size, and snags that were created following tree harvest.

Now consider a stand managed to enhance stand complexity. The stand is clearcut with 25 trees/ha (10 trees/acre) retained from the previous stand and representing five species (Douglas-fir, western hemlock, bigleaf maple, western redcedar, and black cottonwood). The site is then prepared for planting taking care to retain advance regeneration of as many species as possible and planting a mixture of tree seedlings representing the aforementioned species plus red alder (Figure 6.10; see Figure 6.11 for a similar approach from central Washington). At stand age 30, there are approximately 865 trees/ha (350 trees/acre) representing seven different species, and having a total basal area of approximately 41 m²/ha (180 ft²/acre) and greater complexity (Figure 6.12). Approximately 7.5 MBF/ha (3 MBF/acre) are then removed in a commercial thinning that reduces the stand density to 125 trees/ha (50 trees/acre) and the site is underplanted with a mix of five tree species to develop a second age class. At age 50, there is a broad range of tree diameters, multiple species, and approximately 84 MBF/ha (34 MBF/acre) of timber. If this mixed-species sawtimber sold for $300/MBF, then the landowner would gross approximately $25,194/ha ($10,200/acre) from which harvesting, two plantings, and site preparation costs must be deducted. Hence, the bottom-line profit would be less, but there still would be a profit and money to pay for developing habitat complexity.

SUMMARY

Even-aged management typically involves the use of clearcut, seed-tree, or shelterwood regeneration methods to establish a new stand consisting of one dominant tree age. The choice of the regeneration method is dependent on the characteristics of the site and the desired stand composition and structure in the future. Complexity in habitat elements can be manipulated during stand management by retaining legacy from the previous stand and choices in the intensity and arrangement of site preparation, planting, and thinning activities. The combination of choices made throughout a rotation can lead to structurally simple or complex stands at rotation age. Hartley (2002) provided some useful and simple guidelines that should be considered if a forest manager wishes to increase complexity in managed even-aged stands:

1. Retain dead and living trees in the stand or in retention islands or strips where they will not interfere too badly with other stand management activities. Snag and reserve tree management (e.g., leaf strips). These should include mature native trees and understory vegetation which are left unharvested or allowed to regenerate.
2. Polycultures should be favored over monocultures by planting multiple crop species and/or leaving some native trees unharvested.
3. Native species should generally be favored over exotics.
4. Site preparation should favor methods that reflect natural disturbances and conserve dead wood.
5. Retain horizontal complexity in structure and composition.
6. Extend rotations as long as possible.
7. Grow some crop trees through two rather than one rotation.

REFERENCES

Anderson, S.H., and B.J. Crompton. 2002. The effects of shelterwood logging on bird community composition in the Black Hills, Wyoming. *For. Sci.* 48: 365–372.

Annand, E.M., and F.R. Thompson. 1997. Forest bird response to regeneration practices in central hardwood forests. *J. Wildl. Manage.* 61: 159–171.

Ball, J.P., K.L. Danell, and P. Sunesson. 2000. Response of a herbivore community to increased food quality and quantity: an experiment with nitrogen fertilizer in a boreal forest. *J. Appl. Ecol.* 37: 247–255.

Boyer, W.D. 1993. Long-term development of regeneration under longleaf pine seedtree and shelterwood stands. *Southern J. Appl. For.* 17: 10–15.

Bragg, D.C., M. Shelton, and B. Zeide. 2003. Impacts and management implications of ice storms on forests in the southern United States. *For. Ecol. Manage.* 186: 99–123.

Brandeis, T.J., M. Newton, and E.C. Cole. 2002. Biotic injuries on conifer seedlings planted in forest understory environments. *New For.* 24: 1–14.

Bull, E.L., and A.D. Partridge. 1986. Methods of killing trees for use by cavity nesters. *Wildl. Soc. Bull.* 14: 142–146.

Burns, R.M., and Honkala, B.H. (eds.). 1990. *Silvics of North America*, Vol. 1, Conifers. U.S Dep. Agric. For. Serv. Agric. Handbook 654.

Cameron, I.R. 1988. *An Evaluation of the Density Management Diagram for Coastal Douglas-Fir.* B.C. Ministry of Forests and lands, Victoria, B.C.

Canon, S.K., P.J. Urness, and N.V. DeByle. 1987. Habitat selection, foraging behavior, and dietary nutrition of elk in burned aspen forest. *J. Range Manage.* 49: 433–438.

Carey, A.B., and S.M. Wilson. 2001. Induced spatial heterogeneity in forest canopies: responses of small mammals. *J. Wildl. Manage.* 65: 1014–1027.

Carpenter, L.H., and G.L. Williams. 1973. *A Literature Review on the Role of Mineral Fertilizers in Big Game Range Improvement.* Colo. Div. Game Fish and Parks, Denver. Spec. Rep. No. 28.

Chambers, C.L., T. Carrigan, T. Sabin, J. Tappeiner, and W.C. McComb. 1997. Use of artificially created Douglas-fir snags by cavity-nesting birds. *West. J. Appl. For.* 12: 93–97.

Chambers, C.L., W.C. McComb, J.C. Tappeiner II, L.D. Kellogg, R.L. Johnson, and G. Spycher. 1999a. CFIRP: what we learned in the first ten years. *For. Chron.* 74: 431–434.

Chambers, C.L., W.C. McComb, and J.C. Tappeiner II. 1999b. Breeding bird responses to 3 silvicultural treatments in the Oregon Coast Range. *Ecol. Appl.* 9: 171–185.

Cole, E.C., W.C. McComb, M. Newton, C.L. Chambers, and J.P. Leeming. 1997. Response of amphibians to clearcutting, burning, and glyphosate application in the Oregon Coast Range. *J. Wildl. Manage.* 61: 656–664.

Cole, E.C., W.C. McComb, M. Newton, J.P. Leeming, and C.L. Chambers. 1998. Response of small mammals to clearcutting, burning, and glyphosate application in the Oregon Coast Range. *J. Wildl. Manage.* 62: 1207–1216.

Conner, R.N., A.E. Snow, and K.A. Ohalloran. 1991. Red-cockaded woodpecker use of seed-tree and shelterwood cuts in eastern Texas. *Wildl. Soc. Bull.* 19: 67–73.

Crouch, G.L., and M.A. Radwan. 1981. *Effects of Nitrogen and Phosphorus Fertilizers on Deer Browsing and Growth of Young Douglas-Fir.* USDA For. Serv. Res. Note PNW-368.

DellaSala, D.A., J.C. Hagar, K.A. Engel, W.C. McComb, R.L. Fairbanks, and E.G. Campbell. 1996. Effects of silvicultural modifications of temperate rainforest on breeding and wintering bird communities, Prince of Wales Island, Southeast Alaska. *Condor* 98: 706–721.

Dey, D.C. 1995. *Acorn Production in Red Oak.* Ontario Ministry of Natural Resources, Ontario Forest Research Institute, Sault Ste. Marie, Ontario, Forest Research Information Paper No. 127, 22.

Easton, W.E., and K. Martin. 1998. The effect of vegetation management on breeding bird communities in British Columbia. *Ecol. Appl.* 8: 1092–1103.

Gevorkiantz, S.R. 1957. *Site Index Curves for Red Oak in the Lake States.* USDA For. Serv. Tech. Note 495.

Gibbens, R.P., and R.D. Pieper. 1962. The response of browse plants to fertilization. *Calif. Fish Game.* 48: 268–281.

Graham, R.T., and T.B. Jain. 1998. Silviculture's role in managing boreal forests. *Conserv. Ecol.* 2, Article 8 (http://www.consecol.org/Journal/vol2/iss2/art8/), accessed March 24, 2007.

Hagar, J., W.C. McComb, and W.H. Emmingham. 1996. Bird communities in commercially thinned and unthinned Douglas-fir stands of western Oregon. *Wildl. Soc. Bull.* 24: 353–366.

Hansen, E.M., J.K. Stone, B.R. Capitano, P. Rosso, W. Sutton, L. Winton, A. Kanaskie, and M.G. McWilliams. 2000. Incidence and impacts of Swiss needle cast in forest plantations of Douglas-fir in coastal Oregon. *Plant Dis.* 84: 773–778.

Hartley, M.J. 2002. Rationale and methods for conserving biodiversity in plantation forests. *For. Ecol. Manage.* 155: 81–95.

Hayes, J.P., J.M. Weikel, and M.M.P. Huso. 2003. Response of birds to thinning young Douglas-fir forests, *Ecol. Appl.* 13: 1222–1232.

Huff, M.H., and C.M. Raley. 1991. Regional patterns of diurnal breeding bird communities in Oregon and Washington. In Ruggiero, L.F., K.B. Aubry, A.B. Carey, and M.H. Huff (technical coordinators). *Wildlife and Vegetation of Unmanaged Douglas-Fir Forests.* U.S. Forest Service General Technical Report PNW-GTR-285.

Johnson, B., and K. Kirby. 2001. Potential impacts of genetically modified trees on biodiversity of forestry plantations: a global perspective. In *Proceedings of Tree Biotechnology in the New Millennium* July 22–27, 2001, Columbia River Gorge, USA. www.fsl.orst.edu/tgerc/accessed 8 August 2006.

Jules, E.S., M.J. Kauffman, W.D. Ritts, and A.L. Carroll. 2002. Spread of an invasive pathogen over a variable landscape: a nonnative root rot on Port-Orford cedar. *Ecology* 83: 3167–3181.

Kerr, G. 1999. The use of silvicultural systems to enhance the biological diversity of plantation forests in Britain. *Forestry* 72: 191–205.

Marquis, D.A. 1974. *The Impact of Deer Browsing on Alleghany Hardwood Regeneration.* USDA For. Serv. Res. Paper NE-308.

Martin, K.J., and B.C. McComb. 2003. Amphibian habitat associations at patch and landscape scales in western Oregon. *J. Wildl. Manage.* 67: 672–683.

McCarter, J.M., J.S. Wilson, P.J. Baker, J.L. Moffett, and C.D. Oliver. 1998. Landscape management through integration of existing tools and emerging technologies. *J. For.* 96: 17–23.

McComb, W.C. 2001. Management of within-stand features in forested habitats. In Johnson, D.H., and T.A. O'Neill (managing editors), *Wildlife Habitat Relationships in Oregon and Washington.* OSU Press, Corvallis, OR.

McComb, W.C., and D. Lindenmaycr. 1999. Dying, dead and down trees. Pages 335–372 in Hunter, M.L. Jr. (ed). Maintaining Biodiversity in Forest Ecosystems. Cambridge Univ. Press, Cambridge, England.

McGarigal, K., and W.C. McComb. 1995. Relationships between landscape structure and breeding birds in the Oregon Coast Range Ecol. Monogr. 65: 236–260.

Morrison, M.L., and E.C. Meslow. 1984. Effects of the herbicide glyphosate on bird community structure, western Oregon. *For. Sci.* 30: 95–106.

Pierson, D.J., J. Patterson, H.I. Brent, and C. Stoddard. 1967. Elk and forest succession. Job Completion Report, P.R. Proj. W69-R, Wash. Dep. Game. Olympia. 6 p.

Randall-Parker, T., and R. Miller. 2002. Effects of prescribed fire in ponderosa pine on key wildlife habitat components: preliminary results and a method for monitoring. Pages 823–834 *in* W.F. Landenslayer, Jr. B. Valentine, C.P. Weatherspoon, and T.E. Lisle, Tech. Coordin. Proceedings of the Symposium on ecology and management of dead wood in western forests. USDA Gen. Tech Rep. PSW-GTR-181.

Rose, R., and J.S. Ketchum. 2003. Interaction of initial seedling diameter, fertilization and weed control on Douglas-fir growth over the first four years after planting. *Ann. For. Sci.* 60: 1–11.

Santillo, D.J., P.W. Brown, and D.M. Leslie Jr. 1989. Response of song-birds to glyphosate-induced habitat changes on clear-cuts. *J. Wildl. Manage.* 53: 64–71.

Schreiber, B., and D.S. DeCalesta. 1992. The relationship between cavity-nesting birds and snags on clearcuts in western Oregon. *For. Ecol. Manage.* 50: 299–316.

Schurbon, J.M., and J.E. Fauth. 2003. Effects of prescribed burning on amphibian diversity in a southeastern U.S. National Forest. *Conserv. Biol.* 17: 1338–1349.

Scotter, G.W. 1980. Management of wild ungulate habitat in the western United States and Canada: a review. *J. Range Manage.* 33: 16–27.

Smith, D.M., B.C. Larson, M.J. Kelty, and M.S. Ashton. 1997. *The Practice of Silviculture: applied Forest Ecology*, 9th edn. John Wiley & Sons, New York.

Smith, H.C., N.I. Lamson, and G.W. Miller. 1986. An esthetic alternative to clearcutting. *J. For.* 87: 14–18.

South, D.B., J.B. Zwolinski, and H.L. Allen. 1995. Economic returns from enhancing loblolly pine establishment on two upland sites: effects of seedling grade, fertilization, hexazinone, and intensive soil cultivation. *New For.* 10: 239–256.

Sparks, J.C., R.E. Masters, D.M. Engle, M.W. Palmer, and G.A. Bukenhofer. 1998. Effects of late growing-season and late dormant-season prescribed fire on herbaceous vegetation in restored pine-grassland communities. *J. Veg. Sci.* 9: 133–142.

Sullivan, T.P., and D.S. Sullivan. 2001. Influence of variable retention harvests on forest ecosystems. II. Diversity and population dynamics of small mammals. *J. Appl. Ecol.* 38: 1234–1252.

Sullivan, T.P., and D.S. Sullivan. 2003. Vegetation management and ecosystem disturbance: impact of glyphosate herbicide on plant and animal diversity in terrestrial systems. *Environ. Rev.* 11: 37–59.

Suzuki, N., and J.P. Hayes. 2003. Effects of thinning on small mammals in Oregon coastal forests. *J. Wildl. Manage.* 67: 352–371.

Taulman, J.F., and K.G. Smith. 2004. Home range and habitat selection of southern flying squirrel in fragmented forests. *Mammalian Biol.* 69: 11–27.

Thompson III, F.R., and D.R. Dessecker. 1997. *Management of Early-Successional Communities in Central Hardwood Forests*. USDA For. Serv. Gen. Tech. Rep. NC-195.

Tiarks, A.E., and J.D. Haywood. 1996. Site preparation and fertilization effects on growth of slash pine for two rotations. *Soil Sci. Soc. Am. J.* 60: 1654–1663.

Vega, R.M.S. 1993. Bird communities in managed conifer stands in the Oregon Cascades: habitat associations and nest predation. MS Thesis, Oregon State University, Corvallis, OR, USA.

Wendel, G.W., and H.C. Smith. 1990. Eastern white pine. In Burns, R.S., and B.H. Honkala (eds.), *Silvics of North America*, Vol. 1, Conifers. USDA For. Serv. Agric. Handbook 654.

7 Silviculture and Habitat Management: Uneven-Aged Systems

Even-aged management is the most common approach to managing forests for commodity production. But on many of the privately owned forests in the eastern United States and Europe where a single forest owner might manage 2 to 4 ha, uneven-aged systems are more consistent with owner goals. Where certain characteristics of forest structure and composition are desired while still managing a forest for commodities, uneven-aged management represents a useful and practical approach, regardless of landowner.

CHARACTERISTICS OF UNEVEN-AGED STANDS

Uneven-aged stands consist of three or more age classes represented in the same stand. Managing a variety of tree ages in the same stand can be challenging because of the potential for large trees to outcompete smaller trees for growing space. Further, in mixed-species stands, each tree species varies in its tolerance to shade, water, or nutrients, and so management becomes even more challenging. The advantage to managing stands using this system is that there are trees of various ages and sizes in the same stand at all times. So, from the standpoint of providing homogeneity over a large area, or for a small private landowner to always have trees on her property, uneven-aged stands can be an attractive alternative to even-aged stands. Because there should be a variety of tree sizes represented in the same stand, vertical and horizontal complexity and tree size diversity can be high and hence can provide a diverse set of food and cover resources for many, but not all, species in a region. Further, many old-growth stands have structural characteristics typical of uneven-aged stands, but uneven-aged stands should not be considered a substitute for old-growth. Old-growth stands often have much higher stocking levels than managed uneven-aged stands, and most trees in a managed stand are not allowed to get very old. Old trees, regardless of their size, support some habitat elements that can only develop over a long period of time, such as tree hollows, lichen communities, and large dead limbs.

In order to maintain an uneven-aged stand, the manager must be sure that there always are enough small, young trees to replace larger trees that are harvested or die over time. Consequently, there are three primary factors defining an uneven-aged stand and the habitat elements therein:

1. Basal area — How much growing space is occupied?
2. Tree density in each diameter class — Are there enough small diameter trees to replace the larger ones that are cut?
3. Target tree size — What diameter class represents the largest harvestable trees in the stand?

First it is important to understand when cutting occurs. Uneven-aged management is based on a *cutting cycle* or period of time between harvests when some trees of all tree diameters are cut. A typical cutting cycle is once every 15 to 30 years in most North American managed forests.

The following example is for a fully regulated balanced uneven-aged stand — something that rarely exists. But it provides the conceptual basis for approaching uneven-aged management. It may

TABLE 7.1

Examples of Tree Densities by Diameter Classes for Three-Diameter Distributions with a Target Tree Size of 76 cm (30 in.)

	Trees/ha		
dbh (cm)	Q = 2.0	Q = 1.5	Q = 1
76	1.0	1.0	1.0
71	2.0	1.5	1.0
66	4.0	2.3	1.0
61	8.0	3.4	1.0
56	16.0	5.1	1.0
51	32.0	7.6	1.0
46	64.0	11.4	1.0
41	128.0	17.1	1.0
36	256.0	25.6	1.0
31	512.0	38.4	1.0
25	1024.0	57.7	1.0
20	2048.0	86.5	1.0
15	4096.0	129.7	1.0
10	8192.0	194.6	1.0
5	16384.0	291.9	1.0
Basal area (m^2/ha)	407	25	2.5
	Impossible!	Fully stocked	Understocked

be easiest to understand how uneven-aged management works by working backwards from the target tree size. Let us assume that the forest manager wants at least 1 tree/ha (1 tree/2.5 acres) that is 76 cm (30 in.) in dbh (diameter at breast height) in the stand at all times and that the manager is using a 15-year cutting cycle. When the manager decides to cut these 76-cm trees, then there must be at least 1 tree/ha that is 71 cm (28 in.) dbh to grow to be 76 cm during the next 15 years. And there must be at least 1 tree/ha that is 66 cm (26 in.) to grow to be 71 cm, and so on. This means that you need at least 1 tree/ha of all size classes down to the smallest size class, which is the regeneration that you want to establish (Table 7.1). But that would be a perfect world, and we know that unpredictable things happen in forests, and so we always try to have more trees in each smaller size class. In our example, let us say you want twice as many trees in each successively smaller diameter class. So you may want to have 1 tree/ha that is 76 cm dbh, but 2 that are 71 cm, and 4 that are 66 cm, and so on.

But, if you calculate the number of trees that you would need in the smallest diameter class, then you would need to have over 16,000 5-cm (2-in.) trees/ha (Table 7.1)! If you could get that many seedlings started per hectare in a stand, their growth would probably be very slow because the stocking would be impossibly high. It is simply impossible to maintain a stand like this. But if we wished to maintain a reasonable stocking level where there is enough room for all trees to grow then we would want to have 1.5 trees/ha in each successively smaller diameter class in this example.

Note that under ideal circumstances, there is a negative exponential distribution of tree diameters in an uneven-aged stand (Figure 7.1). Also note that the shape of the curve is a function of the factor by which you multiply the number of trees in one diameter class to get the number in the next smallest diameter class. This is known as the *Q-factor*. The higher the *Q*-factor, the steeper is the curve (more small trees). The larger the target tree size, the longer or more protracted is the curve. But because growing space is limited, setting a large target tree size automatically means that there have to be many fewer small trees to allow space for all trees to grow. Consequently, managers will

TABLE 7.2

Example of Trees Harvested (Bold = Merchantable) by Diameter Class in an Idealized Uneven-Aged Stand at the End of a Cutting Cycle (see Figure 7.1)

dbh cm (in.)	Trees cut/ha	Trees cut/acre
5(2)	97.3	39
10(4)	64.9	26
15(6)	43.2	17
20(8)	28.8	12
25(10)	19.2	8
31(12)	**12.8**	**5**
36(14)	**8.5**	**3**
41(16)	**5.7**	**2**
46(18)	**3.8**	**1.5**
51(20)	**2.5**	**1.0**
56(22)	**1.7**	**0.7**
61(24)	**1.1**	**0.4**
66(26)	**0.8**	**0.3**
71(28)	**0.5**	**0.2**
76(30)	**1.0**	**0.4**

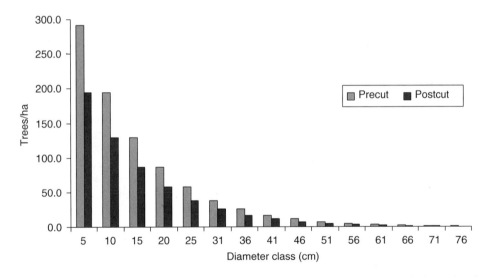

FIGURE 7.1   Idealized distribution of tree sizes in an uneven-aged stand. Note that trees of all tree size classes are harvested during each cutting cycle.

estimate basal area for a stand and see if the stand is overstocked (trees dying from competition), fully stocked (growing without imminent tree mortality), or understocked (not all growing space is being used). In our previous example, let us assume that 25 m²/ha (110 ft²/acre) of basal area is a fully-stocked stand. When this point is reached at the end of a cutting cycle, we may want to reduce the basal area to 16 m²/ha (70 ft²/acre) to provide more growing space for the remaining trees and allow the remaining trees to grow faster. But we want all trees to grow in diameter, not just the big ones, and so we have to provide growing space for all tree size classes. Therefore, a harvest that would reduce the basal area to 16 m²/ha would remove some trees from each size class (Table 17.2).

After harvesting, the remaining trees will grow into the larger size classes and replace those that were cut. In addition, there is a new influx of regeneration established by creating a *seedbed* or growing site for seedlings and sprouts that will replace the trees in the 5 cm (2 in.) dbh class that grew larger. It is important to realize though that if the minimum marketable tree size is 31 cm (12 in.) dbh, then only those trees 31 cm dbh or larger that are cut can be sold for a profit. Trees less than 31 cm dbh are cut to provide growing space at a cost to the landowner and effectively represent a precommercial thin. In addition, the stand is harvested more frequently than an even-aged stand might be, increasing harvesting costs and impacting the site more often. Finally, less timber volume is removed during each harvest than in an even-aged stand, and so the net short-term profit to the landowner may be less, but the landowner will see a more regular income from the property, which will come at the end of every cutting cycle instead of at the end of a rotation.

The aforementioned example is idealized. Managers can never control tree densities by size classes so accurately as we have in this example, and so there is considerable art involved in managing uneven-aged stands to ensure that the resulting diameter distribution after a harvest approximates a negative exponential distribution. For tree species that are intolerant of shade as they regenerate and grow, the ability to maintain the shape of the diameter distribution (and hence the foliage height profile) becomes even more complicated and dictates the type of regeneration harvest that will be used.

CONSIDERING THE SITE POTENTIAL

The choice of which regeneration approach to use, what the target tree size should be, and what stocking level to maintain are influenced by many factors. The ability to manage an area using uneven-aged regeneration methods is often constrained by topography. Because of the high cost of harvesting per income from volume harvested using this system, typically ground-based harvesting equipment is used, such as horses, skidders, and feller bunchers. Cable, skyline, and helicopter logging often cost more than the value of the timber to be removed, and so are cost-prohibitive except in very-high-value stands.

The site index for the tree species being grown will influence the cutting cycle length. Low site-index sites grow trees more slowly and extend cutting cycles. As mentioned earlier, the tree species that you wish to manage influences the basal area removed depending on the shade or moisture tolerance of these species. In addition, the presence of competing vegetation such as shrubs or herbs may represent excellent forage resources, but inhibit establishment of regeneration. Consequently, careful consideration must be given to the restrictions that the characteristics of the site place on your ability to use uneven-aged systems to achieve habitat structure and timber goals.

UNEVEN-AGED REGENERATION METHODS

Uneven-aged management usually involves *group selection* or *individual tree selection* regeneration methods. Individual tree selection is usually used with tree species that are moderately to very shade tolerant because it is the removal of one or a few trees from a location in the stand to create a canopy gap to allow tree regeneration to occur. For many tree species there is simply not enough light entering the forest floor to allow the regeneration to survive and grow if only one tree crown is removed. For these less shade-tolerant species a group selection system may be used, which involves creating small openings in the stand to allow more light and somewhat larger patches of regeneration to become established. These groups are usually less than one tree height in width, but may necessarily exceed that width for very shade-intolerant species. The point at which a large group becomes a small clearcut is somewhat semantic, as is the point at which a small group selection becomes individual tree selection. These uneven-aged regeneration systems cause a fine-scale disturbance, and so within-stand vertical structure and fine-scale horizontal patchiness are usually high compared with even-aged systems (Figure 7.2).

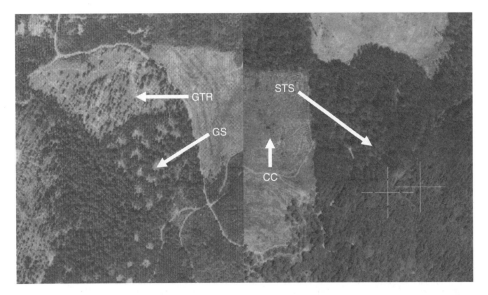

FIGURE 7.2   Single-tree (STS) and group selection (GS) stands, compared to green tree retention (GTR) and clearcut (CC) stands, McDonald Forest, Benton County, Oregon. Note the fine scale patchiness in the single-tree selection stand and the coarser level of patchiness in the group selection stand.

NATURAL REGENERATION AND PLANTING OPTIONS

Site preparation in uneven-aged systems may range from none to mechanical scarification or prescribed burning. If advance regeneration is already established in the existing litter layer, then no site preparation is needed. But if the regeneration needs to be established and the plant species requires bare mineral soil for seed germination, then litter layer disturbance may be done during harvesting by using the blade on the skidder to expose soil. Alternatively, if the remaining trees are reasonably tolerant of fire, a cool burn may be used to expose an ash layer as a seedbed. Mechanical scarification of the harvest groups and compaction of the soils along designated skid trails can significantly affect below-ground habitat by reducing the availability of burrow systems to many animal species and restricting the ability of animals to burrow in the compacted soil. Careful harvest planning and use of designated skid trails is essential on many soil types (Kellogg et al. 1996). Alternatively, fires and litter-layer disruption can lead to increased sprouting of a wide variety of plant species as potential forage, and allow seeds in the soil seed bank to germinate and proliferate. Chemical site preparation is also an option but can be expensive and time consuming because herbicides often must be applied using backpack sprayers.

Artificial regeneration may be established if advance regeneration is not present. Tree species that are somewhat shade intolerant can be regenerated more easily with a group selection than a single-tree selection. Small patch, group selection methods or single-tree selection methods that rely on existing advance regeneration or large planting stock may eliminate or significantly reduce the duration of grass–forb–shrub patches in the stand.

UNEVEN-AGED STAND DEVELOPMENT

During management of uneven-aged stands, the structure is in a continual state of flux. Trees grow until the growing space leads to a fully-stocked condition at the end of a cutting cycle, and then some trees of all size classes are harvested to produce the growing space needed for regeneration and continued growth of residual trees. In unmanaged uneven-aged forests, growing space occurs from competition mortality or disturbances such as insects, disease, fire, or wind. When competition causes mortality and creates growing space, usually shade-tolerant tree species fill the gap unless the gap is large and disturbs the soil. For instance, in an old-growth forest in Great Smoky Mountains National Park, Tennessee, when large (dbh >70 cm) shade-tolerant trees die and fall, they are usually replaced by less shade-tolerant species such as yellow birch, yellow-poplar, and Fraser magnolia (Barden 1979). Uneven-aged management simply imitates this process and selects certain species to favor during the harvesting process, causing the desired species to dominate the stand.

One uneven-aged approach taken by some forest managers is *diameter-limit cutting* or cutting all the trees above some minimum diameter, usually the merchantable tree diameter. Although widely used, this approach is often criticized by silviculturists because the harvest leaves small trees (usually with small crowns), diseased, damaged, or shade-tolerant trees to regenerate the next stand. So from a timber production standpoint, diameter-limit cutting may not produce consistent long-term yields of products in some forest types. Effects on habitat elements and vertebrates are, however, less clear. A study in the eastern United States demonstrated only minor effects on abundances of various bird species using this system (Weakland et al. 2002). Abundances of most songbird species present prior to harvest changed little after the timber removal (Weakland et al. 2002). Two species, the Canada warbler and dark-eyed junco, were more abundant in harvested areas than unharvested forest. Stands that were harvested differed from unharvested ones in only a few structural characteristics. Harvested stands had more snags, more trees of 8 to 14.9 cm (3 to 6 in.) dbh, and more down wood (Weakland et al. 2002). Canopy cover over 24 m (80 ft), density of saplings, and the amount of leaf litter decreased after harvesting. Another study modeled three stand-management options in southern Indiana. The "do-nothing" management provided the best gray squirrel habitat but the worst economic return; the diameter-limit alternative produced poorer squirrel habitat but a better short-term financial return; and intensive management provided the highest long-term economic return but produced the poorest squirrel habitat (Brand et al. 1986). Although the impacts of diameter-limit harvesting may be well accepted from a timber management perspective, the effects on habitat elements and vertebrates is highly variable and may not be problematic at low levels of volume removals.

HABITAT ELEMENTS IN UNEVEN-AGED STANDS

Several factors influence the development of an uneven-aged stand and the resulting habitat elements. Clearly the tree species composition and the ability of the desired species to regenerate in the stand have the greatest effects on stand development. Mixed-species stands can be difficult to manage because of the varying growth rates of the different species, but can also provide the manager with the opportunity to favor some species over others during management. The desired range of basal area can also influence stand development. If a manager wishes to provide growing space for regeneration, browse, soft-mast, and hard-mast production, then the stocking level must be kept quite low from one cutting cycle to the next, much lower than would be expected under most natural disturbance processes unless disturbances are frequent. Nonetheless, uneven-aged stands seem to support more species of birds typical of unmanaged forests than even-aged stands, at least early in stand development. Chambers et al. (1999) reported that many bird species found in clearcut and green-tree retention stands did not occur in stands managed using small group selection (Figure 7.2). Many of the species using the group selection stands were also found in uncut mature forest stands (Chambers and McComb 1997, Chambers et al. 1999, Gram et al. 2001). These patterns likely reflect the distribution of habitat elements in uneven-aged stands. High vertical structural diversity and fine-scale horizontal patchiness tend to be associated with the single-tree selection stands (Kenefic and Nyland 2000; Figure 7.2). Although selection methods decrease total canopy closure, they maintain high vertical structural diversity and an even distribution of foliage among canopy strata. Selection regeneration methods can reduce the number of cavity-bearing trees and dead wood but increase browse (Kenefic and Nyland 2000, McComb and Noble 1980). Cutting cycle length, target tree size, and stocking all affect the structure and composition of uneven-aged stands.

VERTICAL STRUCTURE

Probably the most obvious effect of using uneven-aged approaches is that the vertical structure of the stand is more complex than would typically be found in even-aged systems, particularly in single-species stands. The shape of the diameter distribution, though, can have a significant effect on the distribution of foliage in the stand. A stand with a high Q will have more foliage represented among the smaller trees and less in the larger trees (steeper diameter distribution). A stand with a low Q (flatter diameter distribution) will have proportionally more foliage in the larger trees. Consequently. depending on the species of animals that you wish to manage and which foliage layers they are associated with, you may wish to use different diameter distributions to meet the needs of those species.

The plant species composition of the various foliage layers can also have an effect on the responses of vertebrates to this vertical structure. If the lower foliage layers are manipulated to remove shrubs and allow tree regeneration to become established, then those species of birds and mammals that rely on shrubs more heavily than tree seedlings could be adversely affected. Removal of understory vegetation in uneven-aged management could decrease populations of some ground- and shrub-nesting forest interior species of birds (Rodewald and Smith 1998).

HORIZONTAL DIVERSITY

Horizontal diversity or patchiness is high at a small spatial scale especially using group selection approaches. If the groups are sufficiently large, then early successional species might colonize them, though they may be too small to be of value to some early seral bird species. King et al. (2001) found that gaps served as sinks, not sources, for many of these bird species. Chambers et al. (1999) found that small gaps of 0.2 ha (0.5 acres) were not colonized by early-successional bird species in western Oregon. As you would expect, the responses of various species to these small gaps varies with species. Small gaps and single-tree selection systems tend to support a species assemblage more similar to that of a mid- to late-successional forest, especially if snags, logs, hardwoods, and shrubs are allowed to persist. Large gaps (small clearcuts) allow the colonization of some early seral associates.

FORAGE AND BROWSE

Group selection methods can also provide patches of browse and forage adjacent to cover for ungulates, hares, and other herbivores. The interspersion of forage and cover can be an excellent management strategy for these species if the openings are large enough to produce browse of the correct species and quality. Creating small gaps often leads to increased levels of shade in the gaps and reduces the production of browse, but this apparently is a problem only with very small gaps (80 to 100 m^2) in northern hardwood stands (Webster and Lorimer 2002). But quantity may not be as important as quality for many herbivores. Gap sizes of 100 m^2 or larger were needed to allow dominance of more palatable browse species (Webster and Lorimer 2002). This may be particularly important where edges of gaps decrease plant growth (York et al. 2004). On the other hand, plants may allocate more energy to growth than to defense under low-light conditions, allowing the plants growing in partial shade to be higher-quality browse than plants grown in full sunlight (Dudt and Shure 1994). Selection method harvests to various densities in loblolly–shortleaf pine stands, indicated that herbage and browse production were generally related to residual pine basal area and site quality (Wolters et al. 1977). Browse made up about one-fourth of the forage under stands having high residual pine basal area but represented considerably lower proportions in clearings (Wolters et al. 1977). Stands with lower basal areas tend to have higher browse production, denser and higher vertical structure, more woody vine and fern biomass, and higher plant species diversity and richness (Miller et al. 1999).

In addition, Hanley and Barnard (1998) suggested that patches of hardwoods, specifically red alder in conifer forests of southeast Alaska, offer significant food resources to herbivores beyond simply browse. These patches allow more sunlight to the forest floor and provide a diversity of forage species for Sitka black-tailed deer in this region.

DEAD AND DYING TREES

Maintaining stands at low stocking levels means that competition mortality is kept to a minimum. Snag and fallen log availability in uneven-aged stands is often lower than in unmanaged old-growth stands (Goodburn and Lorimer 1998). If competition mortality is occurring in a stand, then the trees most likely to die are the smaller ones in the stand, and in an uneven-aged system these trees are the regeneration and browse resources. So providing dead wood in uneven-aged stands often requires active management through killing trees or retaining patches of forest that are allowed to remain dense while giving up the opportunity to recruit regeneration and browse in those patches. In addition, legacy trees can be retained from one cutting cycle to another to ensure that some of the elements of old trees are present in the managed stand. But remember that these legacy trees will often grow to a size larger than the target tree size and take up growing space that could be occupied by regeneration if they were not retained.

MAST

Soft-mast production in shelterwood stands and clearcuts is often greater than in single-tree selection, group selection, and unharvested stands (Perry et al. 1999), and we would expect that soft-mast production increases as gap sizes increase to allow more full sunlight to strike the shrubs. Hard-mast production is generally associated with crown size and tree age, and consequently can respond to silvicultural treatments that provide more sunlight to the crowns of mast-producing trees (Perry and Thill 2003). Mast production in many oak species is highly variable from year to year and seems to be heavily influenced by weather and time since the last heavy mast crop. Nonetheless, trees with large crowns should periodically produce an abundant crop of mast. Generally, open crowns are capable of producing many more fruits than closed crowns (Johnson 1994). Larger stem diameters (and consequently larger crowns) also produce greater crops of acorns than smaller diameter stems, and so uneven-aged management methods that use large target tree sizes and keep stocking levels near or below crown closure should produce more abundant mast during years of high-mast production (Desmarais 1998).

As a rule of thumb, the shape of the diameter distribution and the target tree size will influence your ability to provide vertical and horizontal complexity, forage, and mast. If the diameter distribution is steep with very many small trees and only a few large ones, then it will probably function similar to an early- to mid-successional even-aged stand for most species; browse availability may be greater in these stands. If the diameter distribution is somewhat flat or if the target tree size is large, with very few small trees and more large trees, then it may function more similarly to a late-successional even-aged stand; hard-mast production may be greater in these stands. In both cases, uneven-aged regeneration systems cause a more fine-scale of disturbance than even-aged systems, and so within-stand vertical structure and fine-scale horizontal patchiness is usually high compared with even-aged systems.

CHALLENGES TO USING AN UNEVEN-AGED SYSTEM

Achieving timber and habitat goals using uneven-aged regeneration methods presents a few challenges that should be understood before accepting this technique as a way of meeting these goals. First, if stocking levels are not kept low enough, many species of shade intolerant plants will likely decline in abundance in the stand. These may be important plant species as food resources for herbivores or valuable timber species. Hence, cutting cycles may need to be more frequent on highly productive sites, or volume removals heavier than one might wish, to achieve goals related to vegetative cover.

In addition, each entry will require that trees of a wide range of tree diameters (and often tree species) be harvested. These various tree sizes and species have different market values. So it is quite likely that sawtimber, pulpwood, firewood, and perhaps veneer logs could all be removed in one harvest. Ensuring that logs are sorted and that markets are available for each tree size and species can present challenges to the manager. Harvesting these various sized trees can also be a challenge especially where advance regeneration occurs. Felling large trees onto existing regeneration can damage the smaller trees and reduce the ability to maintain the desired diameter distribution. Use of directional felling and designated skid trails can help to reduce these problems, but may increase harvesting costs (Kellogg et al. 1996).

Finally keeping unmerchantable trees and shrubs in the stand as legacy trees or shrub patches is feasible but must be taken into consideration during each cutting cycle to maintain these structures or plan for their replacement as they age and die. These residual plants also occupy growing space and consequently represent a trade-off between timber production and habitat availability for desired species.

NONTRADITIONAL MANAGEMENT APPROACHES

The uneven-aged systems described in this chapter and the even-aged systems described in the previous chapter represent only a few examples of stand-management approaches that span a spectrum of possibilities (Figure 7.3). The opportunities to develop stand structure and composition to meet land manager objectives is endless and can be crafted to each site to meet those specific objectives.

FIGURE 7.3 Several silvicultural approaches to increase complexity in 50-year-old Douglas-fir stands in the Oregon Cascades. (a) is an unthinned stand, (b) is a stand with 0.2-ha gaps, and (c) is a heavily thinned stand with shade-tolerant trees and shrubs in the understory.

McComb et al. (1993) used the structure and composition of unmanaged stands that were meeting habitat objectives for late-seral species as models for proposed managed stands. This is one approach, though certainly not the only one, to defining a *desired future condition* — a description of the structure and composition of a stand that you would hope to achieve through active management. Defining the desired future condition, or specific goals for the stand, is the first step in stand management.

One type of management approach described by McComb et al. (1993) is a many-storied stand that uses small-group selection cutting to create a stand that is composed of more than three layers of canopy trees in a mosaic of gaps while retaining large legacy trees and snags in the stand. The approach contains elements of a forest found in and produced by gap-phase forest dynamics and may be applicable to many forest types. The many-storied system is patterned after fine-scale natural disturbances. Cut gaps may have to be larger than most natural canopy gaps to allow successful natural regeneration of shade-intolerant species and to make harvesting more efficient. This system would have high within-stand variability in tree size and vertical complexity. This system might provide acceptable habitat for mature forest species while allowing some small but regular timber removal and as such be attractive for nonindustrial forestland managers.

The choice of which silvicultural system to use is determined by the plant community, site conditions, logging constraints, and species of vertebrates of highest interest. Uneven-aged management strategies that could improve habitat quality for species that inhabit late-seral-stage conditions include establishing a large target tree size, lengthening cutting cycles, minimizing disturbance to the stand during logging with designated skid trails, harvesting with small-group or single-tree selection methods where they are appropriate, managing for shade-tolerant tree species, and maintaining high-density groups of regeneration (Figure 7.4). An allocation of dead or large, living trees also would increase habitat quality for many species typical of late-seral stages.

Altering the scale or frequency of cutting also might influence habitat quality for forest vertebrates. Imposing a single-tree selection system in a forest with a cutting cycle of 10 to 15 years and target tree sizes of more than 50 cm dbh, for example, would result in small, widely scattered openings. On the other extreme, locating 60-ha clearcuts side-by-side within a watershed would create huge areas of early-seral-stage stands. Colonization of parts of this area by relatively less mobile species would be less likely than colonization by larger, more mobile species. Both the silvicultural strategy employed and its arrangement in context with other stands on the landscape, therefore, can have a tremendous influence on the future abundance and distribution of animals in the landscape.

Regeneration

FIGURE 7.4 Vertical complexity arising from regeneration in a single-tree selection stand (a) and a group selection stand (b), in a Douglas-fir-grand fir forest type, Oregon.

CASE STUDY: MANAGING A SMALL PRIVATELY OWNED FOREST

As an example of using uneven-aged management to provide a variety of ecosystem services, we can examine how a family owning 10 ha of forest land in western Massachusetts might approach management. First, it is important to recognize that they have multiple objectives for the forest, and include in order of priority:

1. Always having a forest on their land
2. Firewood to help heat their home
3. Periodic income sufficient to cover taxes
4. A multistory forest to support a diversity of nesting songbirds
5. Enough browse and mast production to attract white-tailed deer and ruffed grouse.

The stand is a mixed oak–pine forest on glacial till. It is approximately 80 years old and established following farmland abandonment in the early 1900s. There is currently 46 m^2/ha (200 ft^2/acre) of basal area and is dominated by northern red oak, red maple, black birch, and eastern white pine, with seven other tree species common in the stand (Figure 7.5). If we reduce the basal area to 16 m^2/ha (70 ft^2/acre), then that results in about 27 MBF/ha or 271 MBF removed from the property. At current stumpage values that would be yield about $27,000. Using a 20-year cutting cycle we can have another harvest that yields about $13,000 in 2026. A third harvest 20 years later is a cordwood sale (no sawtimber in cut). During each harvest, 25 to 50 cords of wood are cut per hectare to provide firewood. At the end of 60 years of management, the stand contains 44 cords of firewood per hectare and 25 MBF of sawtimber per hectare available for future harvests. Openings are sufficient to always have an understory present and large enough to provide browse and soft mast. Red and white oaks average about 30 cm in diameter and should produce regular acorn crops. With careful

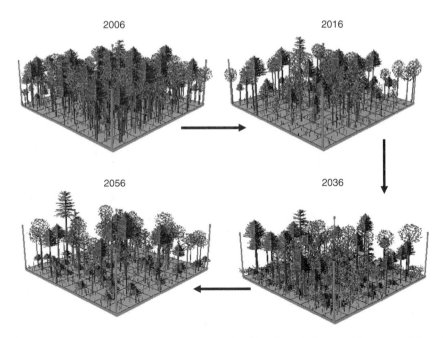

FIGURE 7.5 Example of several cutting cycles in a mixed hardwood-pine stand in western Massachusetts. Data are from the UMass–Amherst Cadwell Forest, and projections are from the Landscape Management System (McCarter et al. 1998). Stand visualization is from the Stand Visualization System (Dixon 2003).

FIGURE 7.6   Abundant regeneration following single-tree selection in a ponderosa pine mixed-conifer forest type, Ochoco National Forest, Oregon.

attention to regeneration of desired tree species, the needs of the landowners should be met for several generations of the family.

SUMMARY

Uneven-aged stands consist of three or more age classes represented in the same stand (Figure 7.6). Uneven-aged stand structure and hence the structure and function of habitat elements is governed largely by basal area, tree density in each diameter class, and target tree size. These characteristics are manipulated to achieve the desired negative exponential diameter distribution to meet the goals for a stand. This structure can be achieved using single-tree selection or group selection regeneration methods. In general, vertical complexity in an uneven-aged stand is high and horizontal complexity is fine-scaled. Browse and mast are less abundant than in early stages of even-aged management unless large gaps are made using group selection methods. Because stocking must be kept low to allow trees to grow, recruitment of dead wood is usually minimal, and so active management is usually needed to ensure adequate dead wood for desired wildlife species. Alternative management approaches usually define a desired future condition and then adapt even-aged and uneven-aged approaches to meet that goal.

REFERENCES

Barden, L.S. 1979. Tree replacement in small canopy gaps of a *Tsuga canadensis* forest in the southern Appalachians, Tennessee. *Oecologia* 44: 141–142.
Brand, G.J., S.R. Shifley, and L.F. Ohmann. 1986. Linking wildlife and vegetation models to forecast the effects of management, pages 383–387. In Verner, J., M.L. Morrison, and C.J. Ralph (eds.), *Wildlife 2000: Modeling Habitat Relationships Terrestrial Vertebrates*. University of Wisconsin Press, Madison, WI.

Chambers, C.L., and W.C. McComb. 1997. Effects of silvicultural treatments on wintering bird communities in the Oregon Coast Range. *Northwest Sci.* 71: 298–304.

Chambers, C.L., W.C. McComb, and J.C. Tappeiner II. 1999. Breeding bird responses to 3 silvicultural treatments in the Oregon Coast Range. *Ecol. Appl.* 9: 171–185.

Desmarais, K.M. 1998. *Northern Red Oak Regeneration: Biology and Silviculture.* University of New Hampshire Cooperative Extension Service, Durham, NH.

Dixon, G.E. (Compiler). 2003. *Essential FVS: A User's Guide to the Forest Vegetation Simulator.* Internal Report. USDA For. Serv. Forest Management Service Center. Fort Collins, CO.

Dudt, J.F., and D.J. Shure. 1994. The influence of light and nutrients on foliar phenolics and insect herbivory. *Ecology* 75: 86–98.

Goodburn, J.M., and C.G. Lorimer. 1998. Cavity trees and coarse woody debris in old-growth and managed northern hardwood forests in Wisconsin and Michigan. *Can. J. For. Res.* 28: 427–438.

Gram, W.K., V.L. Sork, R.J. Marquis, R.B. Renken, R.L. Clawson, J. Faaborg, D.K. Fantz, J. LeCorff, J. Lill, and P.A. Porneluzi. 2001. Evaluating the effects of ecosystem management: a case study in a Missouri Ozark forest. *Ecol. Applic.* 11: 1667–1679.

Hanley, T.A., and J.C. Barnard. 1998. Red alder, *Alnus rubra*, as a potential mitigating factor for wildlife habitat following clearcut logging in southeastern Alaska. *Can. Field-Naturalist.* 112: 647–652.

Johnson, P.S. 1994. The silviculture of northern red oak, pages 33–68. In J.B. Isebrands and R.E. Dickson, eds. *Biology and Silviculture of Northern Red Oak in the North Central Region: A Synopsis*, U.S. Dep. Agric. For. Serv. Gen. Tech. Rep. NC-173.

Kellogg, L.D., P. Bettinger, and R.M. Edwards. 1996. A comparison of logging planning, felling, and skyline yarding between clearcutting and five group-selection harvesting methods. *West. J. Appl. For.* 11: 90–96.

Kenefic, L.S., and R.D. Nyland. 2000. Habitat Diversity in Uneven-Aged Northern Hardwood Stands: A Case Study. U.S. Dep. Agric. For. Serv. Res. Pap. NE-714.

King, D.I., R.M. DeGraaf, and C.R. Griffin. 2001. Productivity of early-successional shrubland birds in clearcuts and groupcuts in an eastern deciduous forest. *J. Wildl. Manage.* 65: 345–350.

McCarter, J.M., J.S. Wilson, P.J. Baker, J.L. Moffett, and C.D. Oliver. 1998. Landscape management through integration of existing tools and emerging technologies. *J. For.* 96: 17–23.

McComb, W.C., and R.E. Noble. 1980. Effects of single-tree selection cutting upon snag and natural cavity characteristics. *Trans. Northeast Fish and Wildl. Conf.* 37: 50–57.

McComb, W.C., T.A. Spies, and W.H. Emmingham. 1993. Stand management for timber and mature-forest wildlife in Douglas-fir forests. *J. For.* 91: 31–42.

Miller, D.A., B.D. Leopold, M.L. Conner, and M.G. Shelton. 1999. Effects of pine and hardwood basal areas after uneven-aged silvicultural treatments on wildlife habitat. *So. J. Appl. For.* 23: 151–157.

Perry, R.W., and R.E. Thill. 2003. Initial effects of reproduction cutting treatments on residual hard mast production in the Ouachita Mountains. *So. J. Appl. For.* 27: 253–258.

Perry, R.W., R.E. Thill, D.G. Peitz, and P.A. Tappe. 1999. Effects of different silvicultural systems on initial soft mast production. *Wildl. Soc. Bull.* 27: 915–923.

Rodewald, P.G., and K.G. Smith. 1998. Short-term effects of understory and overstory management on breeding birds in Arkansas oak-hickory forests. *J. Wildl. Manage.* 2: 1411–1417.

Weakland, C.A., P.B. Wood, and W.M. Ford. 2002. Responses of songbirds to diameter-limit cutting in the central Appalachians of West Virginia, USA. *For. Ecol. Manage.* 155: 115–129.

Webster, C.R., and C.G. Lorimer. 2002. Single-tree versus group selection in hemlock–hardwood forests: are smaller openings less productive? *Can. J. For. Res.* 32: 591–604.

Wolters, G.L., A. Martin Jr., and W.P. Clary. 1977. Timber, Browse, and Herbage on Selected Loblolly-Shortleaf Pine-Hardwood Forest Stands. U.S. Dep. Agric. For. Serv. Res. Note. SO-223.

York, R.A., R.C. Heald, J.J. Battles, and J.D. York. 2004. Group selection management in conifer forests: relationships between opening size and tree growth. *Can. J. For. Res.* 34: 630–641.

8 Desired Future Conditions

To effectively managing habitat for a species, a group of species, or to contribute to biodiversity conservation, we need goals or targets toward which management can be directed. This may involve a condition that will occur on its own in the absence of active management, or it may require intervention to guide the development of the stand or landscape toward your goal. Describing the structure, composition, and scales of a condition that you think will meet the needs for species on your site is one of the first steps in developing a habitat management plan for a stand, forest, or landscape. Landres et al. (1999) described *desired future conditions* as expressions of ecosystem conditions preferred by stakeholders and managers. Kessler et al. (1992) also referred to an articulation of a desired future condition as a goal in ecosystem management. This may be a reference condition, or more appropriately a set of reference conditions that currently achieve some desired objectives, or it may be a sequential set of future conditions that achieve different objectives for different species over time. Given the inherent uncertainty in achieving goals in the face of stochastic disturbances, ecological pathways, and novel stresses on forest dynamics (e.g., climate change and spread of invasive species), monitoring to assess progress toward a desired future condition is probably a reasonable strategy for achieving habitat objectives. Goals for habitat are typically set at large scales (regions) and achieved at small scales (stands). Foresters typically develop plans for managing stands that contribute to some overall forest-level goal.

However the desired future conditions are described, they must be implementable, that is, the site must be capable of producing those conditions. All the factors described in the previous chapters come into play when considering whether current conditions, past actions, and likely future changes will result in achievement of a set of desired future conditions. Models of forest development under alternative management strategies can help guide development of management plans for a stand or landscape.

DEVELOPING THE STAND PRESCRIPTION

Foresters write *prescriptions* or silvicultural management plans for stands to achieve a desired future condition. Personally I find the term "prescription" unfortunate because to the lay public it implies that the stand is unhealthy and needs fixing, which may or may not be the case! Nonetheless, once the desired future condition is deemed achievable, the prescription for the current stand can be developed. It is important to have a written prescription that clearly describes the current condition, the desired future condition, management actions, schedule for anticipated future entries, and a monitoring plan. For many prescriptions the desired future condition may not be attained for several decades. Consequently, it is important that there is a written record for the rationale behind the prescription. Future managers of this stand may be able to make more informed decisions based on written records.

Your prescription should include a description of any of the stand summary information (species, basal area, tree diameter distribution, site index, stocking, etc.) that would be needed to develop marking guides. In addition, to achieve habitat objectives, the prescription should contain:

Species background: What plant and animal species goals are intended for this stand? Are individual species or groups of species to be managed? What are their habitat needs? Over what spatial scale (e.g., nest sites, foraging patches, or home ranges) must they be met? How do these goals complement those for adjacent stands?

Current stand condition: What are the habitat conditions in the stand now for the species that you intend to manage? What factors are limiting habitat suitability for the species? What is the tree and shrub species composition? Site index? Stocking? Existing or possible future regeneration?

Desired future condition: What are the habitat conditions that you would like to produce? Be specific and describe the plant species composition, size classes, basal area, and any predictions of future stand development that you can develop. Stand-growth models such as the Forest Vegetation Simulator (Dixon 2003) are particularly useful for understanding if your desired future condition can be met.

Management actions to achieve the desired future condition: What will you do to the stand now and over time to achieve your desired future conditions? How long do you think it will take to achieve them? How long will they last? How much will it cost?

Monitoring plans: What will you measure and how often will you measure to determine if your management plan was implemented correctly and if the actions were successful? How will you decide if you need to change your management plans?

Budget: What will implementation and monitoring of the plan cost? Regardless of the landowner, cost becomes a factor when implementing a prescription to achieve a goal. There must not be a net loss to most landowners, and there must be maximum profit for some landowners. Understanding the products that can be derived from implementing the prescription, both economic and ecological, can help the manager decide if the trade-offs are acceptable.

Schedule: When will each step of the plan be completed? Of course, the monitoring of stand development may lead to changes in the schedule, but there should be a plan for when actions will occur.

References: Scientific references should be used to support assumptions used in the development of the prescription. Because new information is always becoming available to guide management, it is important to understand why decisions were made at this point in time.

Once a prescription has been developed, it can be implemented and monitored over time. If the monitoring data indicate that the stand is developing over time differently from what was predicted in the prescription, then midcourse corrections can be made. By having a prescription for each stand that you manage, and keeping it on file, future managers can understand how the stand developed and if it is developing as intended to achieve a desired future condition. If a new desired future condition should be developed at some time in the future, then the manager at that time can learn from the past efforts. Documentation of plans, actions, and results is critical.

CASE STUDY: GROWING RED-COCKADED WOODPECKER HABITAT

Red-cockaded woodpeckers are listed as threatened under the U.S. Endangered Species Act. Their numbers are a fraction of what they were historically. The decline in numbers and distribution is considered to be largely a product of past timber liquidation and current short-rotation forestry for pulp and small sawtimber production in the southern yellow pine region of the United States. This example prescription is developed for a 7 ha pine stand on the Kisatchie National Forest in Louisiana. The stand data were provided by Jo Ann Smith, Forest Silviculturist on the Kisatchie, but I use them merely as an example of managing this stand to achieve a desired future condition for red-cockaded woodpeckers. Managers of the Kisatchie quite likely would have a different prescription based on their goals, their personal knowledge of the stand, its history, and its context on the forest. I simply use the data to demonstrate how a prescription could be developed for red-cockaded woodpeckers and several associated species, Bachman's sparrows and brown-headed nuthatches.

SPECIES BACKGROUND AND MANAGEMENT OPTIONS

Red-cockaded woodpeckers have long been associated with mature southern yellow pine forests containing old trees with red heart disease, a fungal decay softening the heartwood of living pines. Red-cockaded woodpeckers have a broad geographic range throughout the Gulf Coastal Plain and Piedmont north into southern Kentucky. Home range sizes vary from 25 to over 100 ha (DeLotelle et al. 1987, Doster and James 1998). Because our stand is only 7 ha, any plans for habitat improvement must consider the conditions of the surrounding stands that should managed to complement the conditions in our stand.

Within their home range, these woodpeckers defend a somewhat smaller territory around a central nest tree. Home ranges must be large enough to provide resources for young in the nest as well as additional helpers at the nest. Red-cockaded woodpeckers, like acorn woodpeckers, have a social structure known as a clan, where nonreproducing members of the clan assist the reproducing pair with raising young. They feed primarily on insects and fruits and forage on trees more than 60 years of age and on trees of more than 25 cm dbh (diameter at breast height), more than expected by chance (Zwicker and Walters 1999). The nest tree is usually older than 100 years and is more than 40 cm dbh (Conner and O'Halloran 1987). Cavities are excavated in live trees, unlike most other woodpecker species, and they typically make shallow excavations around the cavity entrance into resin wells, causing a flow of sticky resin around the cavity entrance and down the tree. This behavior is presumably to reduce risks of predation by rat snakes (Conner et al. 1998).

The stand around the nest tree typically is dominated by pines (basal area averages 16 m^2/ha) with a few midstory pines (average $= 1.1$ m^2/ha) and very few, if any, midstory hardwoods (Conner and O'Halloran 1987). James et al. (1997) found that variation in the size, density, and productivity of clans was related to the ground cover composition and the extent of natural pine regeneration, both of which are indirect indicators of local fire history.

Bachman's sparrows also are associated with mature pine woodlands characterized by well-spaced pines, an open midstory, and a dense understory of grasses and forbs (Plentovich et al. 1998). They have territory sizes of 2 to 3 ha (Dunning 2006). Populations of the Bachman's sparrow began declining in the 1930s, with both a dramatic retraction in geographic distribution and the extinction of many local populations (Plentovich et al. 1998). Areas suitable for red-cockaded woodpeckers were not always suitable for Bachman's sparrows. Red-cockaded woodpeckers appear more tolerant of a hardwood midstory, and although they do not require a dense cover of grasses and forbs, they seem to be more successful in forests with frequent fires that produce this understory condition. Prescribed burning is the key for development and maintenance of the dense herbaceous understory preferred by Bachman's sparrow (Dunning 2006). Bachman's sparrow populations disappear 4 to 5 years after a burn (Dunning 2006). In areas managed for red-cockaded woodpeckers, frequent (3 to 5 years' interval) burning early in the growing season appears the best way to increase habitat suitability for Bachman's sparrows.

Brown-headed nuthatches are associated with loblolly and shortleaf pines in the Upper Coastal Plain and longleaf and slash pines in the Lower Coastal Plain (Withgott and Smith 1998). They have a territory size of about 3 ha and create nest cavities in pine snags typically more than 20 cm dbh (Harlow and Guynn 1983), but most often forage on live pines (Withgott and Smith 1998). This species is most often found in open, mature, old-growth pine forest, particularly where natural fire patterns have been maintained and where snags are present for nesting and roosting (Withgott and Smith 1998). This combination of vegetation characteristics occurs in mature pine forest in which fire has kept understory open and created snags (Withgott and Smith 1998).

Based on the life history characteristics of these three species, several management options seem possible. James et al. (2001) suggested that smaller size classes of trees in closed-canopy stands should be thinned, creating patchy openings in the forest that will promote natural pine regeneration. This can be achieved by uneven-aged management such as group selection regeneration methods in conjunction with maintaining a low basal area or through irregular shelterwoods (Conner et al. 1991). Because longleaf pine was once a dominant species on the Kisatchie and red-cockaded woodpeckers are often associated with longleaf pine savannah forests, management actions that allow recovery of longleaf into managed stands might also be beneficial to red-cockaded woodpeckers and other species.

Management approaches suggested for recovery of red-cockaded woodpeckers have included both short- and long-term strategies (Rudolph et al. 2004). Nearly all populations require immediate attention in the short term, including manipulation of midstory and understory conditions (Rudolph et al. 2004). Management techniques including prescribed fire and mechanical and chemical control of woody vegetation are often used to achieve these needs. In the long term, cost-effective management of red-cockaded woodpecker populations requires a timber management program and prescribed fire regime that will produce and maintain the stand-structure characteristic of high-quality nesting and foraging habitat so that additional intensive management specific to the woodpeckers is no longer necessary (Rudolph et al. 2004). Management that achieves this goal and still allows substantial timber harvest is feasible. With some attention to understory conditions and availability of snags, these approaches would benefit Bachman's sparrows and brown-headed nuthatches as well. Effects on other species probably would be negative for ground-nesting birds (Wilson et al. 1995) and possibly those associated with midstory conditions, though when done at a small scale negative effects on these other species may be minimal (Lang et al. 2002).

CURRENT STAND CONDITION

The stand is currently dominated by loblolly and shortleaf pines both in basal area and tree density (Table 8.1). Approximately 8% of the basal area is in hardwoods, primarily oaks. Tree diameters range from 30 to 60 cm (12 to 24 in.) dbh, averaging 38 cm dbh. There are about 200 trees/ha and the stand basal area is approximately 23 m²/ha (101 ft²/acre). The stand has understory shrub and hardwood cover but little regeneration (Figure 8.1) and little grass or herbaceous cover. The presence of hardwoods in the stand suggests that it has not been burned for some time and that there may not be adequate fine fuels to carry a burn through the stand. Or worse, suspended fuels (needles in shrub crowns) may allow fires to scorch the crowns of the overstory trees. And if a thick duff layer exists, then the slow-burning hot fire may kill pines and release hardwoods. The stand currently could serve as foraging habitat for red-cockaded woodpeckers, but would not likely be adequate nesting habitat. The stocking is too high, there are hardwoods in the midstory, and the trees are likely not old enough to have heart rot. Nor is the stand likely habitat for Bachman's sparrows, because there would be

TABLE 8.1
Summary Statistics for an Example 7-ha Pine Stand from the Kisatchie National Forest, Louisiana, 2005

Species	Average dbh (cm)	Trees/ha	Basal area (m²/ha)
Loblolly pine	40.6	126	16.5
Shortleaf pine	33.9	54	6.0
Oaks	38.5	15	1.8
Total	38.6	195	23.3

FIGURE 8.1 Schematic of the current stand condition dominated by loblolly and shortleaf pine, Kisatchie National Forest, Louisiana. Based on simulations from the Landscape Management System. (From McCarter, J.M. et al. 1998. *J. For.* 96: 17–23.)

insufficient sunlight to support a dense grass–forb condition in the understory. The area could support brown-headed nuthatches now, especially if intertree competition has created snags (no snag data were available).

DESIRED FUTURE CONDITION

To provide nesting habitat for red-cockaded woodpeckers, brown-headed nuthatches, and Bachman's sparrows, we would need to have a stand in which the dominant trees were more than 100 years of age (increasing the probability that some had red heart disease), had a basal area closer to 16 m^2/ha or less to provide grass–forb understory conditions, a few midstory pines, and no hardwoods. Because the area once supported longleaf pine, we would want to reestablish longleaf as a functional component in the stand. We will also want to ensure that tree mortality is likely to occur at some times during stand development to provide snags as potential nesting sites for brown-headed nuthatches. Because pine basal area will be kept somewhat low and dominant trees well-spaced for Bachman's sparrows, snags may be created either from fire or they may need to be created by killing a few dominant trees per hectare periodically depending on tree mortality and snag fall rates. Once the desired future condition is reached we would like to maintain the condition for the foreseeable future to contribute to red-cockaded woodpecker population recovery and maintenance.

MANAGEMENT ACTIONS TO ACHIEVE THE DESIRED FUTURE CONDITION

Although the specifics of management approaches for these species varies from one author to another, the general theme is that stands be managed using either group selection (McConnell 2002) in longleaf systems or irregular shelterwood regeneration methods in loblolly–shortleaf systems (Hedrick et al. 1998, Rudolph and Conner 1996). Use of an irregular shelterwood approach displaces red-cockaded woodpeckers that might use the site for foraging, suggesting that group selection regeneration, planting of longleaf pine and frequent prescribed fire could be used to manage the site (personal communication, Dr. Robert Mitchell, Jones Center, Georgia). But to illustrate how both even- and uneven-aged systems can be used to manage habitat, I will initiate management with an irregular shelterwood. Irregular shelterwood methods retain all or a part of the overwood well into the next rotation, and can provide older and larger trees as habitat from one rotation to another. Management practices that reduce litter, maintain relatively low tree and shrub densities, and encourage the growth of forbs and grasses are recommended for Bachman's sparrows (Haggerty 1998). Haggerty (1998) suggested that a combination of thinning and burning would contribute to the most suitable habitat for this species. Similarly, brown-headed nuthatches also find suitable habitat in mature pine forests, but with snags present. For this example, I propose initiating management with an irregular shelterwood to remove hardwoods and establish longleaf pine, then manage stand density and hardwood competition through a combination of thinning and prescribed burning. The series of management actions needed to produce a desired future condition that follows is based on use of the Southern version of the Forest Vegetation Simulator (version SN) (Dixon 2003) and the Landscape Management System (McCarter et al. 1998).

Year 2015: We will initiate stand management with an irregular shelterwood retaining 25 pines/ha (8 m^2/ha; Boyer and Peterson 1983) and cutting all hardwoods. Following harvest, we will prepare the site for planting using a cool winter burn to remove fine fuels and then plant 250 longleaf pine seedlings per hectare. Because the overwood is dominated by loblolly and shortleaf pines, I will also assume that natural regeneration of these two species will become established following harvest. There should be no prescribed burning for the first few years after longleaf pine establishment (Boyer and Peterson 1983), and then early spring burns (prior to bird nesting) should be initiated to control hardwoods. Once longleaf regeneration is well established, we can continue with prescribed burning every 2 to 3 years to control hardwoods, shortleaf pine regeneration, and any brownspot disease that might become established on the longleaf pine seedlings. Longleaf persists in a "grass stage" for several years in which it puts on root growth but not shoot growth. Dense long needles protect the terminal bud from fires at this time and fires remove needles infected with brownspot fungus.

Year 2025: By 2025, I would anticipate that the new stand would be much lower in basal area with a high density of pine regeneration (Table 8.2; Figure 8.2). Note that although we thinned from below, the average dbh declined. This is because the stand is no longer even-aged. By 2025, there will be 20 trees/ha that are more than 56 cm dbh along with many that are small seedlings. Hence, average dbh may not be a good indicator of tree sizes in the stand. Also note that the basal area is far below our target. This low basal area is necessary to allow longleaf pine to become established while still retaining some trees that can become potential nest trees in the future. However, with the initiation of prescribed burning, habitat for Bachman's sparrows has likely improved tremendously. Finally, note that although we retained 25 trees/ha in the overwood, only 20/ha were predicted to survive the following 10 years. We will assume that 5/ha have become snags, some of which may remain after burning and provide potential nesting habitat for brown-headed nuthatches.

TABLE 8.2

Summary Statistics for a 7-ha Pine Stand Projected to the Year 2025, 10 Years after Initiating an Irregular Shelterwood, Kisatchie National Forest, Louisiana

Species	Average dbh (cm)	Trees/ha	Basal area (m^2/ha)
Longleaf pine	8.3	250	1.3
Loblolly pine	23.9	71	6.3
Shortleaf pine	6.9	50	0.1
Total	11.2	371	7.9

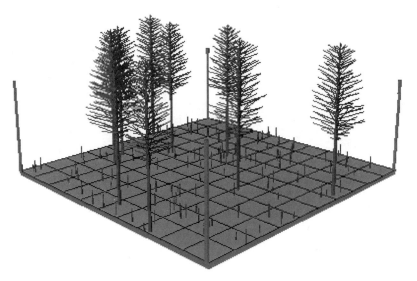

FIGURE 8.2 Schematic of the stand condition at year 2025, 10 years after initiation of management, Kisatchie National Forest, Louisiana. Based on simulations from the Landscape Management System. (From McCarter, J.M. et al. 1998. *J. For.* 96: 17–23.)

TABLE 8.3

Summary Statistics for a 7-ha Pine Stand Projected to the Year 2045, 30 Years after Initiating Management, Kisatchie National Forest, Louisiana

Species	Average dbh (cm)	Trees/ha	Basal area (m^2/ha)
Longleaf pine	12.2	120	1.4
Loblolly pine	37.8	59	9.1
Shortleaf pine	20.8	25	0.3
Total	20.6	203	10.8

Year 2035: With continued prescribed burning, hardwoods should be kept under control and longleaf pine will initiate height growth. The overstory trees, now released from intertree competition, will also continue to grow. Although red-cockaded woodpecker nesting habitat has likely not yet developed, habitat for Bachman's sparrows and brown-headed nuthatches is available. However, as basal area increases, overstory cover will reduce the production of grasses and forbs following burning. If frequent fire does not allow adequate production of grasses and forbs, then we may need to initiate a light thinning of the overstory by creating snags and a precommercial thin of the seedlings and saplings to ensure continued tree growth while providing growing space for grasses and forbs.

Year 2045: If regeneration of longleaf was not adequate, we can create gaps in the stand and replant longleaf pines, adding to the uneven-aged character of the stand. Or else we will let the stand grow, with early spring burns every 3 to 4 years (Table 8.3). Note though that regeneration may again reduce the availability of grasses and forbs and that some precommercial thinning may be necessary simply to maintain Bachman's sparrow habitat (Figure 8.3).

Year 2055: In addition to prescribed burning to reduce hardwoods and to maintain a grass–forb understory, density management will now be necessary to ensure that growing space is available for regenerating trees as well as the grass–forb layers. We will initiate a precommercial thin reducing the density of the regeneration by 50% if necessary to ensure continued stand development.

FIGURE 8.3 Schematic of the stand condition at year 2045, 30 years after initiation of management, Kisatchie National Forest, Louisiana. Based on simulations from the Landscape Management System. (From McCarter, J.M. et al. 1998. *J. For.* 96: 17–23.)

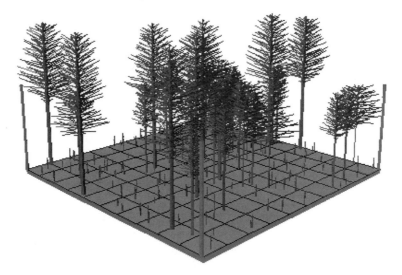

FIGURE 8.4 Schematic of the stand condition at year 2065, 50 years after initiation of management, Kisatchie National Forest, Louisiana. Based on simulations from the Landscape Management System. (From McCarter, J.M. et al. 1998. *J. For.* 96: 17–23.)

Year 2065: With the precommercial thin in the previous decade, and the site maintained with fire, additional regeneration may become established in the understory (Figure 8.4). Stand basal area has increased to nearly 15 m^2/ha (Table 8.4).

Year 2075: We will continue to manipulate the density of all tree sizes and by this time have entered a group selection regeneration system. Group sizes will vary, and cutting cycles of 10 to15 years will be needed to ensure that all size classes of longleaf pine continue to develop. We also will retain some of the large loblolly pine in the stand as potential nest trees. The stand will now be at the stage that red-cockaded woodpeckers may begin using the stand for nesting if adjacent stands provide adequate foraging habitat.

TABLE 8.4
Summary Statistics for a 7-ha Pine Stand Projected to the Year 2065, 50 Years after Initiating Management, Kisatchie National Forest, Louisiana

Species	Average dbh (cm)	Trees/ha	Basal area (m²/ha)
Longleaf pine	9.0	274	1.8
Loblolly pine	30.1	108	12.8
Shortleaf pine	6.9	50	0.2
Total	14.0	431	14.9

TABLE 8.5
Summary Statistics for a 7-ha Pine Stand Projected to the Year 2105, 90 Years after Initiating Management, Kisatchie National Forest, Louisiana

Species	Average dbh (cm)	Trees/ha	Basal area (m²/ha)
Longleaf pine	24.1	58	2.8
Loblolly pine	56.1	25	6.2
Total	33.5	83	9.0

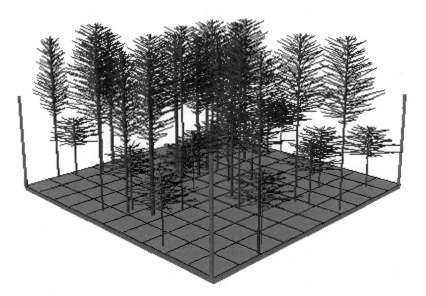

FIGURE 8.5 Schematic of the stand condition at year 2105, 90 years after initiation of management, Kisatchie National Forest, Louisiana. Based on simulations from the Landscape Management System. (From McCarter, J.M. et al. 1998. *J. For.* 96: 17–23.)

Years 2085–2105: Management will continue as needed in a maintenance mode, by managing stand density to ensure rapid tree growth, retaining potential nest trees, and burning to maintain a grass–forb understory (Table 8.5; Figure 8.5).

Year 2115: Additional density management may or may not be needed at this point. Monitoring data should be used to ensure that habitat elements needed by the three species are persisting.

FIGURE 8.6 Schematic of the stand condition at year 2125, 110 years after initiation of management, Kisatchie National Forest, Louisiana. Based on simulations from the Landscape Management System. (From McCarter, J.M. et al. 1998. *J. For.* 96: 17–23.)

TABLE 8.6
Summary Statistics for a 7-ha Pine Stand Projected to the Year 2125, 110 Years after Initiating Management, Kisatchie National Forest, Louisiana

Species	Average dbh (cm)	Trees/ha	Basal area (m²/ha)
Longleaf pine	22.4	101	5.3
Loblolly pine	64.1	25	8.2
Total	30.6	126	13.5

Year 2125: By this time the desired future condition (DFC) should be reached or nearly so (Figure 8.6). Monitoring data will help to determine if these stand growth projections produced a reasonable schedule toward development of the desired future condition. Basal area will now be just below our target of 16 m²/ha (Table 8.6), the stand will be 80% longleaf by density, and comprise nearly half the basal area. Loblolly pines are now over 60 cm dbh and over 150 years old. Continued burning has allowed persistence of a grass–forb understory to provide habitat for Bachman's sparrows. Either through natural mortality or by actively killing trees, loblolly pines can be a source of snags for brown-headed nuthatches. Stand density can be reduced even further by cutting some, but not all, of the large loblolly pines to release longleaf pines.

MONITORING PLANS

This plan is a hypothesis. We cannot be sure that the stand will respond as intended, but based on the species, sizes, and site conditions, the steps outlined here represent a reasonable approximation of how we might achieve our desired future condition. As we initiate management, however, we also should initiate a monitoring plan to see if the stand is developing as predicted. Because the stand prescription has a number of steps including regeneration cuts, thinning, and burning, it is important that each step be implemented correctly and in a timely manner. Hence, *implementation monitoring* would be conducted to ensure that the plan is implemented correctly. Questions that you might address in your implementation monitoring might include: Were any unmarked trees harvested? Were burns conducted in the correct seasons? Was the fire return interval as prescribed?

In addition to implementing the prescription correctly, we would want to know if we were effective in achieving both the stand structure as well as function. *Effectiveness monitoring* may take two parts. First we would monitor the survival, growth, and development of the vegetation to ensure that the habitat structure goals were being met. Regeneration surveys to assess survival can be done at 10 to 20 randomly established circular plots in the stand for the first 5 years after planting. Once seedlings are established and free to grow, tree growth and survival should be monitored. Tree growth and survival is typically monitored on continuous forest inventory (CFI) plots and may be a fixed plot or a variable radius (wedge prism) sample. Measuring tree growth and survival every 5 to 10 years would be adequate to see if the stand is developing as predicted. These samples will provide data on growth, basal area, density, and species composition, and should include information on snag densities, and understory cover. Understory-cover estimates can be used to assess if hardwoods are being effectively controlled by burning and if the grass–forb cover needed by Bachman's sparrows is developing (see Appendix 3 for approaches to measuring habitat structure).

We also want to know if the stand is functioning as we intended it. Hence, we will want some indication of species use of the stand over time. Conducting a census of red-cockaded woodpecker nest trees in the stand every 5 years can be quite easily done by searching for resin flows on trees. In addition, 3–5 variable circular plots can be established in association with the CFI plots to sample bird density. Plots would be visited four to six times during the breeding season during the early morning. All birds seen or heard are recorded at each plot, and the distance from plot center to the bird is estimated. These data not only provide evidence of birds using the plot, but also can estimate densities. If our focal species are using the stand, then more intensive nest searching may be needed to document reproductive success in the stand (and more closely estimate potential fitness). Samples taken every 5 years would allow estimates of trends in use of the stand over time. If the structure is not developing as intended or if the species are not using the stand as predicted, then the prescription can be revised to increase the probability of producing functional habitat for these species.

BUDGET

Regardless of the landowner, cost becomes a factor when implementing a prescription to achieve a goal. For most landowners, there must not be a net loss, and for some landowners there must be a maximum profit. Understanding the products that can be derived from implementing the prescription, both economic and ecological, can help the manager to decide if the trade-offs are acceptable.

I developed a budget for our prescription based on a number of assumptions. I assumed that:

1. Pine sawtimber would sell for at least $200/MBF (thousand board feet) and pine pulpwood for $20/cord as stumpage.
2. Site preparation burning would cost $425 and that in-stand burning would cost $175 each time the stand is burned.
3. A fire line can be established for $100.
4. Planting stock, planting and herbicide applications would collectively cost approximately $250/ha.
5. Labor costs associated with marking tallying, sale oversight and monitoring cost $20/h.

TABLE 8.7

Estimated Income and Expenses Associated with Managing a 7-ha Pine Stand on the Kisatchie National Forest to Improve Habitat for Red-Cockaded Woodpeckers, Bachman's Sparrows, and Brown-Headed Nuthatches

Decade	Activity	Income	Expenditure[1]
2005	Tree marking and tally		$(2,700.00)
	Fire line		$(100.00)
	Prescribed burn		$(187.00)
2015	49 MBF/ha sawtimber	$68,000.00	
	66 cords pine pulpwood/ha	$9,225.00	
	Site prep burn		$(425.00)
	Planting weed control		$(1,700.00)
	Implementation monitoring		$(200.00)
	Monitoring regeneration success		$(200.00)
	Prescribed burn (3)		$(561.00)
	Monitoring tree survival and growth		$(200.00)
	Monitoring animal use		$(700.00)
	Snag creation (if necessary)		$(1,000.00)
2025	Prescribed burn (3)		$(561.00)
	Monitoring tree growth and snag recruitment		$(200.00)
	Monitoring animal use		$(700.00)
2035	Prescribed burn (3)		$(561.00)
	Precommercial thin		$(1,700.00)
	Snag creation (if necessary)		$(1,000.00)
	Monitoring tree growth and snag recruitment		$(200.00)
	Monitoring animal use		$(700.00)
2045	Replant longleaf as necessary		$(1,700.00)
	Prescribed burn (3)		$(561.00)
	Monitoring tree growth and snag recruitment		$(200.00)
	Monitoring animal use		$(700.00)
2055	Prescribed burn (3)		$(561.00)
	Monitoring tree growth and snag recruitment		$(200.00)
	Monitoring animal use		$(700.00)
	Precommercial thin		$(1,700.00)
2065	Prescribed burn (3)		$(561.00)
	Monitoring tree growth and snag recruitment		$(200.00)
	Monitoring animal use		$(700.00)
2075	Prescribed burn (3)		$(561.00)
	Monitoring tree growth and snag recruitment		$(200.00)
	Monitoring animal use		$(700.00)
2085	Prescribed burn (3)		$(561.00)
	Monitoring tree growth and snag recruitment		$(200.00)
	Monitoring animal use		$(700.00)
2095	49 MBF/ha sawtimber	$68,000.00	
	47 cords pulpwood/ha	$6,401.00	
	Tree marking and tally		$(2,700.00)
	Implementation monitoring		$(200.00)
	Prescribed burn (3)		$(561.00)
2105	Prescribed burn (3)		$(561.00)
	Monitoring tree growth and snag recruitment		$(200.00)
	Monitoring animal use		$(700.00)
2115	Prescribed burn (3)		$(561.00)
	Monitoring tree growth and snag recruitment		$(200.00)
	Monitoring animal use		$(700.00)
Total		$151,626.00	$(30,683.00)

[1] Numbers in parentheses indicates expenditures.

Based on these assumptions and our prescription we would anticipate two sales that could be sold as stumpage. An irregular shelterwood in 2015 may produce 49 MBF per hectare of pine sawtimber and 66 cords of pulpwood per hectare. Group selection and thinning in 2095 would yield similar volumes (49 MBF/ha and 47 cords/ha). Income and expenses are outlined in Table 8.7 and clearly indicate the potential for having incomes exceed expenses while managing habitat for these three animal species. And I did not include interest earned on income over the life of the plan, which would make the plan even more feasible for private landowners. Are profits maximized? Clearly not. But for a nonfederal forest landowner, both habitat objectives and timber income can be realized in this example. For federal managers there is a positive net return on the investment of taxpayer dollars.

SUMMARY

Developing a prescription or plan for a stand or landscape entails articulating a set of goals and objectives. These objectives usually come in the form of describing a desired future condition, or set of conditions, that will likely produce both stand structure and function in the future. The challenge for the forest manager is to understand the potential of a site to achieve the desired future condition, the opportunities and constraints imposed by the current stand conditions, and the silvicultural approaches that should logically be applied to guide stand development toward one or more desired future conditions. Use of stand-growth models can aid in visualizing future stand conditions and may help in developing a schedule and budget to help ensure that a desired future condition is attainable. Finally, because any plan is a hypothesis, monitoring of stand structure and function must be conducted and data used to refine future management actions.

REFERENCES

Boyer, W.D., and D.W. Peterson. 1983. Longleaf pine. In Burns, R.M. (technical compiler), *Silvicultural Systems for the Major Forest Types of the United States.* USDA For. Serv. Agric. Handb. No. 445.

Conner, R.N., and K.A. O'Halloran. 1987. Cavity tree selection by red-cockaded woodpeckers as related to growth dynamics of southern pines. *Wilson Bull.* 99: 392–412.

Conner, R.N., A.E. Snow, and K. O'Halloran. 1991. Red-cockaded woodpecker use of seed-tree/shelterwood cuts in eastern Texas. *Wildl. Soc. Bull.* 19: 67–73.

Conner, R.N., D. Saenz, D.C. Rudolph, W.G. Ross, and D.L. Kulhavy. 1998. Red-cockaded woodpecker nest-cavity selection: relationships with cavity age and resin production. *Auk.* 115: 447–454.

DeLotelle, R.S., R.J. Epting, and J.R. Newman. 1987. Habitat use and territory characteristics of red-cockaded woodpeckers in central Florida. *Wilson Bull.* 99: 202–217.

Dixon, G.E. (compiler). 2003. *Essential FVS: A User's Guide to the Forest Vegetation Simulator.* USDA For. Serv. Forest Management Service Center, Fort Collins, CO.

Doster, R.H., and D.A. James. 1998. Home range size and foraging habitat of red-cockaded woodpeckers in the Ouachita Mountains of Arkansas. *Wilson Bull.* 110: 110–117.

Dunning, J.B. 2006. Bachman's sparrow (*Aimophila aestivalis*). In Poole, A. (ed.), *The Birds of North America.* Cornell Laboratory of Ornithology, Ithaca, NY.

Haggerty, T.M. 1998. Vegetation structure of Bachman's Sparrow breeding habitat and its relationship to home-range. *J. Field Ornithol.* 69: 45–50.

Harlow, R.F., and D.C. Guynn Jr. 1983. Snag densities in managed stands of the South Carolina coastal plain. *So. J. Appl. For.* 7: 224–229.

Hedrick, L.D., R.G. Hooper, D.L. Krusac, and J.M. Dabney. 1998. Silvicultural systems and red-cockaded woodpecker management: another perspective. *Wildl. Soc. Bull.* 26: 138–147.

James, F.C., C.A. Hess, and D. Kufrin. 1997. Species-centered environmental analysis: indirect effects of fire history on red-cockaded woodpeckers. *Ecol. Appl.* 7: 118–129.

James, F.C., C.A. Hess, B.C. Kicklighter, and R.A. Thum. 2001. Ecosystem management and the niche gestalt of the red-cockaded woodpecker in longleaf pine forests. *Ecol. Appl.* 11: 854–870.

Kessler, W.B., H. Salwasser, C. Cartwright Jr., and J. Caplan. 1992. New perspective for sustainable natural resources management. *Ecol. Appl.* 2: 221–225.

Landres, P.B., P. Morgan, and F.J. Swanson. 1999. Overview of the use of natural variability concepts in managing ecological systems. *Ecol. Appl.* 9: 1179–1188.

Lang, J.D., L.A. Powell, D.G. Krementz, and M.J. Conroy. 2002. Wood thrush movements and habitat use: effects of forest management for red-cockaded woodpeckers. *Auk* 119: 109–124.

McCarter, J.M., J.S. Wilson, P.J. Baker, J.L. Moffett, and C.D. Oliver. 1998. Landscape management through integration of existing tools and emerging technologies. *J. For.* 96: 17–23.

McConnell, W.V. 2002. Initiating uneven-aged management in longleaf pine stands: impacts on red-cockaded woodpecker habitat. *Wildl. Soc. Bull.* 30: 1276–1280.

Plentovich, S., J.W. Tucker Jr., N.R. Holler, and G.E. Hill. 1998. Enhancing Bachman's Sparrow habitat via management of red-cockaded woodpeckers. *J. Wildl. Manage.* 62: 347–354.

Rudolph, D.C., and R.N. Conner. 1996. Red-cockaded woodpeckers and silvicultural practice: is uneven-aged silviculture preferable to even-aged? *Wildl. Soc. Bull.* 24: 330–333.

Rudolph, D.C., R.N. Conner, and J.R. Walters. 2004. Red-cockaded woodpecker recovery: an integrated strategy. In Costa, R., and S.J. Daniels (eds.), *Red-Cockaded Woodpecker: Road to Recovery*. Hancock House Publishing, Blaine, WA.

Wilson, C.W., R.E. Masters, and G.A. Bukenhofer. 1995. Breeding bird response to pine-grassland community restoration for red-cockaded woodpeckers. *J. Wildl. Manage.* 59: 56–67.

Withgott, J.H., and K.G. Smith. 1998. Brown-headed nuthatch (*Sitta pusilla*). In Poole, A., and F. Gill (eds.), *The Birds of North America*, No. 349. The Birds of North America, Inc., Philadelphia, PA.

Zwicker, S.M., and J.R. Walters. 1999. Foraging habitat guidelines for the endangered red-cockaded woodpecker (*Picoides borealis*). *J. Wildl. Manage.* 63: 843–852.

9 Riparian Area Management

Water. A requirement for most if not all life. And free-flowing or standing water is indeed required by many species. Many amphibian species need free water during at least part of their life to lay eggs that will hatch hopefully before the water evaporates (Figure 9.1). Others spend all of their lives in water, such as aquatic forms of Pacific giant salamanders. Waterfowl use lakes, ponds, streams, marshes, and swamps as places to feed and raise young. River otters, snapping turtles, and bullfrogs are all aquatic predators that use resources in both aquatic and terrestrial environments (Figure 9.1). Because so many species are associated with free water, the interface between the water and the land, that area known as a *riparian area*, is usually given special consideration during forest management.

Riparian areas are perceived by many ecologists as being particularly important to animal species for several conditions:

- Free water is available near food and cover for many species.
- Because of the high latent heat of water, riparian, and wetland areas can have more humid and stable microclimates than adjacent uplands.
- Edges are often formed between riparian and upslope vegetative communities, enhancing the number of niches for vertebrate species.
- Streams and riparian areas can provide corridors (or barriers) for animals moving across the landscape.
- Net primary production is often higher in riparian and wetland areas and hence food quantity and quality tends to be higher there for many species.

The availability of free water near the food and cover resources in the terrestrial environment is critical for a large number of vertebrate species. For example, mountain beaver occur near riparian areas (or in very moist forests) because their primitive uretic system requires a humid environment (Schmidt-Neilson and Pfieffer 1970). White-tailed deer use water as a means of escaping predators in bottomland forests. Elk and deer will often use riparian areas as a cooler environment during warm weather. Because of the high latent heat of water (it takes 1 calorie to raise 1 g of water 1°C), riparian and wetland areas can have more stable microclimates than adjacent drier uplands. Some species are able to take advantage of these more stable riparian conditions.

Edges are often formed between riparian and upslope vegetative communities, enhancing the plant species richness and vertical complexity of the forest (Naiman et al. 2000). Horizontal complexity in vegetation is also often high because of the dynamic nature of stream systems creating canopy gaps and patches of shrubs and trees of various species and sizes, some of which are soft-mast producers (Naiman et al. 2000). This increased complexity provides a wide range of habitat element types, abundances, and distributions, thereby providing greater opportunities for more species to occur than in simpler systems.

Although riparian areas can provide many of the elements needed by some species, they also can provide corridors or barriers for animals moving across the watershed. A water shrew may find a fast moving river an excellent corridor, but a short-tailed shrew or Pacific shrew may find the same river to be a significant barrier. Again the function of the stream as a connector or barrier is species specific (Henein and Merriam 1990, Savidge 1973).

FIGURE 9.1 Species typical of riparian areas clockwise from top left: Snapping turtles are found in slow-moving streams and ponds (photo by Mike Jones and used with his permission), spotted salamanders reproduce in vernal pools in the northeastern United States (photo by Mike Jones and used with his permission), cottonmouths are an aquatic pit viper of southeastern swamps and marshes, and beavers create ponds used by many other species.

Net primary production can be quite high, especially in slow-moving rivers and side channels, ponds, and lakes. Marshes represent one of the most productive ecosystems on earth (Waide et al. 1999). Aquatic productivity is dependent on nutrient availability and sunlight in the water column. Nutrient-rich aquatic systems, referred to as *eutrophic systems*, can have very high levels of primary production through algal blooms, submerged vegetation, and emergent vegetation. But there is a price to pay for too much productivity. For instance, enriched streams with high levels of nitrogen and phosphorus from fertilizers can lead to algal blooms, which provide plant food for herbivores, but these plants can also deplete the water of oxygen. As plants photosynthesize, they return oxygen to the water, but in the evening they respire and use oxygen to survive while releasing carbon dioxide. Without adequate oxygen during the night many fish die from oxygen deprivation, reducing not only the fish fauna but also the food resources for their predators. So, is reducing nutrient loads good? Perhaps, but again it depends on the species. Species that rely on a grazing-based energy acquisition by eating aquatic vegetation (e.g., carp) need nutrient-rich systems, while species that feed on aquatic insects associated with decaying leaves or with trees over the stream (e.g., trout) often do better in nutrient-poor (*oligotrophic*) streams.

ANIMAL ASSOCIATIONS WITH RIPARIAN AREAS

Species richness is often higher in riparian areas than in adjacent upslopes (Knopf et al. 1988), especially in arid environments. In arid environments the availability of water not only provides the opportunity for evaporative cooling but also supports a richer and more palatable vegetative community than adjacent upslope areas. In moist environments the importance of riparian areas to many species is significantly reduced (McGarigal and McComb 1992).

Nonetheless, in every forest system in North America, there is a suite of species that require free water. These species are referred to as *riparian obligates* — you only find them near water. River otters, American dippers, beavers, and wood ducks are examples of riparian obligate species. Other species tend to be found more commonly near water but do not require the water directly for some aspect of their life. Yellow warblers, jumping mice, and several species of myotis bats and bald eagles are examples of *riparian associates*.

One group of riparian obligate species, pond-breeding amphibians, preferentially breed in isolated ponds and wetlands that hold water for only a part of the year. These *vernal pools* provide a predator-free (no fish) environment where the pond breeders can lay eggs and the larvae can feed on decaying vegetation, metamorphose into adults, and leave the pond before the pond dries out (Semlitsch and Bodie 1998). In some years ponds dry out too soon, stranding the immature amphibians, while in other years the water lasts until after they have metamorphosed. Apparently, pond-breeding species that reproduce in vernal pools have a selective advantage over those that use permanent fish-bearing water bodies where they face greater risks of predation.

There are also some species that occur in forests that are not associated with riparian areas. They may obtain their water from condensation, rain, or their food. So although riparian areas are an important component of the forested landscape for many species, other upslope parts of the landscape are also important for additional species. Often forest management regulations focus on riparian areas because of the requirements to provide clean water, habitat for fish species, as well as habitat for terrestrial species, but may not consider the condition of the forest for other species some distance away from the riparian area. Such a focus on riparian areas can be the detriment of other species not associated with riparian areas.

GRADIENTS WITHIN RIPARIAN ZONES

There are two dominant gradients of habitat elements associated with riparian areas, the intrariparian and transriparian gradients. *Intrariparian gradients* refer to the continuum of conditions from the headwaters to the confluence with larger water bodies and eventually with the ocean at an estuary. Riparian areas are hierarchical systems constrained by a *watershed*, the area of the land that captures and routes water down hill (Figure 9.2). The upper reaches of a stream system are usually intermittent

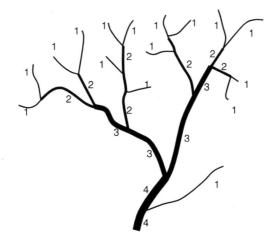

FIGURE 9.2 Hierarchical system of stream orders along an intrariparian gradient. Note that two first-order streams form a second-order stream and two second-order streams form a third-order stream, and so forth.

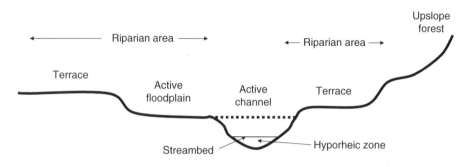

FIGURE 9.3   A schematic of a transriparian gradient from streamside to upslope.

streams that flow only following rain or during snowmelt. Long considered somewhat irrelevant to the function of the permanent stream system, intermittent and headwater streams can provide important refugia for many species of amphibians (Sheridan and Olson 2003, Stoddard and Hayes 2005). These areas can also serve as conduits for transport of nutrients, sediment, and pathogens from upslope areas into the stream system (Naiman et al. 2000). If water temperature, sediment loads, or other pollutants are of concern downstream, then these intermittent streams deserve attention, in addition to meeting habitat needs for species associated with these areas (Wigington et al. 2005). Roadside ditches also often serve as conduits of sediments and chemicals into the stream system. Indeed, road systems are the primary cause of excessive sediment loads in many forested stream systems (Reid and Dunne 1984).

Depending on the geology of the area, the upper portions of mountain watersheds usually support streams that have high gradients and deeply incised channels. The opportunity for the stream to move from side to side and develop a complex floodplain with side channels, pools, riffles, and glides are limited until the gradient declines somewhat and sufficient time has elapsed to lead to stream-bank erosion, deposits of sediments, and expansion of the active channel (Figure 9.3). Within the middle part of watersheds, adequate stream volume usually leads to the development of a floodplain following successive flood events that not only erode stream banks but also deposit sediments from farther upstream. These *alluvial* sediments often form terraces that represent different flood intensities and frequencies (Naiman et al. 2000). Intense floods that carry sediments and wood from the headwaters into the midwatershed area spill over the banks of the active channel into the active floodplain. As the water slows along the flooded edges, it cannot carry the same sediment load, and so it deposits the sediments in this floodplain. Frequent floods such as this allow the development of an active floodplain where tree species such as cottonwoods and willows may become established. Species such as these are well adapted to colonizing the exposed sediments when the water levels recede. They have seeds that are carried on wind and water and deposited on the sediments, and they grow rapidly to claim the site. These midwatershed areas often have braided channels as the stream velocity slows, depositing sediments (Naiman et al. 2000). Depending on the stream volume under high flows, the midwatershed is also the portion of the stream system where beavers most often build dams, creating a staircase of ponds. The active floodplain also provides an opportunity for the stream channel to begin to meander and create a more complex stream channel system. Side cutting of the channel allows the formation of steep stream banks with bars, or areas, of deposition on the opposite side of the stream (Figure 9.4). Steep stream banks provide places for belted kingfishers to nest, and the bars provide nesting sites for spotted sandpipers and other shorebird species. Finally, as the stream approaches an estuary, water velocity reduces further and is affected by tides causing sediment loads to further decline, forming a delta. These delta conditions can allow the formation of *marshes* (wetlands dominated by nonwoody vegetation) and *swamps* (wetlands dominated by woody vegetation). As the water flows into estuarine conditions, many tree and shrub species cannot tolerate the saline conditions and are replaced by grasses, sedges, and rushes to form a marsh. Swamps that

(a)   (b)   (c)

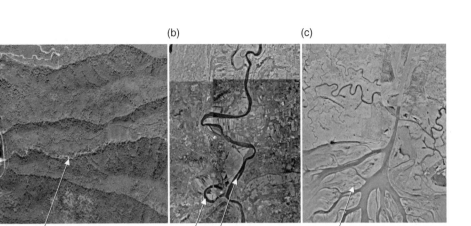

Mainstem Cummins   Oxbow lake and   Distributary pattern
Creek with trellis   sediment bars on   of channels in the
pattern of tributaries   the Connecticut   Wax River Delta
                          River

FIGURE 9.4   Three types of stream morphology in Cummins Creek: Coastal Oregon (a) with steep gradient and narrow valley floors; the Connecticut River Valley, in Massachusetts (b) with broad floodplains and oxbow lakes; and the Wax River Delta in Bayou Vista Louisiana (c) with high sinuosity and delta formation. Note as floodplain width increases and gradient decreases, sinuosity and braided channels increase leading to greater riparian complexity. (Photos from Terraserver.com and used with permission from U.S. Geological Survey and Microsoft Corp.)

often occur at the lower end of a watershed are frequently flooded forests dominated by species tolerant of being partially submerged for prolonged periods, such as baldcypress and water tupelo in the Deep South of the United States. Because of the nutrient-rich environments often associated with these swamps, they tend to be highly productive areas for growth of trees and shrubs and for the animals associated with them. Louisiana's Atchafalaya Basin is the largest swamp in the United States at 241,000 ha (595,000 acres). Because this area and others like it along the Mississippi river and other large rivers produce nutrient-rich floodplain soils, many have been cleared for agriculture and more recently for urban expansion, despite the risks of repeated flooding.

Along the intrariparian gradient, changes in the geomorphology of the watershed give rise to areas of different stream velocities and stream substrates. Continuously rushing water over boulders produces a cascade, which is excellent habitat for Harlequin ducks and tailed frogs. Where an erosion-resistant rock or a log partially dams the stream, a pool can form both above and below the obstacle (Figure 9.5). The upstream pool collects water and allows sediment deposits to build up above the obstacle aggrading the stream channel. As the water flows over the object it forms a plunge pool. These areas provide places where beaver often initiate a dam (Leidholdt-Bruner 1990). Both areas can add to stream channel complexity and hence provide conditions suitable to more species of aquatic and semiaquatic animals.

Large logs falling into a stream or carried downstream from headwaters (often from landslides), can add significantly to channel complexity in many forested areas. Indeed, the transport of wood from the headwaters to the estuary provides opportunities for use by many species from salamanders to otters, and salmonids to mollusks along its journey to the sea (Maser et al. 1988). This stream complexity contributes habitat elements and increases the potential number of species that use the variety of conditions that are created. The variability in flood frequency and severity adds complexity such as fine-scaled features of topography which are strongly related to plant species richness and species composition (Naiman et al. 2000).

Transriparian gradients refer to the changes in conditions as you move from the edge of the stream into upslope forests, perpendicular to the gradient (Figure 9.3). Along the edge of a stream

FIGURE 9.5 Dead wood in a stream in the Berkshires of Massachusetts reduced stream flow and caused sediment deposition upstream and a plunge-pool downstream, adding to channel complexity.

there is often the opportunity for emergent plants to become established, especially if there is adequate sunlight and the gradient is not too steep. Along the edge of a steep active channel, erosion and deposition often prevent trees from becoming established; so water-tolerant herbs and occasionally shrubs dominate. Farther upslope, species of trees that are water tolerant begin to dominate in those areas that are flooded less frequently. Species such as red alder, cottonwoods, silver maple, water oak, and pin oak can be found here (depending on what region of North America you are in). As you move farther upslope, these species give way to others that are not so tolerant of saturated soils and are more drought tolerant; eventually you are in an upslope tree community. But we can see changes in plant species along the moisture gradient all the way to the ridgetop (Figure 4.5). As these plant species change along this transriparian gradient, many animal species also change accordingly. In the northwest, we might find Pacific jumping mice and Dunn's salamanders in the floodplains near the stream and not find them farther upslope (McComb et al. 1993). On the other hand, we might find western red-backed voles upslope but rarely along the stream. Many species distribute themselves throughout a watershed in response to the moisture, soil, and vegetation seen along the transriparian gradient.

RIPARIAN FUNCTIONS

The function of a riparian area in providing energy, nutrients, and other resources to animals is heavily influenced by a set of processes linking streams to adjacent forests. Forested streams often produce cool, clean water; so it is not surprising that most public water supplies are from forested watersheds. Forests are natural filters. But because they are clean, they tend to be low in nutrients. Low nutrient availability can limit the productivity of fish and invertebrates in the stream and hence can limit

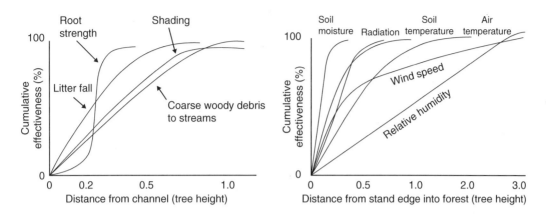

FIGURE 9.6 Hypothesized functional relationships from streamside forests. (From Forest Ecosystem Management Assessment Team (FEMAT). 1993. *Forest Ecosystem Management: An Ecological, Economic, and Social Assessment.* Portland, OR. US Forest Service, US Department of Commerce, National Oceanic and Atmospheric Administration, National Marine Fisheries Service, US Bureau of Land Management, Fish and Wildlife Service, National Park Service, Environmental Protection Agency.)

the abundance of their predators. Indeed, low availability of nitrogen, phosphorus, and carbon in streams of the Pacific Northwest has been suggested as a reason for poor productivity of salmon that once thrived in these streams. Because millions of salmon once spawned and died in streams, the decomposing bodies provided a rich food source for native fish, including young salmon. Now that salmon runs are at 5% or less of the historic levels, this source of nutrients for juvenile salmon is no longer present. In addition, there is evidence that these salmon-derived nutrients influence the function of the entire riparian area (Compton et al. 2006). As a way of giving these streams a boost in nutrients, some fisheries biologists have begun stocking dead salmon to enrich the stream (Compton et al. 2006). But not all forested streams have a natural source of anadromous fish carcasses and so they receive nutrients in other ways. Leaves from trees and shrubs are nutrient rich. Most nutrients in a tree are not stored in wood (except for carbon) but rather in leaves. When leaves and needles fall into a stream as *allochtonous* material, they decompose and provide an energy source for decay organisms. This material then serves as a source of detrital-based energy in nutrient-poor (oligotrophic) streams. Consequently, having a streamside forest ensures that the leaves that fall into a stream provide an important energy pathway (Figure 9.6). Litter falling into the stream provides a source of nutrients and food for aquatic invertebrates, which in turn are food for vertebrates. Streamside forests also provide root systems along the stream margin, stabilizing the streambanks and minimizing erosion and hence sediment loads. Sediment loads that are too high can cover cobble and gravel substrates important to many species of spawning fish (Kondolf 2000), as well as reducing habitat quality for species of salamanders that occur in well-aerated gravels. When trees die or are carried downstream, the wood increases channel complexity. Large pieces of wood are particularly important in large streams because they tend to move through the system more slowly than smaller logs. When a log falls into a stream, water flowing over and under it increases in velocity. Increased water velocity under the log leads to scour pools and the log forms a source of cover for fish and amphibians, as well as a place for otters to feed, muskrats to mark territories, and salamanders to survive in moist conditions.

Dead wood decay rates in water are usually less than on dry land because aerobic decomposition is decreased in water. Indeed, to store logs for long periods they are often submerged in log ponds prior to milling. Nonetheless, logs will eventually decay or move through the stream system; so a continual supply of logs is needed to ensure that these processes of developing channel complexity are maintained.

Tree canopies also provide shade to water and are especially important during the summer. Direct solar radiation striking a water surface can raise the temperature (Sinokrot and Stefan 1993) and increased temperatures can be deadly for some species. Tailed frogs in the western United States usually only occur in streams with water temperatures of 5 to 17°C. Many other species of amphibians and fish also are associated with cold water, and even American dippers, which feed on invertebrates in cascading mountain streams, are often associated with cold water. But some sunlight should enter streams. Sunlight allows photosynthesis to occur in the stream allowing the formation of periphyton on rocks, as well as other aquatic vegetation to grow (Welnitz and Rinne 1999). Periphyton, which grows best under fluctuating light levels, along with decaying plants, provides energy to other species. So a continuous riparian leaf cover, especially conifer cover, may not be ideal for most stream species. The manager often will strive to find a balance between providing enough leaf cover to maintain stream temperature but not so much that it inhibits photosynthesis in aquatic plants throughout too much of the system. The forest structure and composition along riparian areas all contribute to a set of microclimatic conditions that influence habitat quality for some species up to three tree heights from the stream edge, depending on the upslope forest condition (Figure 9.6). If the upslope forest is a recent clearcut or open rangeland, then the advantage of the microclimatic buffering to the riparian area may be critical to allowing some species to persist. On the other hand, if the upslope forest is similar in structure and to the streamside forest, then the stable microclimate can probably be realized over less than three tree heights. The height of mature trees, rather than a fixed distance, is often used as a guideline for maintaining riparian functions because many factors (e.g., shade and dead tree fall) relate more to tree height than to a fixed distance.

RIPARIAN BUFFERS

Most states and provinces in North America have regulations regarding stream protection (Figure 9.7). Often the rationale for these streamside management areas include protecting water quality (especially temperature and sedimentation) and providing habitat for "fish and wildlife." But by now you should be asking, which fish and which wildlife? Therein lies the conundrum.

FIGURE 9.7   Streamside buffer in a managed Oregon forest. These buffers are required by the Oregon Forest Practices Act. (Photo by Dave Vesely and used with his permission.)

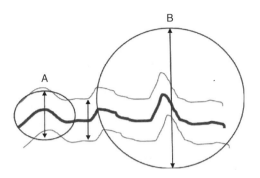

FIGURE 9.8 The width of a riparian management area is different for different species. Species A has a territory size that fits within the proposed buffer; species B has a territory size that would not be accommodated by the buffer.

These streamside management areas may be no-harvest zones, or they may allow harvest down to some minimum basal area, or they may be designed to meet specific needs. Typically, a linear distance from the active channel is used to delineate the buffer strip width (Figure 9.6). Habitat for selected species of vertebrates will be provided dependent on the width and length of the buffer (riparian management area). Buffers that are narrow will likely be used only by species with small home ranges or those that have naturally linear home ranges (e.g., belted kingfishers and mink). The appropriate width will depend on the species and what it is being used for a nesting territory, movement corridor, feeding, or resting. There is not a "one size fits all" prescription for these riparian management areas.

As a first approximation of the necessary width of a riparian buffer for a species, consider its home range and the habitat elements that it needs. If it is a riparian associate and has a home range of 20 ha (50 acres), then the radius of that home range is 142 m (470 ft; $A = \pi r^2$). If the species is mobile and the stream does not represent a barrier to movement, then the width of the buffer should be 142 m on each side of the stream (Figure 9.8). If the stream is a barrier to movement then the width would need to be two times as large or 284 m (940 ft) on each side of the stream. Managing for multiple species requires assessing buffer width for each species. The species requiring the largest buffer width sets the width for the others. So the appropriate buffer width for a small amphibian with a home range of 0.2 ha would be 14 m (47 ft). Correct? Well, maybe. If the species is associated with the riparian area because of the moisture needed for survival (a salamander or frog), then designing a buffer to support a suitable microclimate may be more appropriate. Vesely and McComb (2002) found that buffers probably need to be far wider than would be expected for salamanders based on what is assumed to be a small home range for these species.

Another aspect of buffer width to be considered is the landownership pattern. Buffer strips may be very wide on some federal ownerships or where public drinking water supply is a primary use of the water. The adjacent private landowner may be required only to provide a buffer of 15 m (50 ft) and be able to harvest half of the basal area from it. And the next landowner along the stream may be a dairy farmer who is not required to leave any buffer such that cows graze to the stream bank. The ability for the stream to meet goals of the various landowners is inherently compromised by who is upstream of whom. Should the farmer own the upstream parcel and allow cows to use the stream, then the south end of a north-facing cow (doing what cows will do) will contribute significantly to poor water quality on the downstream ownerships. It takes a cooperative community to manage clean water in a watershed.

Managing within Streamside Management Areas

Many landowners approach riparian areas as set-asides — areas not to be managed. And regulations certainly restrict management options (Figure 9.9). But within the limitations of what is allowed by

FIGURE 9.9 Two species of aquatic amphibians captured in a forest stream in Coastal Oregon, larval forms of Pacific giant salamander and tailed frog tadpoles, in addition to a cutthroat trout. State and federal riparian rules are designed to maintain water quality and protect species such as these. (Photo by Joan Hagar and used with her permission.)

law, the principles of managing habitat elements described in the previous chapters also apply to riparian areas. Depending on the species that you would like to provide habitat for in a riparian area, you can manipulate the vegetation to provide or produce those habitat elements. But as mentioned many times already, managing for one species or a set of species is managing against other species. Over the past 10 to 20 years in the Pacific Northwest there has been an effort to restore conifer dominance to many riparian areas so that there is a continual supply of large dead wood to the stream (MacCracken 2002). This dead wood adds to channel complexity and habitat quality for salmonids and other species. But at least two things must be kept in mind when trying to achieve this goal. First, some species such as white-footed voles (Manning et al. 2003) and several species of myotis bats (Holly Ober, Dep. of Forest Science, Oregon State Univ. personal communication) select hardwood riparian areas over conifer-dominated riparian areas as places to feed. So, some hardwood riparian areas should be retained. Second, dead wood not only comes from the streamside but also from the upslope areas during landslides and hence streamside trees are only part of the source of dead wood to these streams. The conclusion drawn from this information is that there is not a single riparian management strategy that will meet the needs for all species, just as there is no stand or landscape management strategy that will meet the needs for all species.

For instance, if you wished to improve soft-mast production along streamsides, then providing openings in the canopy would provide sunlight to fruit-producing plants and enhance production of soft and/or hard mast. But too much sunlight can raise water temperature and reduce habitat quality for some species. Balancing these conflicting goals is a social dilemma — what are the goals for the stream system? But when setting goals for riparian areas it is important to consider the context within which the riparian area resides. Is it an agricultural area with abrupt edges, a clearcut (Figure 9.7), a thinned stand, or an old-growth stand? How the riparian area will function as habitat for a suite of species will be greatly affected by these adjacent conditions. And although the goals for riparian areas may be different than for upland systems, the two systems are connected and should be considered part of a larger landscape or watershed and not managed in isolation.

Several important principles have been provided by others for maintaining or restoring riparian conditions to meet a variety of riparian goals (Naiman et al. 2000, National Research Council 1999):

- Restoring biophysical properties of riparian zones improves other natural resource values. Riparian zones allowed to respond to disturbance and regrowth may maintain a high level of complexity in plant species composition and structure used by a variety of animal species.
- Protecting interactions between surface flows and groundwater is essential to aquatic–riparian ecosystem integrity. This is particularly important relative to the *hyporheic zone*, the subsurface saturated sediments along the stream bottom (Stanford and Ward 1988).
- Allowing streams and rivers to migrate laterally is necessary for development of riparian habitat elements. This continual disturbance creates a mosaic of substrates and vegetation used by a wide variety of species.
- Incorporating natural flow regimes in regulated rivers promotes aquatic and riparian diversity and resilience. Many species are well adapted and indeed rely on the variability in flow rates in rivers and streams that have occurred for centuries prior to use of dams and levees.
- Modifying human-imposed disturbance regimes can create and maintain a range of habitat conditions in space and time within and among watersheds that reflects the range of conditions to which desired species are well-adapted (Reeves et al. 1995). Humans like stability. Unexpected large-scale disturbances are considered catastrophes by many humans, but it is these events, over entire watersheds, that can influence changes in habitat availability for the full suite of terrestrial and aquatic species found in a region.
- Controlling invasive species that can simplify vegetation structure and composition. Aquatic and streamside vegetation that is invasive can exclude other *phreatophytic* (water-associated) vegetation from the site decreasing the vegetative structure and composition of the streamside area.

BEAVERS — THE STREAM MANAGERS?

Busy beavers just love to stop water from flowing. Although not all beaver populations build dams (some live in dens in streambanks), most build quite impressive dams that flood large areas for long periods. In many respects they are the streamside managers or destroyers. Much depends on your perspective (see Chapter 1). Beavers cut trees and shrubs and use them to build dams, lodges (where they raise young), and as a source of food when they eat the bark from these plants. The openings that they create in streamside forests provide a flush of early-successional plants along streams and provide habitat for a wide variety of early-successional associated species (e.g., yellow warblers, jumping mice, and Carolina wrens; Figure 9.10). Their dams create a *lentic* (lake-like) environment out of a *lotic* (flowing water) system, providing brood habitat for wood ducks and places for pond-breeding amphibians to reproduce (e.g., newts). These changes in vegetation and pool conditions give rise to different assemblages of animal species in the vicinity of beaver ponds than where beaver ponds are not present (Suzuki and McComb 2004). The dams cause ponds to capture sediments and if the pond persists long enough, then it fills in with sediments over time and forms a wet meadow, which is also habitat for a completely different set of organisms.

To a forest manager trying to raise a commercial timber species near a stream or trying to develop forested riparian conditions for other species, beaver can be a problem (Bhat et al. 1993). They preferentially cut hardwoods over conifers and small trees over large ones (Basey et al. 1988). But they eventually cut nearly all trees around their pond. So planted seedlings are often cut and

FIGURE 9.10   Beavers flood bottomland forests creating snags, meadows, and pools, dramatically changing the composition of the riparian animal community in a stream reach.

mature trees may be cut as well, especially those close to the stream. In addition, their dam leads to flooding of low-lying areas, causing tree death. And because culverts are easy places to block and flood a large area upstream of the culvert, damage to culverts and roads also becomes problematic (Jensen et al. 2001). Use of beaver deceivers, devices to allow water to flow through a dam, can help to reduce the damage caused by flooding (Nolte et al. 2001). Tree cutting however can only be reduced if trees are protected by wire mesh — an expensive proposition for a forest manager.

So, at what point do beavers become a problem vs. a natural part of riparian area dynamics? Again that is a social decision. And as beavers do not respect property lines, decisions made by one landowner clearly influence the riparian conditions of his or her neighbor.

CASE STUDY: RIPARIAN AREA MANAGEMENT IN A PATCHWORK OWNERSHIP

To illustrate the conundrum associated with riparian area management, consider the pattern of riparian management areas along streams in western Oregon, where the Bureau of Land Management (BLM) manages public forestland. During 1866, Congress created the Oregon and California railroad lands as alternating square miles of land in western Oregon as an incentive for the railroads to build infrastructure into the region. Congress revested or pulled these lands back into public ownership in 1916 and eventually gave the BLM the responsibility for managing them, mandating that a portion of the timber receipts go to the counties to support schools. Hence, the BLM now manages a checkerboard of lands across much of western Oregon. Intervening lands are privately owned, many of which are forested, but some of which are agricultural.

For years, the BLM followed the Oregon regulations with regard to streamside protection, but when concerns arose regarding endangerment of spotted owls, marbled murrelets, and coho salmon, a new management strategy for all federal lands in the region emerged as the Northwest Forest Plan

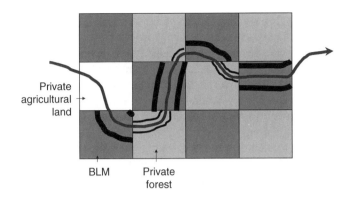

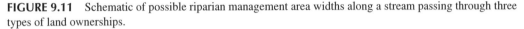

FIGURE 9.11   Schematic of possible riparian management area widths along a stream passing through three types of land ownerships.

(NWFP; FEMAT 1993). So now a stream passing across this landscape has three predominant land uses: federal forests designed to provide habitat for late-seral species, state and private forest lands that are largely timber producers; and private agricultural lands for crop and livestock production. Each land use has its own standards and guidelines for riparian management (Figure 9.11).

Under the NWFP, buffers along non-fish-bearing streams are one site tree height in width and along fish-bearing streams they are two site tree heights in width. A site tree is the height of the dominant trees in the region — 45 m (150 ft) in many of these watersheds. So, in this example, buffers on fish-bearing streams would be 90 m (300 ft wide) and on non-fish-bearing streams they would be 45 m. These buffers could be adjusted up or down depending on the results of a watershed analysis, a process where the functions of the riparian area are more completely considered before additional management is allowed (Montgomery et al. 1995).

On private forested lands, the Oregon Forest Practices Act (OFPA) prescribes streamside management areas that vary depending on stream width and if they are fish-bearing streams (Figure 9.11). Typical OFPA buffers might be 8 to 33 m (25 to 100 ft) in width, with some timber removal allowed within them if canopy cover and basal area guidelines are met. On private agricultural lands, buffers or other management actions are not required, unless the stream is listed as "impaired" under the Clean Water Act (Hill and Blair 2005). Impairment may be caused by many things, but most often by increased temperature, sediment, nutrients, or pesticides compared with standards set by state environmental quality offices.

Now consider a stream passing through a set of these ownerships (Figure 9.11). What imprint does land ownership have on the ability of a stream to maintain certain functions? If the stream passes through agricultural land causing the water to become too warm, and the stream reach to be listed as impaired, and that warm water flows through an adjacent forest owned by someone else, then the downstream neighbor inherits the problem. And we all live downstream. We see similar inconsistencies in streamside protection in forested regions across the country. So is there a more thoughtful design? Clearly there is, but a more thoughtful design makes regulation of policies very difficult. The challenge is to provide incentives to encourage a more thoughtful and cooperative streamside protection strategy that allows streams and other water bodies to meet habitat, clean water, recreation, and other goals.

SUMMARY

Riparian areas are adjacent to and influenced by a body of water, typically a stream, lake, pond, or wetland. Some species are riparian obligates, those that only occur in or near water and the associated streamside conditions. Other species may not require water but are riparian associates and are found in riparian areas more frequently than in upslope areas. The distribution of habitat elements is influenced by the intrariparian gradient, from the headwaters to the ocean, and the transriparian gradient, from the streamside to the ridge top. Contributions of matter and energy to stream systems is often a function of the distance from the stream edge to several tree heights upslope. Delineating and managing streamside areas as habitat for a species can be constrained by regulations or influenced by desired ecosystem services and functions. Buffer strip width appropriate to meet a species needs is dependent not only on the home range size of the species being managed but also the energy and microclimatic conditions needed by the species. Buffer regulations are highly variable, and current state and federal policies in the United States can lead to highly inconsistent riparian area management strategies across a mixed-ownership landscape.

REFERENCES

Basey, J.M., S.H. Jenkins, and P.E. Busher. 1988. Optimal central-place foraging by beavers: tree-size selection in relation to defensive chemicals of quaking aspen. *Oecologia* 76: 278–282.

Bhat, M.G., R.G. Huffaker, and S.M. Lenhart. 1993. Controlling forest damage by dispersive beaver populations: centralized optimal management strategy. *Ecol. Appl.* 3: 518–530.

Compton, J.E., C.P. Andersen, J.R. Brooks et al. 2006. Ecological and water quality consequences of nutrient additions for salmon restoration in the Oregon Coast Range. *Frontiers Ecol. Environ.* 4: 18–26.

Forest Ecosystem Management Assessment Team (FEMAT). 1993. *Forest Ecosystem Management: An Ecological, Economic, and Social Assessment.* Portland, OR. US Forest Service, US Department of Commerce, National Oceanic and Atmospheric Administration, National Marine Fisheries Service, US Bureau of Land Management, Fish and Wildlife Service, National Park Service, Environmental Protection Agency.

Henein, K., and G. Merriam. 1990. The elements of connectivity where corridor quality is variable. *Landsc. Ecol.* 4: 157–170.

Hill, B.H., and R. Blair. 2005. Monitoring the condition of our nation's streams and rivers: from the mountains to the coast. Introduction to the proceedings of the 2002 EMAP symposium. *Environ. Monitoring Assess.* 103: 1–4. WED-05-192.

Jensen, P.G., P.D. Curtis, M.E. Lehnert, and D.L. Hamelin. 2001. Habitat and structural factors influencing beaver interference with highway culverts. *Wildl. Soc. Bull.* 29: 654–664.

Knopf, F.L., R.R. Johnson, T. Rich, F.B. Samson, and R.C. Szaro. 1988. Conservation of riparian ecosystems in the United States. *Wilson Bull.* 100: 272–284.

Kondolf, G.M. 2000. Assessing salmonid spawning gravel quality. *Trans. Am. Fish. Soc.* 129: 262–281.

Leidholt-Bruner, K. 1990. Effects of beaver on streams, streamside habitat, and coho salmon fingerling populations in two coastal Oregon streams. MS Thesis, Oregon State University, Corvallis, OR.

Manning, T., C.C. Maguire, K.M. Jacobs, and D.L. Luoma. 2003. Additional habitat, diet and range information for the white-footed vole (*Arborimus albipes*). *Am. Midl. Nat.* 150: 116–123.

Maser, C., R.F. Tarrant, J.M. Trappe, and J.F. Franklin (eds.). 1988. *From the Forest to the Sea: A Story of Fallen Trees.* USDA For. Serv. Gen. Tech. Rep. PNW-229.

McComb, W.C., K. McGarigal, and R.G. Anthony. 1993. Small mammal and amphibian abundance in streamside and upslope habitats of mature Douglas-fir stands, western Oregon. *Northwest Sci.* 67: 7–15.

MacCracken, J.G. 2002. Response of forest floor vertebrates to riparian hardwood conversion along the Bear River, Southwest Washington. *For. Sci.* 48: 299–308.

McGarigal, K., and W.C. McComb. 1992. Streamside versus upslope breeding bird communities in the central Oregon Coast Range. *J. Wildl. Manage.* 56: 10–23.

Montgomery, D.R., G.E. Grant, and K. Sullivan. 1995. Watershed analysis as a framework for implementing ecosystem management. *Water Resour. Bull.* 31: 369–386.

Naiman, R.J., R.E. Bilby, and P.A. Bisson. 2000. Riparian ecology and management in the Pacific coastal rain forest. *BioScience* 50: 996–1011.

National Research Council. 1999. *New Strategies for America's Watersheds.* National Academy Press, Washington, D.C.

Nolte, D.L., S.R. Swafford, and C.A. Sloan. 2001. Survey of factors affecting the success of Clemson beaver pond levelers installed in Mississippi by Wildlife Services. Pages 120–125. In Brittingham, M.C., J. Kays, and R. McPeake (eds.), *Proceedings of the Ninth Wildlife Damage Management Conference*. Pennsylvania State University, University Park, PA.

Reeves, G.H., L.E. Benda, K.M. Burnett, P.A. Bisson, and J.R. Sedell. 1995. A disturbance-based ecosystem approach to maintaining and restoring freshwater habitats of evolutionarily significant units of anadromous salmonids in the Pacific Northwest. *Am. Fish. Soc. Sympos.* 17: 334–349.

Reid, L.M., and T. Dunne. 1984. Sediment production from forest road surfaces. *Water Resour. Res.* 20: 1753–1761.

Savidge, I.R. 1973. A stream as a barrier to homing in *Peromyscus leucopus*. *J. Mammal.* 54: 982–984.

Schmidt-Nielson, B., and E. Pfeiffer. 1970. Urea and urinary concentrating ability in the mountain beaver. *Am. J. Physiol.* 218: 1363–1369.

Semlitsch, R.D., and J.R. Bodie. 1998. Are small, isolated wetlands expendable? *Conserv. Biol.* 12: 1129–1133.

Sheridan, C.D., and D.H. Olson. 2003. Amphibian assemblages in zero-order basins in the Oregon Coast Range. *Can. J. For. Res.* 33: 1452–1477.

Sinokrot, B.A., and H.G. Stefan. 1993. Stream temperature dynamics: measurements and modeling. *Water Resour. Res.* 29: 2299–2312.

Stanford J.A., and J.V. Ward. 1988. The hyporheic habitat of river ecosystems. *Nature* 335: 64–66.

Stoddard, M.A., and J.P. Hayes. 2005. The influence of forest management on headwater stream amphibians at multiple spatial scales. *Ecol. Appl.* 15: 811–823.

Suzuki, N., and B.C. McComb. 2004. Association of small mammals and amphibians with beaver-occupied streams in the Oregon Coast Range. *Northwest Sci.* 78: 286–293.

Vesely, D.G., and W.C. McComb. 2002. Salamander abundance and amphibian species richness in riparian buffer strips in the Oregon Coast Range. *Forest Sci.* 48: 291–297.

Waide, R.B., M.R. Willig, C.F. Steiner, G.G. Mittelbach, L. Gough, S.I. Dodson, G.P. Juday, and R. Parmenter. 1999. The relationship between primary productivity and species richness. *Ann. Rev. Ecol. System.* 30: 257–300.

Welnitz, T., and B. Rinne. 1999. Photosynthetic response of stream periphyton to fluctuating light regimes. *J. Phycol.* 35: 667–672.

Wigington Jr., P.J., T.J. Moser, and D.R. Lindeman. 2005. Stream network expansion: a riparian water quality factor. *Hydrologic. Proc.* 19: 1715–1721.

10 Dead Wood Management

Dead trees are the losers in a density-dependent competition and a product of forest disturbance and disease. Considered by many to be a waste of wood fiber and a fire hazard, dead wood provides habitat for many animal species, nursery sites for germination of plants, and pathways for energy in a cellulose-based environment (Harmon et al. 1986). A large western redcedar may live to be 300 years old, and then may take another 300 years or more to decay (Embry 1963). Throughout its life and after its death, a tree can play a role in contributing to habitat quality for a succession of organisms (Maser et al. 1979). Consider the pathway of energy following a natural disturbance that creates an early-successional forest (Figure 10.1). Photosynthesis leads to allocation of energy to leaves, fruits, tree boles, and roots. In later stages of forest succession, most forest energy is stored in cellulose, and cellulose must be broken down into simpler molecules to allow the stored energy to become available to other organisms. Following an intense disturbance, that cellulose is abundant and can be decomposed to provide energy to other life forms. This process is the primary mechanism allowing energy flow through trophic levels in detrital-based systems. Cellulose also is the primary source of stored carbon in forest systems. Carbon is slowly released as CO_2 during wood decomposition (Harmon et al. 1986). The decaying wood also is associated with nitrogen-fixing bacteria that may contribute to the soil nitrogen, thereby influencing soil fertility in some forest types (Sollins et al. 1987).

The fungi and invertebrates responsible for decomposing and fragmenting the wood become the basis for energy flow into other organisms. The organisms responsible for decomposition can differ markedly between aquatic and terrestrial systems, often leading to slower rates of decay in submerged wood vs. wood exposed to air. Further, dead wood can affect the function of terrestrial and aquatic systems. Dead wood adds complexity to forest floors, increasing ground-surface and

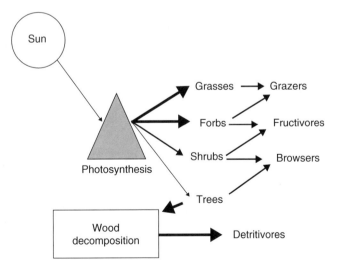

FIGURE 10.1 Early in succession following a natural disturbance, energy is transferred to higher trophic levels through both grazing based systems and detrital systems because of the high levels of dead wood at this successional stage.

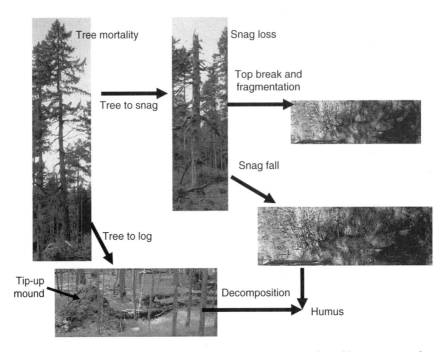

FIGURE 10.2   The fate of trees, snags, and logs in forests. Live trees can die and become snags, fragment and fall and become logs, or fall directly and become logs. Each step in the process provides habitat for a different suite of animals.

below-ground heterogeneity. Trees and snags that fall into streams can have significant impacts on sediment deposition and scouring within the channel, leading to more complex channel structure than would be present without these logs (Bisson et al. 1987).

When a tree dies, it may (1) remain standing, in some cases for decades, (2) be uprooted by wind, or (3) progressively break into pieces from damage or decay (Putz et al. 1983, Tyrrell and Crow 1994; Figure 10.2). As trees die and decay, the species that can use the tree change as well. Further, changes in the structure of forest through ecological succession influence the function of the dead and dying trees. Many cavity-nesting species rely on dead and dying wood as a source for their nest or roost cavity. Scott et al. (1977) listed 85 species of cavity-nesting birds (CNB) that occur in North American forests. In New England forests, 41 species of birds and mammals use standing trees with decay present (DeGraaf and Shigo 1985). Forest management decisions have become increasingly based on habitat relationships of animals dependent on dead wood in forests around the world. Often, these relationships are summarized for large functional groups of species, such as primary and secondary cavity users, and log users.

PRIMARY CAVITY EXCAVATORS

Up to 40% of the bird species in North American forests are cavity nesters (Evans and Conner 1979, Scott et al. 1977). In many forest systems, primary CNB (species such as woodpeckers that excavate their own cavities) play a key role by providing cavities that are used by secondary cavity nesters (species that use cavities excavated by primary cavity nesters or natural cavities created by decay). Much attention has been given to management of primary cavity nesters based on the assumption that if they are present and excavating cavities, then secondary cavity nesters will have the habitat that they need to survive (Neitro et al. 1985). Although some species of primary cavity nesters can excavate cavities in living wood, most excavate cavities in either dead wood or through live wood into decaying heartwood (Conner et al. 1976). Because most hardwoods and some conifers compartmentalize heart rot (Shigo 1984), excavation through sapwood into softened heartwood may allow organisms to create cavities in tree sections that are only two to three times the diameter of a bird's body. However, in many conifers and some hardwoods, decay of sapwood must occur to a sufficient depth toward the heartwood to allow excavation of the sapwood alone (Miller and Miller 1980). For instance, pileated woodpeckers may excavate a cavity in a tree of only 55 cm (22 in.) in diameter in eastern hardwood forests of the United States (Evans and Conner 1979), but often select much larger conifer snags for nesting in the Pacific northwest of the United States (Nelson 1988). Generally, snags or dead limbs <10 cm (4 in.) in diameter are of little or no value as nest sites for primary cavity-nesting vertebrates. Small pieces of dead wood may become important feeding substrates for some species, but foraging probably is more energy efficient on larger stems than on smaller ones, leading to selection of large stems for foraging by most species (Brawn et al. 1982, Weikel and Hayes 1999).

Most species of primary CNB use only one nest cavity per year, although a few species may use different cavities if they raise more than one brood of young in a year (Bent 1939). The excavation of a cavity is a required part of the nesting ritual for most primary cavity-nesting species (Nilsson 1984). Additional cavities often are created and used by CNB as roost and rest sites (Bent 1939). A pair of CNB may use one to ten or more cavities within a territory for nesting and roosting each year. For instance, species such as acorn woodpeckers and red-cockaded woodpeckers have nesting clans that include helpers to help raise the young (Lennartz and Harlow 1979, Neitro et al. 1985). Roost sites must also be available for the breeding pair as well as the helpers. Consequently, primary cavity excavators create many cavities in a pair's nesting territory over time.

Many species of primary CNB feed on wood-boring insect larvae, and so require dead wood as a foraging substrate within a territory (Otvos and Stark 1985). Consequently, there must be a continual replacement of feeding sites as well as nest sites within territories to allow them to remain occupied. Other species, such as common flickers, feed primarily on insects found on the ground or in understory vegetation; dead substrates are not as important as foraging sites for these species (Brawn et al. 1982). In summary, the need for dead trees or limbs as feeding sites varies considerably among different species of primary CNB occupying any given tract of forest.

SECONDARY CAVITY USERS

Secondary cavity nesters can be conveniently placed into one of two groups: (1) obligate cavity users (those species that must have a cavity for nesting or breeding) and (2) opportunistic cavity users (those that use cavities but do not require them). There are many species in the second group ranging from invertebrates to black bears (McComb and Lindenmayer 1999) that opportunistically use dead or dying trees as cover. But we will focus on obligate cavity users.

Secondary cavity nesters can use cavities created by primary cavity nesters or by wood decay following damage to a tree. Trees that sustain physical damage from wind or fire often become infected with fungal decay (Shigo 1965). The death of branches by self-pruning, incomplete branch shedding and wound occlusion, or mechanical damage usually provide avenues for decay microbes to enter live trees. Compartmentalization of decay can lead to isolated columns of decay, commonly producing a cavity (Shigo 1984). If the tree remains alive, then compartmentalization of the wound may allow cavity formation, or subsequent healing may preclude development of a cavity (Sedgwick and Knopf 1991). Tree cavities provide a very secure and microclimatically stable den, nest, or roost site (McComb and Noble 1981b).

The number of cavities used by an individual varies widely among species. Some secondary CNB change nest sites between broods presumably to avoid parasite burdens (Mason 1944); some mammals also move among den sites in response to high ectoparasite loads (Muul 1968). For example, house wrens and bluebirds may use one to three nest cavities each year and defend each from other species. Cavity-using mammals also tend to use many den sites. In North America, northern flying squirrels use multiple cavities as well as external nests within their home range (Martin 1994).

Some species use communal nest and roost sites. Swifts and bats may roost communally, with hundreds of individuals occupying one site.

There are many more species of secondary cavity nesters than of primary cavity nesters, and each species has its own requirements for the type of cavity or roost site used (Balda 1973). Long-legged bats and brown creepers use spaces behind loose bark on snags (Ormsbee and McComb 1998, Scott et al. 1977). Species such as wood ducks have more specific requirements and occupy large cavities usually near water (Lowney and Hill 1989).

Cavities may be particularly important roost sites during the winter for species in temperate climates (Haftorn 1988). Energy savings of cavity-roosting species can be significant where ambient temperatures drop below freezing over long winter nights (Weigl and Osgood 1974).

LOG USERS

Logs are used by many species of vertebrates and invertebrates as cover (e.g., red-backed voles), foraging sites (e.g., shrews and moles), and sites for attracting mates (e.g., ruffed grouse). Logs in streams provide cover for fish and influence the scouring and deposition of sediments in streams, thereby increasing stream complexity for many fish species (Bisson et al. 1987). In terrestrial environments, the interior of hollow logs, or the spaces beneath a log, provide a stable and often moist microenvironment that is especially important to the survival of some species of amphibians and reptiles (deMaynadier and Hunter 1995). Other species use the space between the bark and the wood (e.g., scarlet kingsnakes) and some use the interior of well-decayed logs (e.g., clouded salamanders; Stelmock and Harestad 1979).

Log size dictates the area or volume of space available to be occupied (Maser et al. 1979). Logs smaller than 10 cm (4 in.) in diameter are probably of little value to most vertebrates; large logs seem to be used by more species than small ones. Moreover, large logs persist longer than small logs. Decay status also affects log use by organisms. Few species are capable of using undecayed logs (e.g., ruffed grouse; Figure 10.3); most use well-decayed logs (e.g., clouded salamanders and California red-backed voles). Ideally, the habitat requirements of each species must be considered when deciding where logs should be retained and what log characteristics are sufficient to meet their needs. Obviously, with species representing a range of organism sizes from microbes, mites, and tardigrades to salamanders, fishers, and bears, managing the spatial distribution of logs must consider a wide range of home-range sizes. Realistically, the needs of most species will probably best be met if large logs are retained in clumps of various sizes in numbers that are representative of the range of conditions one might expect following a natural disturbance (Landres et al. 1999).

There are several key attributes of logs that influence their value to vertebrates: piece size and condition (decay stage), biomass or areal cover, and the successional stage in which it occurs. Piece size can be important to vertebrates for a number of reasons. Large-diameter logs provide more cover per piece than those of small diameter. Western red-backed voles select large logs as cover (Hayes and Cross 1979), and logs provide cover and a source of fungi for food for southern red-backed voles (Buckmaster et al. 1996). A wide variety of other species also reportedly use logs as cover: shrews, weasels, mink, and northern river otters, among many others (Maser et al. 1981). Further, long logs provide more connectivity across the forest floor than short ones. Connectivity throughout a home range can, theoretically, influence animal fitness because an individual can remain under cover during movements, thereby reducing the risk of predation, while also possibly providing microclimatic advantages to the organism.

The distribution of log sizes in a forest generally reflects the site quality for tree growth, stage of stand development, and sources of mortality that in the past that led to tree death. Trees dying from suppression mortality are typically 50% the diameter (but similar in length) of dominant and codominant trees in the stand (McComb and Lindenmayer 1999). Because large-diameter pieces take longer to fully decay than small diameter pieces, large piece sizes may last longer as functional

FIGURE 10.3 Ruffed grouse use logs in dense patches of forest as drumming sites where males attract females during the spring. Ermine use hollow logs as den sites. (Photos by Michele Woodford. With permission.)

habitat for more species. The desired size class distribution for the suite of species being managed in a stand or landscape should be determined by the species requiring the largest piece size.

The areal cover or biomass of logs may influence the function of the wood as cover to some mammal and amphibian species (McComb 2003). The physical structure of the log is also important to some species. Maser et al. (1979) described stages of log decay that are similar to that used to describe snag decay stages. Each stage of decomposition can provide different resources to a suite of organisms (Maser et al. 1979). Early in the decay process, sloughing bark and infestation by bark beetles, carpenter ants, and termites provide food and cover resources for small mammals, bears, and woodpeckers (Maser et al. 1979, Torgersen and Bull 1995). Once the wood has softened and fragmented, vertebrates can begin to excavate the wood to extract insects and build nests. Red-backed voles and shrews use very decayed logs as nest sites (Tallmon and Mills 1994, Zeiner et al. 1990) that provide cryptic, dry, and thermally stable environments for their young. Eventually the structural integrity of the log is so severely compromised by the fungal infection that the log loses value as a potential nest site or feeding site.

Some species such as wood rats, foxes, black bears, skunks, and ermine also use hollow logs as dens (Figure 10.3). Hollow logs become available after a hollow tree falls to the ground. Hollow trees form because a column of decay develops following top breakage that extends up and down the bole of the tree from the wound (Shigo 1984). Recruiting hollow logs into managed stands requires the identification and retention of injured and decaying trees, allowing them to grow to sufficient size or to decay to an acceptable extent, then allowing or promoting their death. Black bears use hollow

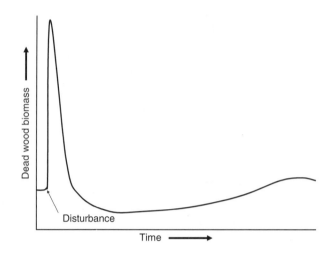

FIGURE 10.4   Generalized pattern of changes in dead wood biomass over time following a natural forest disturbance which creates a pulse of dead wood followed decompositional losses and a slow recovery from tree mortality. (From Spies, T.A. et al. 1988. *Ecology* 69: 1689–1702.)

logs averaging 106 cm (42 in.) in diameter for winter denning in British Columbia (Davis 1996), so recruitment of potential den sites for bears may take centuries. It is apparent that logs can function as a habitat element for many species in all successional stages of forests in North America.

PATTERNS OF DEAD WOOD FOLLOWING DISTURBANCE

Two processes contribute to dead wood recruitment in a stand over time: (1) the number of trees dying increases rapidly shortly after stand establishment, then declines in a negative exponential manner through the period of "self-thinning" (Oliver and Larson 1996), and (2) the biomass of dead wood increases immediately after an intense disturbance (unless biomass is removed during logging), declines slowly over time, then recovers as large trees die late in stand development (Spies et al. 1988; Figure 10.4). Dead wood biomass accumulates when inputs of dead wood are greater than decomposition losses. Inputs (suppression mortality or exogenous disturbance) and losses (decomposition or fire) interact to produce a "U"-shaped trend in dead wood biomass over time seen in forest types throughout North America (D'Amato 2007, Gore and Patterson 1986, Spies et al. 1988, van Lear and Waldrop 1994; Figure 10.4).

Natural old forests contain high volumes of large pieces of dead wood, but not to the level found following intense disturbances such as fires or hurricanes. Infrequent but severe disturbances create pulses of dead wood (Spies et al. 1988). High levels of dead wood produced following a disturbance also may represent a fuel source for subsequent fires in fire-prone systems (Spies et al. 1988). Fear of recurring fire led to salvage logging and snag removal several decades ago in the Pacific Northwest of the United States (McWilliams 1940). Now managers often try to recruit dead wood to stands that were salvaged in past years.

Changes in Dead Wood over Time

Dead wood changes over time through decomposition (Miller and Miller 1980). When a tree dies, fungal decay usually begins. Fungal decay facilitates wood fragmentation when combined with the activities of invertebrates, such as termites (Atkinson et al. 1992). Tree mortality and wood decomposition rates interact to dictate dead wood biomass on a site. The size and species composition of the live trees influence the potential dead wood production on the site. Hardwood forests generally have less dead wood than conifer forests (Harmon and Hua 1991, Harmon et al. 1986). Eastern hardwood forests may support 11 to 220 m³/ha of dead wood (D'Amato 2007, Tyrell and Crow 1994), but western coniferous forests may have 376 to 1421 m³/ha of dead wood (Huff 1984). Variability in the amounts and distribution of both standing and fallen dead wood is considerable (Everett et al. 1999). Indeed, managing dead wood to reflect variability among sites over a landscape may be a more meaningful approach than mandating a minimum retention level in managed stands or trying to manage for individual animal species (Everett et al. 1999).

The size of the dead wood influences the rate of decomposition and its value to organisms. Large pieces of dead wood provide habitat for a large number of species in various seral stages. These large remnant snags and logs can last for centuries before becoming an unrecognizable part of the forest humus (Tyrrell and Crow 1994). Trees that die and remain standing provide habitat for species as snags until they fall over. Fall rates of live trees and snags vary among tree species (McComb and Lindenmayer 1999). Ten-year fall rates (the proportion of trees expected to fall in a 10-year period) for pine and fir snags in the western United States and many hardwoods in the eastern United States exceed 50% (Morrison and Raphael 1993, Wilson and McComb 2005). Fall rates of large-diameter Douglas-fir snags may be <20% per decade (Cline et al. 1980).

The combination of a tree's size and the variability among species in their resistance to decay leads to considerable variation among trees in rate of decay and fragmentation (Harmon et al. 1986). As fragmentation of the tree bole advances, the diameter and height of snags (or length of logs) decreases (Figure 10.2). Tree species and size also influence other characteristics of dead and dying trees. Decomposition rates are generally described using decay rate constants (Olson 1963):

$$D_t = D_0 e^{-kt},$$

where D is the wood density, t the time (years), and k the decay rate constant.

Decay rates vary among species, with conifers generally being more decay resistant than hardwoods (Harmon and Hua 1991; Table 10.1). As decay proceeds within a bole of wood, the bole becomes subject to fragmentation (Harmon et al. 1986, Tyrell and Crow 1994). Consequently, the dead wood biomass on a site at any one time will be dependent on a number of factors. These include site quality and tree species composition, the disturbance regime for the site, and the climatic factors that influence tree growth and decomposition (Muller and Liu 1991).

Dead Wood during Stand Development

Stand establishment following a disturbance often results in over 1,000 tree seedlings per hectare in some western U.S. coniferous stands and over 40,000 seedlings per hectare in eastern hardwood stands. In plantations, stand density is controlled. In both cases, as stands develop, intertree competition results in mortality among those trees that are intolerant of shade or drought (Oliver and Larson 1996). It is not uncommon to see over 90% of the trees in a stand die during the first few decades following natural stand establishment. High-density stands produce many small dead stems early in stand development. Competition mortality commences later in stand development in lower-density stands, allowing trees to grow rapidly for many years prior to competition (Figure 6.5A). Thinning that reduces stand density may benefit the production of dead wood of large sizes later in stand development. Hence, manipulating stand density allows the manager to influence the size and numbers of dead trees throughout even-aged stand development. Manipulating stocking rates in uneven-aged stands can produce similar results.

MANAGEMENT OF TREE CAVITIES AND DEAD WOOD

There are two general approaches to dead wood and tree cavity management, and they represent two complementary philosophies. The first is based on the concept of a historical range of

TABLE 10.1

Comparison of Decay Constants (k) among Tree Species in Various Parts of North America (Listed from Slowest to Fastest Decay Rates)

Taxon	Location	k (per year)[a]	References
Douglas-fir	Oregon	0.005–0.10	Harmon and Hua (1991)
Douglas-fir	Oregon	0.0063	Means et al. (1985)
Balsam-fir	New Hampshire	0.011	Lambert et al. (1980)
Western hemlock	Oregon	0.012	Grier (1978)
Western hemlock	Oregon	0.016–0.018	Harmon and Hua (1991)
Mixed oaks	Indiana	0.018	MacMillan (1988)
Western hemlock	Oregon	0.021	Graham (1982)
Eastern hemlock	Wisconsin	0.021	Tyrell and Crow (1994)
Red Spruce	New Hampshire	0.033	Foster and Lang (1982)
Jack pine	Minnesota	0.042	Alban and Pastor (1993)
Mixed maples	Indiana	0.045	MacMillan (1988)
Red pine	Minnesota	0.055	Alban and Pastor (1993)
White spruce	Minnesota	0.071	Alban and Pastor (1993)
Trembling aspen	Minnesota	0.080	Alban and Pastor (1993)
Mixed hardwoods	New Hampshire	0.096	Arthur et al. (1993)
Mixed hardwoods	Tennessee	0.110	Onega and Eickmeier (1991)

[a] k = a decay rate constant when calculating decay rates as $D_t = D_0 e^{-kt}$, where D = wood density, t = time (years).

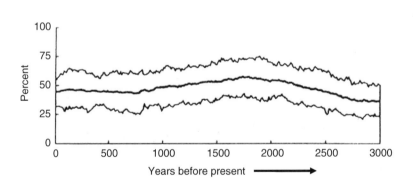

FIGURE 10.5 Expected range of variability of old-growth forests in the Oregon Coast Range. As seral stages change in abundance over time so do patterns of dead wood associated with them. (Redrafted from Wimberly, M.C. et al. 2000. *Conserv. Biol.* 14: 167–180. With permission from Blackwell Press.)

variability (HRV) or the range conditions produced through natural disturbances over an area (Figure 10.5). So a manager might ask, "Do the levels of dead wood biomass, piece size, and condition over large areas fall within the range of conditions that the species should be adapted, the range that might be represented under natural disturbances (HRV)?" If the answer is "no," then you might question which species might be at risk (if any) based on this departure from the HRV, and if management actions should be taken to address these risks. If impacted species and processes are adequately addressed elsewhere in the landscape, then allowing some stands or landscapes to fall outside the HRV may be an acceptable risk. If, however, the addition of another stand or landscape to areas that already fall outside the HRV means that there is a likelihood of cumulative risks on species or processes over space and time, then the manager may wish to take actions that contribute to goals related to the HRV (Table 10.2; Landres et al. 1999).

TABLE 10.2

Likely Ranges of Dead wood (CWD) Variability among Forest Age Classes in the Oregon Coast Range under the Historical Range of Variability

Age class (years)	Percent of region	Area of region (ha × 100,000)	Number of patches	CWD range (m³/ha)
0–30	4–11	0.9–2.5	1–4	376–1421[a]
31–80	6–19	1.4–4.3	1–6	163–305[b]
80–200	15–45	3.4–10.1	2–14	93–165[b]
>200	25–75	5.6–16.9	4–24	219–324[b]

[a] From Huff (1984).
[b] From Spies and Franklin (1991).

Source: Based on Wimberly, M.C. et al. 2000. *Conserv. Biol.* 14: 167–180 and McComb, W.C. 2003. Ecology of coarse woody debris and its role as habitat for mammals. In Zabel, C.J., and R.G. Anthony (eds.), *Mammal Community Dynamics: Management and Conservation in the Coniferous Forests of Western North America*, Chapter 11. Cambridge University Press, Cambridge, England.

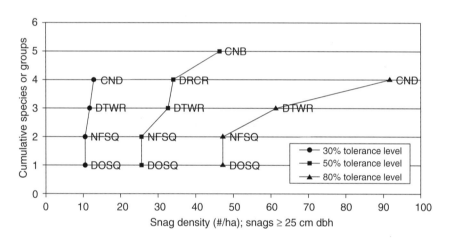

FIGURE 10.6 Cumulative species curves of snags (number per hectare) supporting species use of areas for nesting, roosting, and occurrence for 30, 50, and 80% tolerance levels, Westside Lowland Conifer-Hardwood habitat type in the small tree structural class. (Acronyms are various species and groups of species described by Mellen, K. et al. 2005. *DecAID, the Decayed Wood Advisor for Managing Snags, Partially Dead Trees, and Down Wood for Biodiversity in Forests of Washington and Oregon.* Version 2.0. USDA For. Serv. Pac. Northwest Res. Sta. and USDI Fish and Wildl. Serv., Oregon State Office, Portland, OR.)

Alternatively, the manager can assess functional relationships between animals and dead wood and manage for these conditions as part of a desired future condition (McComb and Lindenmayer 1999, Mellen et al. 2005). These functional relationships are not clear for most species, but they can be hypothesized and tested in an adaptive management approach. Indeed, some habitat relationships models already include estimates of the abundance of dead wood as a component contributing to habitat quality for a species (e.g., Allen 1983). The compilations of these relationships for desired species in future landscapes would dictate the dead wood goals (Mellen et al. 2005; Figures 10.6 and 10.7).

Regardless of the dead wood management approach chosen, managers should identify high-priority sites for dead wood management. Intensively managed plantations might fall within this group because they have dead wood and tree-cavity levels that often fall outside of the HRV (Butts and McComb 2000). Modest inputs of dead wood to these stands may make a greater impact on animal habitat or ecological processes than a similar treatment in stands that already contain dead wood.

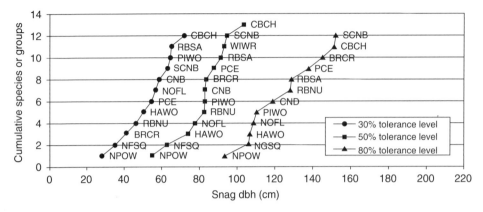

FIGURE 10.7 Cumulative species curves of snag dbh (cm) supporting species use of areas for nesting and denning for 30, 50, and 80% tolerance levels, Westside Lowland Conifer-Hardwood habitat type in the small tree structural class. (Acronyms are various species and groups of species described by Mellen, K. et al. 2005. *DecAID, the Decayed Wood Advisor for Managing Snags, Partially Dead Trees, and Down Wood for Biodiversity in Forests of Washington and Oregon.* Version 2.0. USDA For. Serv. Pac. Northwest Res. Sta. and USDI Fish and Wildl. Serv., Oregon State Office, Portland, OR.)

The steps in the management process that I recommend are given below:

1. Inventory dead wood at the desired scale at time 0 (see Bull et al. [1990] and Harmon and Sexton [1996] for inventory techniques). The chosen spatial scale should be biologically meaningful to the species of interest.
2. Compare dead wood levels to the HRV estimates for the region and compare estimates to your habitat goals for species. In western U.S. forests, DecAID can provide a useful reference for comparison (Mellen et al. 2005; Figures 10.6 and 10.7).
3. Conduct this analysis across your planning area, and prioritize stands for dead wood management based on the risk of not meeting future dead wood goals.
4. Beginning with the highest-priority stands, determine if there are trees of sufficient size that could be felled or killed now to fulfill the snag or log goals or that could be retained into the future to replace those snags that fall over time.
5. If trees in the current stand are not appropriate for meeting dead wood goals, then silvicultural actions should be considered to achieve goals. Thinning from below to allow dominant and codominant trees to grow more rapidly may be preferable to allowing an overstocked stand to grow slowly and to contribute small amounts of dead wood to the stand.
6. Monitor species of highest concern prior to and following active management and assess population changes. Given the long-term nature of wood decay and the habitat functions that develop throughout decay processes, monitoring may need to occur periodically for decades.
7. Assess monitoring results and decide if changes should be made to the dead wood goals for the area.

Many states and provinces have either regulations, or standards (you shall do them) and guidelines (you should do them) regarding dead wood retention levels. Several factors come into play when regulating dead wood levels in managed stands. First, the minimum level of the range chosen for regulation is usually the level that managers will strive to retain in stands. Providing one dead wood level in all managed stands homogenizes that condition over managed landscapes. Although current dead wood guidelines could be rewritten to ensure that dead wood levels fall *within* the HRV, it is much more difficult to develop regulations that will lead to dead wood levels that *represent* the HRV for the region. Incentives such as dead wood credits provided to landowners by local, state, or federal agencies may allow better representation of the HRV in dead wood conditions across landscapes than mandating it by law (McComb 2003).

Clearly such management actions will require a commitment of time and money to providing dead wood. Costs can be modest if management is for one or a few species, but much higher if dead wood is managed to represent goals for multiple species or the HRV. Dead wood guidelines should, however, be scale-dependent. Dead wood biomass among many stands should collectively contribute to landscape goals. Landscapes should also represent variability in dead wood levels, but collectively contribute to regional goals.

A delay in dead wood management in a stand with low levels of dead wood now may result in a gap in dead wood availability in the future. Certainly a few stands with low dead wood levels in an area with otherwise high levels may be relatively unimportant, unless overall dead wood levels decline over time and no action is taken now to ensure that advanced decay class (class 5) logs will occur in the stands 50 years from now.

LIVE CAVITY-TREE MANAGEMENT IN MANAGED STANDS

Managing forests to achieve goals for secondary cavity nesters can partially be achieved by managing habitat for primary cavity nesters. But not entirely (Figure 10.8). Clearly there are some secondary cavity users that are larger in body mass than the largest primary cavity nesters. And there are more species of secondary cavity nesters than primary one. Based on nest box studies, they can occur at much higher densities than primary cavity nesters. So providing natural cavities can be an important supplement to the cavities created by primary cavity nesters.

FIGURE 10.8   Dead limbs on live hardwoods and cavities in live hardwoods both contribute to cavity resources for secondary cavity nesters. Hence, providing some hardwoods in conifer stands can add to snags as a source of cavities for cavity-nesting species. Natural cavities also provide nest sites for species such as barred owls (owlet shown here), which cannot use cavities created by primary cavity nesters.

Nest boxes are one alternative to providing natural cavities, but nest boxes are expensive to build and maintain and they are likely to last only a fraction of the time that a natural cavity would last in a live tree (McComb and Lindenmayer 1999). Nonetheless, nest box programs have been very successful for some species such as bluebirds and wood ducks. Nest boxes are widely used to increase nesting and roosting site availability for a number of species, and the proportion of nest boxes used by animals can be higher than use of natural cavities for many species (McComb and Noble 1981a). However, maintenance costs for nest boxes are high, microclimates are less stable than natural cavities (McComb and Noble 1981b), and primary cavity nesters rarely use them unless they are filled with a substance that can be excavated. Nest boxes should only be considered a temporary solution to a shortage of nest cavities and one that usually can only be used in a relatively small area for a small number of species.

Managing natural cavity abundance in forests is a bit more challenging than managing dead wood because they are more difficult to inventory and the rate of gain and loss in a forest is very slow, and somewhat unpredictable. Estimating cavity abundance is challenging. Sampling trees for cavities often is complicated by inadequate access to or visibility of cavities in standing trees. Cavities judged to be suitable from the ground may not be useable by a given species (Healy et al. 1989). Typically, sampling for cavities is conducted during the leafless period in temperate climates if hardwoods are present in the stand. The size and number of plots used to sample for cavities will be largely a function of the density and among-plot variability in cavity density. To adequately predict the prevalence of trees with cavities, a very large number of plots may be required to sample cavity abundance (Healy et al. 1989). DeGraaf and Shigo (1985) provided guidelines for managing natural cavities in eastern U.S. forests.

Predicting cavity availability in a stand from tree size and species information is even more problematic. Cavity occurrence in a tree is a function of tree size and tree age, as well as the often highly stochastic disturbance factors that initiate cavity formation. Nonetheless, it seems that there are relationships that can be developed for some hardwood species in North America (Allen and Corn 1990, McComb et al. 1986). In general, large-diameter trees with some past injury are more cavity-prone than small-diameter trees that lack obvious signs of past injury (Figure 10.9). Assumptions made regarding the processes of cavity formation, such as the continued role of insects and fire, must be monitored carefully throughout prescription development and implementation in order to ensure that cavities will be available over time in a stand.

DEAD WOOD RETENTION AND HARVEST SYSTEM CONSIDERATIONS

Because of the logistics of harvesting around dead and green trees reserved from harvest, snags and replacement green trees often are left in clumps between cable corridors or between skid trails, and soft snags are left opportunistically between the clumps. But clumping snags can have adverse effects on snag use. Location matters. In clumps, a territorial individual can exclude other individuals of the same species from a clump. If the same number of snags were distributed at a spacing consistent with the territory size of the species being managed, then snag use can be optimized. Where human safety issues occur, then some balance must be achieved between the optimum distribution for animal use of snags and reducing risk to forest workers. In the United States, harvest operations must be coordinated with retention of snags, logs, and cavity trees to avoid interference with harvest systems (e.g., skid trails and cable corridors) and to ensure worker safety during the operations (Hope and McComb 1994; Figure 10.9). In the United States, the Occupational Safety and Health Administration (OSHA) places restrictions on loggers working around dead or dying trees limiting options when managing dead wood in stands.

CREATING SNAGS AND LOGS FOR WILDLIFE

The goals for dead wood abundance in a stand should be compared with the levels of dead and dying trees predicted to occur in the stand over time. These estimates can be developed using a

FIGURE 10.9   Leaving snags and logs must be balanced with worker safety and fuels management in managed stands to meet the needs for species such as woodpeckers and alligator lizards (shown here).

forest growth model that includes a tree mortality function (e.g., Forest Vegetation Simulator; Dixon 2003). Once we know how many trees are likely to die each decade and what size they are likely to be, then we can predict additions of dead wood over time. We can also estimate dead wood loss through decomposition (Table 10.1) or snag fall rates. If the predicted recruitment of dead wood does not balance the losses to meet or maintain dead wood goals, then some trees can be killed to meet the goals.

The process for deciding which trees to retain as replacement snags during management activities have largely been driven by tree species, tree size, and costs associated with forgoing timber value (Washington DNR et al. 1992). Generally, large trees with some timber defect have the potential to provide tree cavities and dead and dying wood (Healy et al. 1989). In intensively managed stands, defective or diseased trees may be thinned early in stand development. In these stands, dominant and codominant trees may provide habitat for cavity-using species early in the rotation if some of these large trees are retained and killed (Bull and Partridge 1986) or injected with fungal spores (Parks et al. 1995). Indeed, thinning can accelerate tree diameter growth tremendously in some forest types, providing an opportunity to kill some large trees much sooner than would occur in the absence of management.

There is a range of methods available for killing trees to produce snags or cavity trees for vertebrates (Bull and Partridge 1986). Topping the trees with a chain saw or explosives is effective for both Douglas-fir and ponderosa pine (Bull and Partridge 1986, Chambers et al. 1997; Figure 10.10). Herbicides also have been shown to be an effective method for killing trees that are then used by primary cavity nesters (McComb and Rumsey 1983). Girdling, although potentially effective, may be less cost-effective than other techniques simply because trees often break at the point of girdling, creating short snags of limited value to some species (Figure 10.10). Hardwoods have been killed to increase invertebrate food resources for woodpeckers in Europe (Aulen 1991), but live hardwoods

FIGURE 10.10   (a) Snags created by girdling often will break at the point where the girdling occurred, (b) whereas topping trees creates a longer lasting snag, and (c) one more typical of a snag that develops following natural death and decay.

may be used by more species for a longer period of time than dead hardwoods. Killing trees as habitat management for selected vertebrates must be done based on needs for primary cavity excavators and the potential for subsequent use of these cavities by secondary users. Generally, killing trees as a remedial measure is most appropriate in managed conifer forests.

Other techniques are available, but rarely used. Wood-decaying fungi has been experimentally injected into live trees to create a pocket of rot that can be excavated by cavity nesters at some later date (Parks et al. 1995). Artificial cavities also have been created by excavating holes in live trees in eastern hardwood forests (Carey and Sanderson 1981), and cavity inserts have been used to create artificial nest sites for red-cockaded woodpeckers in pine trees without heart rot.

MONITORING CAVITY TREES, SNAGS, AND LOGS

Most goals for dead wood management in managed forests are based on a number of assumptions. These include, but are not limited to, the following: estimates of the number of snags required by each individual or breeding pair; distribution of trees, snags and logs within territories; estimates of fall rates and decay rates of snags; and persistence of populations that may become isolated over time. Monitoring of management effectiveness becomes a key part of the management process, especially given the uncertainties associated with requirements for each species, stand projection estimates, and estimates of snag decay and fall rates. Effective management of dead wood habitat will require consideration of not only the primary cavity nesters (Neitro et al. 1985) but also foraging and nesting sites for those secondary cavity nesters that do not use nest sites abandoned by the primary cavity nesters (e.g., bats, wood ducks, and some invertebrates; Figure 10.8). Secondary cavity nesters are generally dependent on the activities of primary cavity nesters and on cavities formed by wood decay processes. Consequently, secondary cavity nesters may be better candidates to monitor the effects of forest management on dead wood dependent species.

CASE STUDY: MANAGING DEAD WOOD IN OREGON FORESTS

To illustrate the process of managing dead wood in a managed stand, let us consider a 100-year-old stand in the Oregon Coast Range:

1. *Inventory dead wood at the desired scale now.* The stand is fully stocked at 75 m^2 of basal area per hectare (250 ft^2/acre), and is dominated by Douglas-fir with minor components of grand fir, western hemlock, bigleaf maple, and red alder. A stand examination (a systematic or random sample of trees and habitat elements in the stand) revealed an estimate of 10 snags/ha (4 snags/acre) >76 cm dbh (30 in. dbh) and 7/ha (3/acre) are larger than 80 cm dbh (32 in. dbh). The remainder of dead trees in the stand (12/ha; 5/acre) are <5 cm dbh (2 in. dbh). There are 106 trees/ha (43 trees/acre) >76 cm dbh (30 in. dbh).

2. *Compare dead wood levels to your goals.* Using the Dead wood Adviser (DecAID) developed by Mellen et al. (2005), we chose to manage for snag levels that represent a 50% tolerance level, or a likelihood of providing ecosystem functions intermediate between management to provide primarily ecosystem function goals (80% tolerance level) and providing primarily timber production (30% tolerance level). DecAID uses empirical relations from dead wood–species relationships to develop these curves (Figures 10.6 and 10.7). Because we intend to use a clearcut regeneration system with legacy to regenerate the stand yet provide habitat elements, we selected the early-successional condition of the Westside Lowland Conifer Hardwood habitat type to best represent the stand that will result from our management (Mellen et al. 2005). Our species goal is to manage to provide snags at a level that will meet the needs for CNB as a group at the 50% tolerance level, or approximately 42 snags/ha of >25 cm dbh (17/acre of >10 in. dbh; Figure 10.6). But not all cavity nesters can use such small snags, so we also need to set snag size goals. DecAID indicates that 80-cm-dbh snags (32 in. dbh) are needed to meet our size goals (Figure 10.7), so at least some of the snags in the stand should be greater than this diameter. Hence, we want >42 snags/ha that are >25 cm dbh, and as many as possible of these should be >80 cm dbh.

3. *Prioritize stands for dead wood management based on the risk of not meeting future dead wood or species goals.* Because there are only 10 snags/ha (4 snags/acre) in the stand now and we want to have 42 snags/ha (17/acre) following harvest, this stand becomes a high-priority stand for increasing dead wood availability.

4. *Determine if there are trees of sufficient size that could be felled or killed now to fulfill the snag or log goals.* Because we have 106 trees/ha of >76 cm dbh, there are sufficient live trees that can be killed to provide these snags.

5. *If trees in the current stand are not appropriate for meeting dead wood goals, then silvicultural actions should be considered to achieve goals, including retaining trees for future snags, and killing trees to create snags and logs.* We know that we will need more than our 43 snags/ha (17/acre) goal because some snags will fall during the future development of the stand. But how many more? We may chose to mark and retain 50 trees/ha (20/acre) of which we will kill 32/ha to supplement the 10/ha that are on the site now to meet our goal. The remaining reserved trees (18/ha) are retained as live trees as a future source of dead wood, if needed, later in stand development. The stand is then harvested and after harvest, 32 of the 50 retained trees per hectare (13 of the 20 retained trees per acre) are killed by topping the tree (cost = $50/tree, or $1600/ha; $650/acre). The tops will be left on the site to add to the log availability. Because the largest trees in the stand were retained as logs for species requiring dead wood on the forest floor; 108 MBF/ha (44 MBF/acre) were harvested and 141 MBF/ha (57 MBF/acre) were retained and killed. If Douglas-fir sold for $500/MBF, then the

gross timber receipts would be $53,340/ha ($22,000/acre); $69,160/ha ($28,000/acre) would have been allocated as timber value forgone to create dead wood.

6. *Monitor species of highest concern prior to and following active management and assess if populations decline.* Following harvest and snag creation we will monitor snag fall rates and populations of CNB every 5 years until the stand moves into another vegetative structural condition (Mellen et al. 2005). Projections of snag loss using the Snag Recruitment Simulator (SRS; Marcot 1992) suggests that of the 42 snags/ha (17 snags/acre) available immediately after harvest that there would be 37/ha (15/acre) available after 10 years, 35/ha (14/acre) after 20 years, 20/ha (8/acre) after 30 years, 10/ha (4/acre) after 40 years, and 5/ha (2/acre) after 50 years because of snag decomposition, decay, and subsequent fall. Of course, these fallen snags add to the log biomass available for other species, but substrates for cavity nesters would decline considerably during the first 50 years of stand development because of snag fall. Monitoring data collected over the 50 years would allow the managers over that time to assess if these projections were correct and if additional trees should be killed.

7. *Assess monitoring results and decide if changes should be made to the dead wood goals for the area.* Recall that we retained 50 trees/ha (20/acre) and only killed 32/ha (13/acre) of them after harvest. So there are still 18 trees/ha (7 trees/acre) carried into the new stand that have grown for 50 years and that could be killed if monitoring indicated that more snags were needed. These retention trees represent the insurance policy for dead wood in the stand so that future goals can be met as the created snags fall over (Figure 10.10).

Approaches such as this that also consider the cavity tree resources and logs on the forest floor can be used to help ensure that the species and ecological processes associated with dead and decaying wood are maintained in stands and across landscapes. But it should be apparent that these activities come at a cost, sometimes a significant cost, to a landowner. On public lands, where timber profits are not a goal, such a dead wood recruitment and maintenance strategy is clearly feasible, but on private lands different goals and approaches may be necessary.

SUMMARY

Forest management activities that influence the frequency, severity, and pattern of disturbances in forest systems can have marked effects on the abundance of cavities and dead and dying trees in the system. Dead and dying trees function differently in each stage of forest succession and the trees themselves progress though a succession of decay stages. Decay stages provide opportunities for use by vertebrates, with a different suite of species selecting each decay stage. If forest managers wish to maintain these functions in managed forests, then they must actively manage the dead wood resource.

Live trees with decay are especially important to animals in hardwood forests. Standing and fallen dead trees are particularly important in conifer forests. Integration of management of dead, dying, and decayed trees in forest management will be key in any management prescription designed to balance biodiversity conservation with commodity production. Delay in initiating active management can have long-term implications because of the time needed to both recruit large trees and for the large wood to decay to a stage suitable for certain organisms. There are seven steps to managing habitat for species that depend on cavities, snags, or logs:

1. Inventory dead wood at the desired scale now.
2. Compare dead wood levels to your goals.
3. Prioritize stands for dead wood management based on the risk of not meeting future dead wood or species goals.
4. Determine if there are trees of sufficient size that could be felled or killed now to fulfill the snag or log goals.

5. If trees in the current stand are not appropriate for meeting dead wood goals, then silvicultural actions should be considered to achieve goals, including retaining trees for future
snags, and killing trees to create snags and logs.
6. Monitor species of highest concern prior to and following active management and assess
if populations decline.
7. Assess monitoring results and decide if changes should be made to the dead wood goals
for the area.

REFERENCES

Alban, D.H., and J. Pastor. 1993. Decomposition of aspen, spruce, and pine boles on two sites in Minnesota.
Can. J. For. Res. 23: 1744–1749.

Allen, A.W. 1983. *Habitat Suitability Index Models: Southern Red-Backed Vole.* U.S.Dep. Interior Fish and
Wildl. Serv., FWS/OBS-82/10.42.

Allen, A.W., and J.G. Corn. 1990. Relationships between live tree diameter and cavity abundance in a Missouri
oak-hickory forest. *No. J. Appl. For.* 7: 179–183.

Arthur, M.A., L.M. Tritton, and T.J. Fahey. 1993. Dead bole mass and nutrients remaining 23 years after
clear-felling of a northern hardwood forest. *Can. J. For. Res.* 23: 1298–1305.

Atkinson, P.R., K.M. Nixon, and M.J.P. Shaw. 1992. On the susceptibility of Eucalyptus species and clones to
attack by *Macrotermes natalensis* Haviland (Isoptera: Termitidae). *For. Ecol. Manage.* 48: 15–30.

Aulen, G. 1991. Increasing insect abundance by killing deciduous trees: a method of improving the food situation
for endangered woodpeckers. *Holarctic Ecol.* 14: 68–80.

Balda, R.P. 1973. *The Relationship of Secondary Cavity Nesters to Snag Densities in Western Coniferous Forests.*
USDA For. Serv., Southwestern Region. Wildl. Habitat Tech. Bull. No. 1.

Bent, A.C. 1939. *Life Histories of North American Woodpeckers.* U.S. National Muzeum Bulletin 174.
Smithsonian Institute, Washington, D.C.

Bisson, P.A., R.E. Bilby, M.D. Bryant, C.A. Dolloff, G.B. Grette, R.A. House, M.L. Murphy, K.V. Koski,
and J.R. Sedell. 1987. Large woody debris in forested streams in the Pacific Northwest: past, present,
and future. In Salo, E.O., and T.W. Cundy (eds.), *Streamside Management: Forestry and Fishery
Interactions.* Contrib. No. 57. University of Washington, Institute of Forest Resources, Seattle, WA.

Brawn, J.D., W.H. Elder, and K.E. Evans. 1982. Winter foraging by cavity nesting birds in an oak-hickory
forest. *Wildl. Soc. Bull.* 10: 271–275.

Buckmaster, G., W. Bessie, B. Beck, J. Beck, M. Todd, R. Bonar, and R. Quinlan. 1996. Southern red-backed vole
(*Clethrionomys gapperi*) year-round habitat. Draft habitat suitability index (HSI) model. In Beck, B.,
J. Beck, W. Bessie, R. Bonar, and M. Todd (eds.), *Habitat Suitability Index Models for 35 Wildlife Spe-
cies in the Foothills Model Forest: Draft Report.* Canadian Forest Service, Edmonton, Alberta, Canada.

Bull, E.L., and A.D. Partridge. 1986. Methods of killing trees for use by cavity nesters. *Wildl. Soc. Bull.* 14:
142–146.

Bull, E.L., R.S. Holthausen, and D.B. Marx. 1990. How to determine snag density. *West. J. Appl. For.* 5: 56–58.

Butts, S.R., and W.C. McComb. 2000. Associations of forest-floor vertebrates with coarse woody debris in
managed forests of western Oregon. *J. Wildl. Manage.* 64: 95–104.

Carey, A.B., and H.R. Sanderson. 1981. Routine to accelerate tree-cavity formation. *Wildl. Soc. Bull.* 9: 14–21.

Chambers, C.L., T. Carrigan, T. Sabin, J. Tappeiner II, and W.C. McComb. 1997. Use of artificially created
Douglas-fir snags by cavity-nesting birds. *West. J. Appl. For.* 12: 93–97.

Cline, S.P., A.B. Berg, and H.M. Wight. 1980. Snag characteristics and dynamics in Douglas-fir forests, western
Oregon. *J. Wildl. Manage.* 44: 773–786.

Conner, R.N., O.K. Miller Jr., and C.S. Adkisson. 1976. Woodpecker dependence on trees infected by fungal
heart rots. *Wilson Bull.* 88: 575–581.

D'Amato, A.W. 2007. Structural attributes, disturbance dynamics, and ecosystem properties of old-growth
forests in western Massahchusetts. Ph.D. dissertation, University of Massachusetts-Amherst.

Davis, H. 1996. Characteristics and selection of winter dens by black bears in coastal British Columbia. Thesis,
Simon Fraser University, Burnaby, British Columbia, Canada.

DeGraaf, R.M., and A.L. Shigo. 1985. *Managing Cavity Trees for Wildlife in the Northeast.* USDA For. Ser.
Gen. Tech. Rep. NE-101.

DeMaynadier, P.G., and M.L. Hunter Jr. 1995. The relationship between forest management and amphibian ecology: a review of the North American literature. *Environ. Rev.* 3: 230–261.

Dixon, G.E. (compiler). 2003. *Essential FVS: A User's Guide to the Forest Vegetation Simulator.* USDA Forest Service. Forest Management Service Center. Fort Collins, Co. Internal Report.

Embry, R.S. 1963. Estimating how long western hemlock and western redcedar trees have been dead. USDA Forest Service Res. Note NOR-2.

Evans, K.E., and R.N. Conner. 1979. Snag management. In DeGraaf, R.M. (technical coordinator), *Proceedings of the Workshop, Management of North Central and Northeastern Forests for Nongame Birds.* USDA For. Serv. Gen. Tech. Rep. NC-51.

Everett, R., J. Lehmkuhl, R. Schellhaas, P. Ohlson, D. Keenum, H. Reisterer, and D. Spurbeck. 1999. Snag dynamics in a chronosequence of 26 wildfires on the east slope of the Cascade Range in Washington State, USA. *Int. J. Wildfire* 9: 223–234.

Foster, F.R., and G.E. Lang. 1982. Decomposition of red spruce and balsam fir boles in the White Mountains of New Hampshire. *Can. J. For. Res.* 12: 617–626.

Gore, J.A., and W.A. Patterson III. 1986. Mass of downed wood in northern hardwood forests in New Hampshire: potential effects of forest management. *Can. J. For. Res.* 16: 335–339.

Graham, R.L. 1982. Biomass dynamics of dead Douglas-fir and western hemlock boles in mid elevation forests of the Cascade Range. Ph.D. Dissertation. Oregon State University, Corvallis.

Greir, C.C. 1978. A *Tsuga heterophylla* — *Picea stichensis* ecosystem of coastal Oregon: decomposition and nutrient balances of fallen logs. *Can. J. For. Res.* 8: 198–206.

Haftorn, S. 1988. Survival strategies of small birds during winter. In Oullet, H. (ed.), *Acta XIX Congressus Internationalis Ornithologica.* Vol. II. Ottawa, Canada.

Harmon, M.E., and C. Hua. 1991. Coarse woody debris dynamics in two old-growth ecosystems. *BioScience* 41: 604–610.

Harmon, M.E., and J. Sexton. 1996. *Guidelines for Measurements of Woody Detritus in Forest Ecosystems.* U.S. Long Term Ecological Research Publication No. 20. University of Washington, Seattle, Washington, USA.

Harmon, M.E., J.F. Franklin, F.J. Swanson et al. 1986. Ecology of coarse woody debris in temperate ecosystems. *Adv. Ecol. Res.*,15: 133–302.

Hayes, J.P., and S.P. Cross. 1987. Characteristics of logs used by western red-backed voles, *Clethrionomys californicus,* and deer mice, *Peromyscus maniculatus. Can. Field-Naturalist.* 101: 543–546.

Healy, W.M., R.T. Brooks, and R.M. DeGraaf. 1989. Cavity trees in sawtimber-size oak stands in central Massachusetts. *No. J. Appl. For.* 6: 61–65.

Hope, S., and W.C. McComb. 1994. Perceptions of implementation and monitoring of wildlife tree prescriptions on National Forests in western Washington and Oregon. *Wildl. Soc. Bull.* 22: 383–392.

Huff, M.H. 1984. Post-fire succession in the Olympic Mountains, Washington: forest vegetation, fuels, avifauna. Ph.D. dissertation, University of Washington, Seattle, Washington, USA.

Landres, P.B., P. Morgan, and F.J. Swanson. 1999. Overview of the use of natural variability concepts in managing ecological systems. *Ecol. Applic.* 9: 1179–1188.

Lennartz, M.R., and R.F. Harlow. 1979. The role of parent and helper red-cockaded woodpeckers at the nest. *Wilson Bull.* 91: 331–335.

Lowney, M.S., and E.P. Hill. 1989. Wood duck nest sites in bottomland hardwood forests of Mississippi. *J. Wildl. Manage.* 53: 378–382.

MacMillan, P.C. 1988. Decomposition of coarse woody debris in an old-growth Indiana forest. *Can. J. For. Res.* 18: 1353–1362.

Marcot, B.G. 1992. *Snag Recruitment Simulator, Rel. 3.1* [Computer program]. USDA For. Serv. Pacific Northwest Research Station, Portland, OR.

Martin, K.J. 1994. Movements and habitat associations of northern flying squirrels in the central Oregon Cascades. M.S. Thesis, Oregon State Univ., Corvallis. p. 44.

Maser, C., R.G. Anderson, and K. Cromack Jr. 1979. Dead and down woody material. In Thomas, J.W. (technical editor), *Wildlife Habitats in Managed Forests: The Blue Mountains of Oregon and Washington.* USDA For. Serv. Agric. Handb. No. 553.

Maser, C., B.R. Mate, J.F. Franklin, and C.T. Dyrness. 1981. *Natural History of Oregon Coast Mammals.* USDA For. Serv. Gen. Tech. Rep. PNW-133.

Mason, E.A. 1944. Parasitism by *Protocalliphora* and management of cavity-nesting birds. *J. Wildl. Manage.* 8: 232–247.

McComb, W.C. 2003. Ecology of coarse woody debris and its role as habitat for mammals. In Zabel, C.J. and R.G. Anthony (eds.), *Mammal Community Dynamics: Management and Conservation in the Coniferous Forests of Western North America*, Chapter 11. Cambridge University Press, Cambridge, England.

McComb, W.C., and D. Lindenmayer. 1999. Dying, dead, and down trees. In Hunter Jr., M.L. (ed.), *Maintaining Biodiversity in Forest Ecosystems*. Cambridge University Press, Cambridge, England.

McComb, W.C., and R.E. Noble. 1981a. Nest box and natural cavity use in three mid-South forest habitats. *J. Wildl. Manage.* 45: 92–101.

McComb, W.C., and R.E. Noble. 1981b. Microclimates of nest boxes and natural cavities in bottomland hardwoods. *J. Wildl. Manage.* 45: 284–289.

McComb, W.C., and R.L. Rumsey. 1983. Characteristics and cavity-nesting bird use of picloram created snags in the central Appalachians. *So. J. Appl. For.* 7: 34–37.

McComb, W.C., S.A. Bonney, R.M. Sheffield, and N.D. Cost. 1986. Den tree characteristics and abundances in Florida and South Carolina. *J. Wildl. Manage.* 50: 584–591.

McWilliams, H.G. 1940. Cost of snag falling on reforested areas. British Columbia For. Serv. Res. Note. No. 7.

Means, J.E., K. Cromack, and P.C. MacMillan. 1985. Comparison of decomposition models using wood density of Douglas-fir logs. *Can. J. For. Res.* 15: 1092–1098.

Mellen, K., B.G. Marcot, J.L. Ohmann, K. Waddell, S.A. Livingston, E.A. Willhite, B.B. Hostetler, C. Ogden, and T. Dreisbach. 2005. *DecAID, the Decayed Wood Advisor for Managing Snags, Partially Dead Trees, and Down Wood for Biodiversity in Forests of Washington and Oregon*. Version 2.0. USDA For. Serv. Pac. Northwest Res. Sta. and U.S. Dep. Interior Fish and Wildl. Serv., Oregon State Office, Portland, OR.

Miller, E., and D.R. Miller. 1980. Snag use by birds. In DeGraaf, R.M. (technical editor), *Management of Western Forests and Grasslands for Nongame Birds*. USDA For. Serv. Gen. Tech. Rep. INT-86.

Morrison, M.L., and M.G. Raphael. 1993. Modeling the dynamics of snags. *Ecol. Applic.* 3: 322–330.

Muller, R.N., and Y. Liu. 1991. Coarse woody debris in an old-growth deciduous forest on the Cumberland Plateau, southeastern Kentucky. *Can. J. For. Res.* 21: 1567–1572.

Muul, I. 1968. *Behavioural and Physiological Influences on the Flying Squirrel* Glaucomys volans. University of Michigan Museum of Zoology Misc. Publ. No. 134.

Neitro, W.A., V.W. Binkley, S.P. Cline, R.W. Mannan, B.G. Marcot, D. Taylor, and F.F. Wagner. 1985. Snags. In Brown, E.R. (technical editor), *Management of Wildlife and Fish Habitats in Forests of Western Oregon and Washington*. USDA. Forest Serv. Publ. No. R6-F&WL-192-1985.

Nelson, S.K. 1988. Habitat use and densities of cavity nesting birds in the Oregon Coast Range. M.S. Thesis, Oregon State University, Corvallis.

Nilsson, S.G. 1984. The evolution of nest-site selection among hole-nesting birds: the importance of nest predation and competition. *Ornis Scandinavica* 15: 167–175.

Oliver, C.D., and B.C. Larson. 1996. *Forest Stand Dynamics*. Update Edition. John Wiley and Sons, Inc. New York.

Olson, J.S. 1963. Energy storage and the balance of producers and decomposers in ecological systems. *Ecology* 44: 322–331.

Onega, T.L., and W.G. Eickmeier. 1991. Woody detritus inputs and decomposition kinetics in a southern temperate deciduous forest. *Bull. Torrey Bot. Club* 118: 52–57.

Ormsbee, P.C., and W.C. McComb. 1998. Selection of day roosts by female long-legged myotis in the central Oregon Cascade Range. *J. Wildl. Manage.* 62: 596–603.

Otvos, I.S., and R.W. Stark. 1985. Arthropod food of some forest-inhabiting birds. *Can. Entomol.* 117: 971–990.

Parks, C.G., E.L. Bull, and G.M. Filip. 1995. Using artificial inoculated decay fungi to create wildlife habitat. In Aguirre-Bravo, C., L. Eskew, A.B. Vilal-Salas, and C.E. Gonzalez-Vicente (eds.), *Partnerships for Sustainable Forest Ecosystem Management*. USDA For. Serv. Gen. Tech. Rep. RM-GTR-266.

Putz, F.E., P.D. Coley, A. Montalvo, and A. Aiello. 1983. Snapping and uprooting of trees: structural determinants and ecological consequences. *Can. J. For. Res.* 13: 1011–1020.

Scott, V.E., K.E. Evans, D.R. Patton, and C.P. Stone. 1977. *Cavity-Nesting Birds of North American Forests*. USDA For. Serv. Agr. Handb. 511.

Sedgwick, J.A., and F.L. Knopf. 1991. The loss of avian cavities by injury compartmentalization. *Condor* 93: 781–783.

Shigo, A.L. 1965. Pattern of defect associated with stem stubs on northern hardwoods. USDA For. Serv. Res. Note NE-34.

Shigo, A.L. 1984. Compartmentalization: a conceptual framework for understanding how trees defend themselves. *Ann. Rev. Phytopathology* 22: 189–214.

Sollins, P., S.P. Cline, T. Verhoeven, D. Sachs, and G. Spycher. 1987. Patterns of log decay in old-growth Douglas-fir forests. *Can. J. For. Res.* 17: 1585–1595.

Spies, T.A., and J.F. Franklin. 1991. The structure of natural young, mature and old-growth Douglas-fir forests in Oregon and Washington. In Ruggiero, L.F., K.B. Aubry, A.B. Carey, and M.H. Huff (technical coordinator), *Wildlife and Vegetation of Unmanaged Douglas-Fir Forests.* USDA For. Serv. Gen. Tech. Rep. PNW-GTR-285.

Spies, T.A., J.F. Franklin, and T.B. Thomas. 1988. Coarse woody debris in Douglas-fir forests of western Oregon and Washington. *Ecology* 69: 1689–1702.

Stelmock, J.J., and A.S. Harestad. 1979. Food habits and life history of the clouded salamander (*Aneides ferreus*) on northern Vancouver Island, British Columbia. *Syesis* 12: 71–75.

Tallmon, D.A., and L.S. Mills. 1994. Use of logs within home ranges of California red-backed voles on a remnant of forest. *J. Mammal.* 75: 97–101.

Torgersen, T.R., and E.R. Bull. 1995. Down logs as habitat for forest dwelling ants — the primary prey of pileated woodpeckers in northeastern Oregon. *Northwest. Sci.* 69: 294–303.

Tyrrell, L.E., and T.R. Crow. 1994. Dynamics of dead wood in old-growth hemlock-hardwood forests of northern Wisconsin and northern Michigan. *Can. J. For. Res.* 24: 1672–1683.

van Lear, D.H., and T.A. Waldrop. 1994. Coarse woody debris considerations in southern silviculture. In *Proceedings of the Eighth Biennial Southern Silvicultural Research Conference*, Auburn, AL.

Washington DNR, USDA Forest Service, WFPA, Washington Department of Wildlife, Washington contract loggers Association, and State of Washington Department of labor and industries. 1992. *Guidelines for Selecting Reserve Trees.* Allied Printers, Olympia, WA. p. 24.

Weigl, P.D., and D.W. Osgood. 1974. Study of the northern flying squirrel, *Glaucomys sabrinus*, by temperature telemetry. *Amer. Midl. Natur.* 92: 482–486.

Weikel, J.M., and J.P. Hayes. 1999. The foraging ecology of cavity-nesting birds in young forests of the northern Coast Range of Oregon. *Condor* 101: 58–65.

Wilson, B.F., and B.C. McComb. 2005. Dynamics of dead wood over 20 years in a New England forest. *Can. J. For. Res.* 35: 682–692.

Wimberly, M.C., T.A. Spies, C.J. Long, and C. Whitlock. 2000. Simulating the historical variability in the amount of old forests in the Oregon Coast Range. *Conserv. Biol.* 14: 167–180.

Zeiner, D.C., W.F. Laudenslayer Jr., K.E. Mayer, and M. White. 1990. *California's Wildlife. Volume III. Mammals. California Statewide Wildlife Habitat Relationships System.* California Department of Fish and Game, Sacramento, CA.

11 Landscape Structure and Composition

Landscapes: In art, they are paintings of places, usually from a distance, a mosaic of forms interwoven to portray a recognizable place familiar and pleasing to the eye. That is the function of the painting. Ecological landscapes also are a mosaic. They have structure, as portrayed by the pattern of the patches comprising the landscape. They have composition as portrayed by the types of patches on the landscape. And they have function, as defined by the resources of concern, in our case, patches of habitat for a species or collection of species. Forested landscapes as habitat for vertebrates are patterns of habitat patches distributed over forest-dominated areas, with each patch representing a set of resources used or needed by the species. Hence, the pattern of habitat patches across a landscape differs with species. We, as humans, tend to view forested landscapes as collections of forest stands, a mosaic of units of vegetation, each relatively uniform within itself. We do that because, to us, those stands represent different resources, such as economic value, places to hike, or different aesthetics. To some other vertebrates, stands also represent patches of different values, habitat values, which may not be the case with other species. Take the shrub-associated Hermit thrush, which is widely distributed across North America (Figure 11.1). What type of mosaic of patches would a Hermit thrush see? Their search image for a territory likely includes patches of shrubs and small trees with varying quality for foraging on insects in the leaf litter and foliage, as well as nest sites in the shrubs and small trees. The overstory vegetation seems to have little, if any, influence on what Hermit thrushes see as habitat patches except through the effect it has on shrub and small tree development.

FIGURE 11.1 Hermit thrushes select patches of shrubs and small trees as habitat across a landscape mosaic. (Photo from the North Cascades National Park, USDI National Park Service. With permission.)

Old-growth, stand establishment, or understory reinitation conditions are not really as important to this species as the lower vegetation present in the stand. So, if we were to develop a very simple map of habitat with white patches representing habitat and black representing unusable areas, then we would have white patches of shrubs of a certain cover and height for Hermit thrushes and these patches could be quite unrelated to overstory tree species composition or structure.

Mapping of habitat for northern flying squirrels would focus more on snags, tree species, size, and canopy cover, and not so much on shrubs. And to make things even more complicated, consider habitat for cottontail rabbits, which feed in grassy areas but use dense shrubs as cover. They live on the edge between these two types of vegetation and the acceptable habitat (white) on our map would be a linear strip extending perhaps 30 m into the grassy area and 20 m into the shrubs. Everything else would be black. Hence, we have three species and three landscape patterns all in the same forest. Managing stands in this forest could influence the patterns of habitat for each of these species, depending on the types of vegetation and other features provided in each stand.

We know that habitat varies in its quality from place to place and over time. At the very least we might consider sources and sinks for each species, but even these represent a gradient of habitat quality (as expressed through individual fitness). It is more appropriate to think of a mosaic of habitat as represented by shades of gray or perhaps more appropriately a collage of habitat values. A spectrum of resources painted across an area for each species. And if we are concerned about providing habitat for 100 species in our forest, then there are 100 landscapes in our forest, each with its own collage of habitat. Management decisions made at places in our forest represent the building blocks of a habitat mosaic for the species that we manipulate through our forest management activities. So, the structure, composition, and function of a landscape depend entirely on the species or other values that we wish to produce from that landscape.

DEFINING THE LANDSCAPE

There are several attributes of a landscape that must be defined when considering habitat function for a species across a landscape: grain, extent, and context. The *grain* represents the smallest unit of space that we (or the species of interest) use and identify. Recall the hierarchical selection of habitat discussed in Chapter 2. Each species scales its habitat differently at each of four (or more) levels of habitat selection (Johnson 1980). Depending on the scale of habitat selection we wish to understand when managing the species, the smallest unit of space that we would need to consider varies. For many species, the grain might best be an individual tree or shrub where resources are available: browse plants for a herbivore, fruit from a yaupon shrub, or blackgum trees where dens are available. But considering the spatial pattern of all trees and shrubs in a forest is impractical during management. Rather, we typically consider the patches within which these resources occur. The way we define these patches on the ground or on a map characterizes a grain. Some minimum size that is both meaningful to the species of interest and practical to consider during management must be defined. For example, we might define patches of shrubs 0.5 ha or larger of various plant species as resource patches for white-tailed deer. Or we could simply characterize each stand in our forest as having certain levels of food or cover resources within the stand, without explicitly considering where in the stand these resources occur. From a management standpoint we would just know that some stands have more high-quality browse than other stands.

Now let us assume that the forest landscape that we are managing is 20,000 ha; this would define an *extent*, the area over which we are managing resources. Alternatively, we could define the extent as the population of deer that we are managing and the area over which they occur. If we are managing the 20,000 ha landscape with a grain size of 0.5 ha, then we would need to keep track of up to 40,000 patches of browse across the landscape over time. We can do that with the aid of computers and geographic information systems (GIS), but from a forest management standpoint, such a fine level of information may not be practical to monitor or manage over time. Instead, resource patches such as these are often considered as a part of a stand prescription and the resources available in stands and other patches across the landscape are the units that are actually managed.

If we have a 20,000 ha landscape that we are managing, then it resides in a larger area, an area that we are not managing but which affects the function of our landscape. The *context* for the landscape can have a significant effect on what can be achieved through management of the landscape. This is analogous to management of stands in our forest. Manipulating the habitat elements in a stand *may* be effective in managing habitat for a species, depending on the resources available in the surrounding stands. Managing resources across a landscape *may* be effective depending on the resources available in the surrounding landscapes. Managing pileated woodpeckers in Central Park, New York City, is highly constrained by the surrounding urban context. Managing them in a small watershed of similar size in the Catskill Mountains of upstate New York may present far different constraints and opportunities in that context. Manipulating the resources among the patches represented by the grain, over the extent as constrained by the context, is the essence of landscape management for various species.

So how large is a landscape? By now you should know that the answer is, "It depends!" From a functional standpoint, there is no ideal upper or lower bound that describes a landscape. A decaying log on the forest floor represents a landscape with a mosaic of fungal colonies, bryophytes, bacterial colonies, and tree regeneration. The grain is the patch of fungus or the patch of tree seedlings. The extent is the log. The context is the forest in which it resides. Alternatively, the area extending from intermountain rangelands in North America to the pampas of Argentina represent a landscape of appropriate size for Swainson's hawks. They nest and feed in the northern hemisphere and spend the winter in South America (actually Argentina's summer coincides with our winter, so they are never in winter!). Patches representing feeding, nesting, and roosting sites in both hemispheres and along the migratory pathway represent a landscape used as the basis for management of this species.

From a practical standpoint, landscapes are usually considered areas of hundreds to thousands of hectares, areas where consideration of a meaningful grain does not become overwhelming in the landscape extent. One characteristic of a landscape that is particularly important when managing animals, especially mobile animals, is that the extent applies to all owners within the landscape. Recall that in the United States, wildlife are public resources. And most are somewhat mobile. And they do not respect property lines or political boundaries. So a landscape that you define when managing a set of resources that includes multiple owners must not be restricted to just your ownership. Habitat among all owners is what the animals will respond to (Spies et al. 2007). Too often a large-forest owner restricts consideration of impacts of their management on a species to those impacts occurring on their lands, not the multiownership landscape of habitat to which the population responds.

HABITAT QUALITY AT THE LANDSCAPE SCALE

Recall that habitat quality is a function of individual and population fitness. Habitat elements distributed among patches on a landscape change over time. They change as plants grow, soil erodes, and disturbances occur. Hence, the quality of the habitat for a species also changes among patches over time. We can direct those changes to at least some degree by management actions. Over landscapes, we may wish to increase, maintain, or reduce (depending on our goals) the fitness of the individuals in the population of concern. This is usually done in concert with forest management activities as well as natural disturbances and subsequent regrowth. Consequently, we manipulate habitat elements within stands and other management patches on the landscape to influence the direction of the cumulative changes in these resources over time.

Depending on how a species scales its environment, we might expect a population response from these actions. But populations do not respond solely to changes in availability of habitat elements. Disease, competitors, predators, parasites, and other density-dependent factors, which may or may not be mediated by habitat elements, can influence population responses. Altering habitat simply provides the potential for populations to respond in certain ways. And there is always uncertainty in what responses we will see. Disturbances occur, but are usually unpredictable. Climate is changing. Invasive species are having new consequences on some populations. Consequently, tracking the vital rates of the population and understanding the long-term changes in birth rates, death rates, and *lambda* (the population parameter used to define if a population is increasing or declining) allows us to understand the consequences of management actions over large areas. These population parameters are usually used to define habitat quality across a landscape, which reflects the cumulative value of habitat among patches across the landscape. But the location of the patches matters. If a species needs 40% of its home range in high-quality food patches to achieve high levels of fitness and you manage to provide 40% of a landscape in these patch types, then the food patches must be distributed so that each of multiple potential home ranges are distributed appropriately. Locating all high-quality patches on one side of a landscape would leave the other side of no value to this species.

LIVING ON THE EDGE

A feature of landscapes that has an inordinate effect on habitat quality for many species is the boundary between patches. Edges between patches, as well as other linear features such as roads, rivers, and ridgetops, all influence habitat quality for various species even though these linear features occupy only a very small part of a landscape. They can have a disproportionate effect on the function of a landscape for a species. Depending on the species and the type of edge, edges can be beneficial, detrimental, or have no effect whatsoever. Recall our example of the cottontail rabbits earlier in this chapter. Certain types of edges *are* habitat for this species. But other species are adversely affected by certain types of edges. Edges are usually characterized as induced or inherent (Figure 11.2). *Induced edges* are those that occur between two patch types of different successional condition, are usually caused by a disturbance, and represent structural differences between the patches. Edges between early- and late-successional forests, forests and utility right-of-ways, or forests and agricultural lands

FIGURE 11.2 Induced edges reflect structural differences between two patch types, while inherent edges represent floristic differences between patches.

Brown-headed cowbird nestling   Eastern phoebe nestling

FIGURE 11.3   Brown-headed cowbirds are brood parasites that can reduce nest success of other bird species, often neotropical migrants, when they nest near edges. (Photo by Mark Hauber and used with his permission.)

are examples of induced edges, though each of these types of edge functions differently depending on the species. *Inherent edges* are those formed because the composition of the forests differs between the patches. Although this may ultimately be the result of disturbance, it is less a structural interface and more a floristic interface. Induced edges between forests and agricultural lands are ideal conditions for brown-headed cowbirds to nest (Figure 11.3). Cowbirds fed among bison herds historically, and now feed among cattle, where they eat insects disturbed by the grazing animals. They reproduce by laying their eggs in the nests of other birds, a reproductive strategy called *brood parasitism*. Once the female is fertilized she flies into the forests adjacent to the grazing lands and seeks hosts for her eggs. These hosts are often *neotropical migratory birds*, which generally produce one *brood* (cohort of young birds) per year. Once the female finds a host nest, she removes the host's eggs and lays her egg in their place. Should the host continue to lay eggs, then the host eggs hatch after the cowbird egg has hatched. The nestling cowbird grows quickly and the host parent raises it as its own. Being larger and more aggressive it tends to get most of the food provided by the host parents and the host nestlings often die as a result of competition for food. Consequently, the cowbird has a high probability of fledging (successfully leaving the nest) and the host nestlings are unlikely to fledge. Host species within a few hundred meters of agricultural or grassland edges often have depressed levels of nest success due to cowbird parasitism. They can also experience higher levels of predation along edges than in the interior of the forest because raccoons, rat snakes, opossums, and deer mice also often forage near edges. So creating these edges between grasslands and forests can be source habitats for brown-headed cowbirds, cottontail rabbits, and rat snakes, but sinks for neotropical migrant birds that experience high levels of brood parasitism and predation along the edges.

Species can be classified based on their associations with edges. *Edge associates* are those species that find the best quality habitat where there is access to required resources in two or more vegetation patch types. There are also *edge specialists*, those such as cottontail rabbits that are likely to occur only where these edges exist. And there are some species that are found along edges, but do not require the edge; rather, their home ranges simply abut the edge. And there are *forest-interior species*, those that avoid edges and use the interior or core of a patch. Managing these groups together on one landscape often results in a high diversity of species along edges (specialists, associates, and

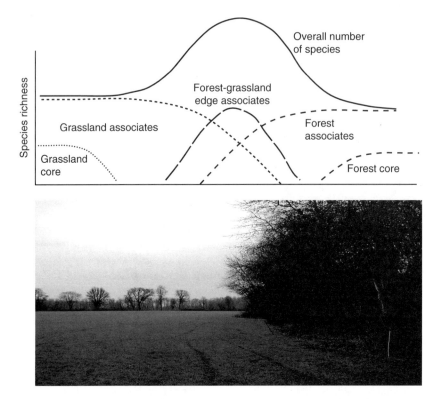

FIGURE 11.4   Species richness is often high along edges due to the contributions of species to edges from both adjoining patches as well as edge specialist species. Note though that patch interior species avoid these edges.

those that occur in each of the adjoining patch types). Hence, you tend to find more species along edges than in the interior of forests or grasslands (Figure 11.4). Since Aldo Leopold's book *Game Management* was published, managers have often focused on providing edges as a way of increasing populations of some game species or increasing species diversity across a landscape. But as in all other management choices, some species are winners and others are losers. Increasing edges can be to the detriment of forest-interior species.

In addition to increased levels of predation and parasitism along edges, some species can be either positively or adversely affected by edges because of changes in the microclimate, probability of natural disturbances, vegetation structure, and a suite of human effects. Microclimatic changes along induced edges are highly influenced by edge contrast, or the degree of difference in forest structure between the two patches. If a forest adjoins an open area, then the edge of that forest receives more sunlight (depending on aspect and slope), is drier, receives more wind, and relative humidity is likely to fluctuate more throughout the day. These conditions can affect habitat quality for some species directly, especially reptiles and amphibians (and other ectotherms), and can result in changes in vegetation structure. Time since edge creation is often an important factor influencing the structure of an edge. An edge formed by a clearcut decreases in contrast to the adjacent forest as the trees in the clearcut grow. And the edge in the forest that receives more sunlight after the clearcut occurs also will fill in with shrubs, trees, and forbs, creating a "softer," less abrupt edge.

Disturbances also can be influenced by the presence of an edge. Fuels for fires that differ between patches can allow intense fire to affect the edges of patches that otherwise would not be so severely affected. Abrupt edges in wind-prone sites can lead to continuous windthrow of the trees along the edge, resulting in accumulations of dead wood (or timber salvage opportunities). Consequently, disturbance intensity can be unusually high along edges, leading to changes in the structure and function of the recovering forest following the disturbance.

Humans not only create edges but use them. Hunters often will follow edges to find game animals. Roads and fences often occur in conjunction with edges. Consequently, domesticated animals tend to use the edges as well; so grazing pressure can be higher there. Free-ranging dogs and feral cats can have a significant impact on native animals and they often hunt near edges. The geographic ranges of nearly all midsized marsupials in Australia have been greatly reduced because of predation by feral cats and foxes. The Wildlife Society produced a position paper on this topic in 2006; excerpts follow:

> The estimated numbers of pet cats in urban and rural regions of the United States have grown … to nearly 65 million in 2000. … domestic cats are a significant factor in the mortality of small mammals, birds, reptiles, and amphibians … . Effects of cat predation are most pronounced in island settings (both actual islands and islands of habitat), where prey populations are already low or stressed by other factors, or in natural areas where cat colonies are established. Competition with native predators, disease implications for wildlife populations, and pet owners' attitudes toward wildlife and wildlife management also are important issues.

Houses in forests perforate if not fragment the forest, proliferating edges. Many of these houses support non-native plants and animals and so the proliferation effects of these non-native species on native species is further exacerbated. Indeed many landowners have planted exotic plants specifically to attract wildlife, which may have long-lasting adverse effects on native plant species.

EDGE GEOMETRY

The effect of an interface between patches on habitat quality for a species often is dependent on the amount of edge occurring within an animal's home range. Size matters, both in the area of the patches constituting the edge as well as in the length of the edge itself. Consider the patches in Figure 11.5. A 10 ha square patch has 1265 m of edge. But two 5 ha patches have 1789 m of edge and five 2 ha patches have 2828 m of edge. So, minimizing edge for forest-interior species can be achieved by having fewer large patches and maximizing edge for edge specialists can be achieved by having many smaller patches.

Edge density (edge length per unit area) can also be influenced by patch shape. Consider the 10 ha patches in Figure 11.6 that all have the same area. The more a patch diverges in shape from a circle, the greater is the edge that is represented per unit area. Patches shaped like amoebas have more edges than regular shapes. So when identifying stand boundaries, stands that are large and approximate a circle or hexagon in shape will minimize edge and maximize *core* (interior) conditions. Patches that

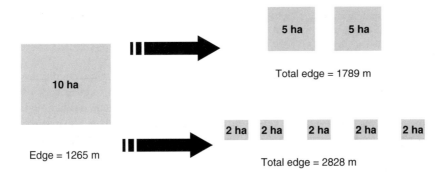

FIGURE 11.5 Decreasing patch size, but keeping total area constant increases the total amount of edge. Increasing patch size per habitat area increases core.

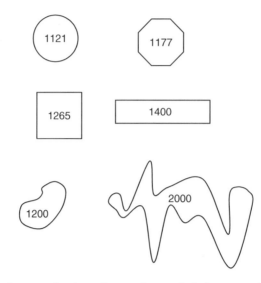

FIGURE 11.6 The more that a regular shape diverges from a circle the greater the edge per unit area and the less core that is available.

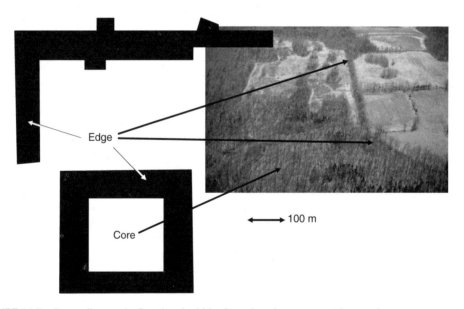

FIGURE 11.7 Depending on the functional width of an edge, then some patches can have no core area. Size, shape, and edge width dictate the amount of core.

are small and irregularly shaped will minimize core and maximize edge. Core conditions are usually identified by considering the area of a patch some distance away from an edge, say 100 m. But it should be clear by now that the distance away from an edge that constitutes a functional core varies from one species to another. For some species it might be 5 m and for others 500 m. For species that are particularly edge sensitive, the amount of core area in small or irregularly shaped patches can easily become zero (Figure 11.7).

The arrangement of patches on landscape can also influence edge conditions. Consider a set of patches in the pattern of a chess board. If all the white squares are acceptable as habitat and the black ones are not, then edges are maximized. Now place all the white squares on one side

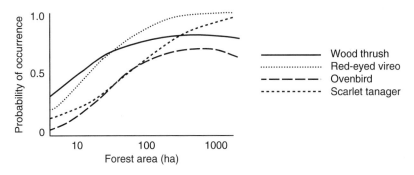

FIGURE 11.8 As the area of a forest patch increases, the probability of detecting forest-interior bird species increases. For some species, very large patches are needed before there is a high probability of occurrence of these species. (Summarized and redrafted from Robbins, C.S., D.K. Dawson, and B.A. Dowell. 1989. *Wildl. Monogr.* 103: 1–34. With permission from The Wildlife Society.)

of the chess board and the black squares on the other side. By blocking patches of similar habitat quality you can increase overall patch size and minimize edge area. Pattern can also influence edge contrast. If a decision is made to minimize edges by systematically harvesting adjacent stands using a clearcut regeneration system across a watershed, then the contrast between recent clearcuts and various ages of regenerating clearcuts will be less abrupt. Alternatively, if we wanted to maximize edges using this system, we would disperse clearcuts throughout a watershed in a "staggered-setting" approach, a technique commonly used on National Forests in the United States for years to improve edge conditions for game species and distribute the clearcut disturbances. The decisions on how to make large patches or small patches during management can be based on the area required by various species or by the area impacted by various disturbances. Even small disturbances can have significant effects on some forest-interior species, because the area requirements for the species extend far beyond the territory size of the species because of adverse edge effects. Consider four species of forest-interior birds from the eastern United States (Figure 11.8). For some species, the patch area must be thousands of hectares before there is certainty that the species will be found in the patch. Because the territory sizes for these species are in the range of 1 to 10 ha, demographic processes associated with avoidance of edges and associations with core conditions are assumed to be largely responsible for these species being associated with such large contiguous forest patches. At least that is the case in landscapes where forests are interspersed among agricultural lands such as in the eastern United States (Robbins et al. 1989). There seem to be less strong associations with large patch sizes for small neotropical birds in forested landscapes where older forests are interspersed with various ages of younger forests (McGarigal and McComb 1995, Welsh and Healy 1993). Agricultural lands provide a very stable *matrix* within which forested patches are embedded and may represent a context for the forests that results in significant adverse edge effects that extend into the forested patches. In forested landscapes interspersed with various-aged forest stands, these edge effects do not seem so significant for many forest-interior species, probably because stands are dynamic. Edge contrast can be ameliorated by quick tree growth on some sites. On high-quality sites, forest structure and composition changes rapidly, minimizing the length of time that high-contrast edges are found on the landscape.

HABITAT FRAGMENTATION

Habitat fragmentation is a landscape-level process in which a specific habitat is progressively sub-divided into smaller, geometrically more complex, and more isolated fragments as a result of both natural and human causes. It involves changes in landscape composition, structure, and function at

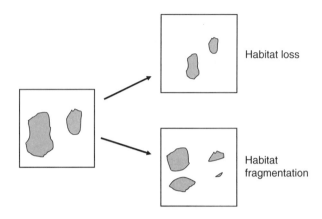

FIGURE 11.9   Habitat loss does not increase the isolation or geometry of patches, while fragmentation is a process that increases patch isolation and exacerbates edge effects. (Based on Fahrig, L. 1999. Forest loss and fragmentation: which has the greater effect on persistence of forest-dwelling animals. In Rochelle, J.P., L.A. Lehman, and J. Wisniewski (eds.), *Forest Fragmentation: Wildlife and Management Implications*. Brill Press, The Netherlands, pp. 87–95. With permission.)

many scales and is overlain on a mosaic created by changing landforms and natural disturbances. Habitat fragmentation is related to but different from habitat loss (Figure 11.9). Habitat can be lost because of natural disturbances, human actions, or ecological succession. That loss can occur in a manner that erodes the edges of a large patch such that habitat area declines, but the patch is still intact and it is not dissected into smaller more isolated pieces. Alternatively, large patches can successively be dissected by a road, utility right-of-way, or clearcut, until there are many smaller patches on the landscape and not just one large patch. Habitat area has declined, but more importantly habitat patches have become subdivided and more isolated from one another. This latter process is termed fragmentation. But the effects of this process are clearly specific to a species. Fragmenting habitat for ovenbirds may be caused by reducing the size and connectedness of mature eastern hardwood forests. Fragmenting habitat for bobolinks (a grassland bird) may be caused by allowing forests to encroach upon grasslands. And fragmentation for both species may be caused by expanding housing developments and roads. The process of fragmentation is species specific, though it is often generalized in the popular literature. Forest fragmentation. How often have you seen that term? But fragmenting a forest can be good, bad, or indifferent depending on the structure and composition of the forest and the species with which you are concerned. Indeed, some forests are fragmented from natural disturbances on a regular basis; others rarely so. The effects of the loss of habitat separate from the fragmentation of habitat must be considered carefully to understand the potential impacts on animal and plant species.

Habitat Area: Species–Area Relationships

Forest fragmentation can occur from natural disturbances or human activities. Indeed, we would expect that many if not all species should be adapted to the frequency, severity, and size of natural disturbances in the landscapes where they occur. We often use this assumption as the basis for designing human disturbances in forests to minimize the risks to species, both known and unknown, through our management actions (Landres et al. 1999). Simply from the standpoint of supporting species richness (number of species in an area), we know that large patches of forest (or grassland) support more species than small patches of forest. Large oceanic islands support more species than small islands, all other things being equal. There is typically an asymptotic relationship between area and species number (Figure 11.10). Hence, the conventional wisdom, when considering identification of reserves designed to meet the needs of many species, is that large areas, areas where an asymptote

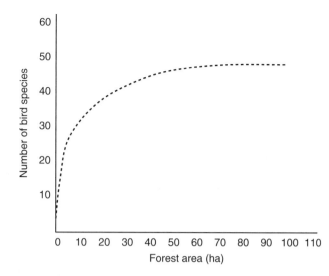

FIGURE 11.10 Example of a typical species–area relationship for forest birds among patches in a landscape. Species richness becomes asymptotic at large patch sizes.

in species richness is reached, represent appropriate areas to "capture" the most species in a region. This approach to conservation of biodiversity is dependent on two spatial scaling properties: (1) the range of spatial scales represented following disturbance and regrowth across a landscape and (2) the range of spatial scales represented by home ranges and populations of organisms occurring on a landscape.

As we have discussed already, natural disturbances of a particular severity typically occur over a range of sizes within a region or forest type. Less frequent disturbances tend to be larger and more severe. The combination of forest regrowth and disturbance produces a range, or domain, of spatial scales that are associated with the severity and frequency of a disturbance. The exact point within this domain that any point on the landscape might experience is dependent on many factors and is not deterministic, but is rather probabilistic. There is some probability that a disturbance will be of a particular size or severity, but we cannot predict exactly what it will be or when it will occur with certainty. Rather, it is this domain of sizes that we can predict and it is the domain of sizes to which most species in the region should be reasonably well adapted to using. How those species react to disturbance severity and size is largely dependent on the spatial domains represented by the suite of species occupying an area. For instance, if we look at the frequency of occurrence of species home-range sizes among all species in a region (Figure 11.11), it is typically a negative exponential distribution, with many species occupying very small portions of a landscape and few species occupying large portions of a landscape. The spatial domain of the cumulative species area requirements defines a domain of spatial scales that would need to be represented across large areas to meet the needs of the full suite of species. The domain of disturbance sizes in a landscape provides the template upon which the species can or cannot find the patch areas and arrangements that meet the cumulative needs of the group of species. The interface between the spatial domains inherent in a landscape (from disturbances and regrowth) and the spatial domain of the animal community (from home ranges of multiple species) causes some landscapes to be better able to support a more full range of species than others. Now consider human impacts on landscape pattern and composition. As land-use decisions such as roads, housing, forest management, and agriculture change the pattern of forests across a landscape, the domain of spatial scales represented in a forested landscape changes and the domain is often narrowed. Forest management occurs over a narrower range of spatial scales (fewer very small patches, fewer very big patches; Figure 11.12) than natural disturbances. The spatial scaling properties of the animals that use a landscape pattern that is constrained by land use

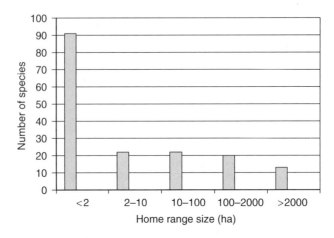

FIGURE 11.11   Frequency distribution of home range sizes for species in the northeastern United States. (Summarized from DeGraaf, R.M. and M. Yamasaki. 2001. *New England Wildlife: Habitat, Natural History, and Distribution*. University of Press of New England, Hanover, NH.)

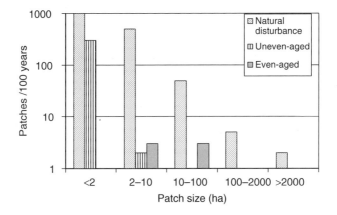

FIGURE 11.12   Range of patch sizes one might expect over a 10,000 ha landscape under natural disturbance regimes and the range of conditions represented by even-aged and uneven-aged forest management approaches. Species associated with patch sizes under-represented by management may face risk unless the range of spatial scales of management is increased or reserves of various sizes are employed within which natural disturbances are allowed to occur.

may change somewhat in response to these new landscape configurations, but unless a species is preadapted to this new narrower range of conditions, it may find that its habitat availability has declined. This reduction in habitat for some species is caused by having a narrower range of patch sizes and arrangements than prior to modern human intervention (Figure 11.12). Small patches may be less common because humans tend to homogenize stand conditions making them easier to manage and producing uniform products within and among stands, unless we take actions to make stands heterogeneous (see Chapters 6 through 8). Large patches are also less likely. Because of social constraints, we are unlikely to manage an entire 10,000-ha watershed using even-aged management to replicate a wildfire, but some species are well adapted to that forest condition at that spatial scale (especially if a complex early-seral condition is provided). Hence, we are often managing forests and the animals that occur in them over a limited spatial domain. Those species adapted to conditions (stand types, edge densities, legacy material, etc.) occurring at smaller and larger spatial scales than those represented by management can be considered most likely to be at risk of having decreased habitat quality (Figure 11.12). If we can identify those species, then we can take management actions to address their needs. Species needing large, complex early-successional patches, for instance, may have their needs met if even-aged systems with legacy retention are employed by clustering harvest units over time that collectively contribute to overall large patches, leaving late-successional forests also in large patches. For those species that require small patches, variable density planting and thinning, or group selection systems using a variety of group sizes, may help to address their needs. These silvicultural and harvest considerations must be planned to ensure that a broader spatial representation is achieved over a landscape than would be achieved simply by maximizing timber profits. That is, in order to achieve the desired patterns and reduce the risk of some species losing habitat availability, planning to represent a broad spatial domain may come at an economic cost when managing a landscape because moving the landscape into an appropriate spatial domain may mean cutting stands earlier or later than would be done to maximize profits.

Just as there are spatial domains of disturbance and regrowth, there are also temporal domains of disturbance and regrowth. And there are temporal domains of population demographics. Some species produce several generations in one year and others produce several generations in decades or centuries. Consequently, the rate of change in forest structure and composition over a landscape, as well as the spatial patterns, can affect species differently. For instance, species adapted to prolonged forest recovery following a disturbance that retains significant legacy from the previous stand may find diminished habitat quality where forest managers are striving to minimize legacy retention and the rate at which growing space is occupied by economically valuable trees. Vegetation management and density management can shorten the period of a diverse early-successional condition and reduce the opportunity for species to find this recently disturbed area. Further, intensive early stand management can minimize the time available for them to produce sufficient young to colonize new areas that grow into habitat elsewhere on the landscape over time. Similarly, species associated with post-rotation-aged forests may have limited opportunities to produce young. This is particularly a problem with species that reproduce late in life, produce few young, and reproduce infrequently. Long-lived species such as tortoises and parrots require some level of habitat stability over long time periods in order to have reproduction and survival requirements met. If forest management or forest land use changes the duration of a particular seral stage, then habitat availability for species associated with habitat elements found in those stages can be reduced. Of particular concern are those species that are not only long-lived but also need to produce multiple generations to ensure population viability within a region.

One way of reducing the impact of changing the temporal domain of scale is to provide legacies from one forest condition to the next over large areas and allow some habitat elements to be represented over large areas over time (e.g., dead wood in managed forests), thereby buffering the diminished domains of scale represented among forest age classes. Hence, attending to the needs of species that are long-lived or have multigenerational requirements to maintain viability may represent a cost to the resource manager not only in silvicultural activities and harvest planning but also in a commitment to legacies of structures and patch types across managed landscapes. It is the intersecting domains of space and time in landscape composition and structure that interfaces with the spatial and temporal domains of a collection of species occurring on the landscape that must be coordinated to minimize the probability that some species will be placed at risk. This approach greatly complicates harvest planning, but it is possible to find harvest planning solutions to spatial problems such as these (Bettinger et al. 2001).

CASE STUDY: HABITAT AREA OR PATTERN?

In the early 1990s, Dr. Kevin McGarigal conducted a study that addressed the relative role of forest area vs. forest pattern in structuring bird communities in the Oregon Coast Range (McGarigal and McComb 1995). Subsequent work by Dr. Karl Martin also addressed associations with mammal and amphibian communities on these sites (Martin and McComb 2002, 2003). The work was set in the

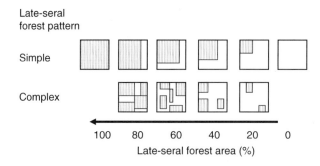

Late-seral
forest pattern

FIGURE 11.13 Experimental design used to assess the contributions of late-seral forest area and pattern to bird, mammal, and amphibian species in the Oregon Coast Range. (Redrafted from McGarigal, K., and W.C. McComb. 1995. *Ecol. Monogr.* 65: 236–260. With permission from the Ecological Society of America.)

dynamic forests of the Oregon Coast Range and followed work by Robbins et al. (1989), Temple (1986), and others, who identified a number of area-sensitive bird species in eastern hardwood forests that seemed to be adversely affected by processes of forest fragmentation resulting from agricultural land uses and urbanization. Few studies had addressed fragmentation effects on vertebrates in managed forest landscapes at that time and many of the findings regarding the effects of fragmentation in eastern hardwood forests were being used to design management strategies for forests in western coniferous forests. McGarigal and Martin wanted to see if these same trends were consistent in forests that were fragmented by forest management while controlling for habitat loss. The experimental design was to sample three replicates of 10 sub-basin (250 to 300 ha) conditions representing two dominant gradients in late-seral forests: 0 to100% late-seral forest area with minimal fragmentation (remaining forest was in large blocks) and also sub-basins with a high level of late-seral forest fragmentation (remaining forest in many blocks) (Figure 11.13). But forests are not structured as simply as is represented in this experimental design. Patch shapes and arrangements vary considerably from one sub-basin to another, and so indicators of landscape structure and function were needed. Consequently, McGarigal and Marks (1995) developed an analytical computer program called FRAGSTATS to provide a huge number of landscape metrics to describe many aspects of patch conditions, sizes, shapes, connectedness, edge conditions, and core areas, among others. Using these metrics in their analyses allowed them to understand if the animals were associated more with the composition of the landscapes (how much late-seral forest was available) or if pattern mattered. Based on over 100,000 observations of birds, mammals, and amphibians, several patterns emerged. For many bird species, habitat area matters. Of 15 species of birds that selected late-seral forests (based on use and availability), 11 were associated more with the *area* of late-seral forest than with its *configuration* (more habitat was better than less). These included species such as brown creepers, pileated woodpeckers, and varied thrushes (McGarigal and McComb 1995). Five species of amphibians also were associated more with the area of a forest condition than its configuration, such as the southern torrent salamander (Martin and McComb 2003). And six species of mammals were more associated with patch area than configuration, including California red-backed voles, deer mice, creeping voles, Pacific shrews, and Pacific jumping mice (Martin and McComb 2002). None of the 14 mammal species captured were negatively associated with edges (Martin and McComb 2002).

But some species were clearly associated with configuration. Most species that exhibited an association with fragmented forests were more abundant in fragmented than unfragmented forests, unlike what was seen in many eastern hardwood forest studies. Species such as olive-sided flycatchers, deer mice, and Pacific jumping mice were positively associated with edges. Very few species were associated with a core area father than 100 m from an edge than with the edges; Pacific giant salamander was one of these core-associated species. Hence, as you would expect, there are species that likely would benefit from fragmentation of late-seral western coniferous forests and those that likely would not. The fact that landscape pattern (fragmented forests) seemed to be only modestly associated with the abundance of many of these species and in most cases was more associated with fragmented than unfragmented forests, may seem counterintuitive, but remember that these forests are dynamic. Harvested stands in these studies were regrowing. Plantations that had regrown even 20 years had reduced edge contrast and provided a matrix condition through which many species could once again disperse. Indeed, in landscapes where the matrix is static (e.g., agricultural lands) and the forest patches are isolated, the results of McGarigal and Martin might have been much different. Further, animals dispersing across fragmented landscapes face different challenges in landscapes with little habitat available than with even marginal habitat available between high-quality patches. So, despite the results reported here, there is reason to consider pattern effects on populations, especially under conditions of low habitat area (With 1999). That is the subject of our next chapter.

SUMMARY

Forested landscapes have structure, composition, and function. Function for each species varies depending on how habitat patches are defined on the landscape and how they are connected across it. The scaling properties of a landscape are defined by the grain (the smallest unit of space that we find useful), the extent or the outer boundaries of the landscape under consideration, and the context, the condition of the area surrounding the extent. Animals integrate patches of varying habitat quality across the complex mosaics of landscapes to meet their habitat needs. As landscape structure, composition, and function change, so will populations and fitness of the species using the landscape.

Edges between successional stages (induced) or plant communities (inherent) represent zones in a landscape where animal species richness can be high, but adverse effects on forest interior species are exacerbated. Predation, brood parasitism, spread of invasives, and altered microclimates along edges can lead to depressed fitness for forest interior species.

Fragmentation is the process of breaking habitat patches into smaller, more complex, and more isolated pieces. This process is different from habitat loss, where habitat reduction may or may not be accompanied by increased isolation. In dynamic forest systems, where the matrix condition among habitat patches is continuously changing, habitat loss is often a more serious issue than fragmentation. But once fragmentation has led to significant isolation of the remaining patches, fragmentation effects may become apparent especially if the matrix is inhospitable to the organism of interest.

REFERENCES

Bettinger, P., K. Boston, J. Sessions, and W.C. McComb. 2001. Integrating wildlife species habitat goals and quantitative land management planning processes. In Johnson D.H. and T.A. O'Neill (managing editors), *Wildlife habitat relationships in Oregon and Washington*. OSU Press, Corvallis, OR, Chapter 23.

DeGraaf, R.M. and M. Yamasaki. 2001. *New England Wildlife: Habitat, Natural History, and Distribution*. University of Press of New England, Hanover, NH.

Fahrig, L. 1999. Forest loss and fragmentation: which has the greater effect on persistence of forest-dwelling animals. In Rochelle, J.P., L.A. Lehman, and J. Wisniewski (eds.), *Forest Fragmentation: Wildlife and Management Implications*. Brill Press, The Netherlands, pp. 87–95.

Johnson, D.H. 1980. The comparison of usage and availability measurements for evaluating resource preference. *Ecology* 61: 65–71.

Landres, P.B., P. Morgan, and F.J. Swanson. 1999. Overview if the use of natural variability concepts in managing ecological systems. *Ecol. Applic.* 9: 1179–1188.

Martin, K.J., and W.C. McComb. 2002. Small mammal habitat associations at patch and landscape scales in Oregon. *For. Sci.* 48: 255–266.

Martin, K.J., and B.C. McComb. 2003. Amphibian habitat associations at patch and landscape scales in western Oregon. *J. Wildl. Manage.* 67: 672–683.

McGarigal, K., and B.J. Marks. 1995. *FRAGSTATS: Spatial Pattern Analysis Program for Quantifying Landscape Structure.* USDA For. Serv. Gen. Tech. Rep. PNW-351.

McGarigal, K., and W.C. McComb. 1995. Relationships between landscape structure and breeding birds in the Oregon Coast Range. *Ecol. Monogr.* 65: 236–260.

Robbins, C.S., D.K. Dawson, and B.A. Dowell. 1989. Habitat area requirements of breeding forest birds of the Middle Atlantic states. *Wildl. Monogr.* 103: 1–34.

Spies, T.A., B.C. McComb, R. Kennedy, M. McGrath, K. Olsen, and R.J. Pabst. 2007. Habitat patterns and trends for focal species under current land policies in the Oregon Coast Range. *Ecol. Applic.* 17: 48–65.

Temple, S.G. 1986. Predicting impacts of habitat fragmentation on forest birds: a comparison of two models. In Verner, J.A., M.L. Morrison, and C.J. Ralph (eds.), *Wildlife 2000: modeling habitat relationships of terrestrial vertebrates.* University of Wisconsin Press, MA, pp. 301–304.

Welsh, C.J.E., and W.M. Healy. 1993. Effect of even-aged timber management on bird species diversity and composition in northern hardwoods of New Hampshire. *Wildl. Soc. Bull.* 21: 143–154.

With, K.A. 1999. Is landscape connectivity necessary and sufficient for wildlife management? In Rochelle, J.P., L.A. Lehman, and J. Wisniewski (eds.), *Forest Fragmentation: Wildlife and Management Implications.* Brill Press, The Netherlands, pp. 97–115.

12 Landscape Connections

Think for a moment about where you were in the past 24 hours. Draw a map and on it locate all those places where you spent more than 5 minutes in the past 24 hours. Now draw a straight line from one place to the other in the order they were visited. Do these straight lines represent your path from one place to another? Probably not. Most likely you used roads, sidewalks, doors, and hallways to get to where you were going. These connections from place to place are similar in some respects to the connections used from patch to patch by some species of animals, but not all. Consider how you would use a compass or GPS to find your way through an unfamiliar forest. You take a bearing, know a distance, and then walk through whatever is between you and your goal (within reasonable limits) more or less in a straight line. Some of the walking through closed canopy forests will be easy, and other places with dense shrubs will be very difficult. So rather than using a connection, you walk in a straight line with intervening patches representing various "risks" to your walking ability. These two types of movement are similar to the types of dispersal that other animal species experience. Some use corridors or connections across a landscape and others disperse across the *matrix* or the intervening conditions between patches of suitable habitat. And the matrix poses various risks to survival during dispersal depending on its habitat quality. This is analogous to a differentially permeable membrane in which some molecules can pass through the membrane easily and others cannot. For dispersing animals, matrix conditions represent differentially permeable conditions for moving organisms. Some intervening conditions allow the animal to move easily during dispersal; others present significant risks to survival (Martin and McComb 2003).

DISPERSAL

Animal movements within a home range often follow paths that are used repeatedly. Deer, elk, and moose, for instance, develop well-worn paths through forests when moving to and from cover and food. Even birds repeatedly use somewhat predictable flight paths in their daily movements (Bélisle and Desrochers 2003). But when an animal is displaced from its home range by its parents (natal dispersal), by dominant individuals, or by a disturbance that alters the quality of habitat in the home range, then the animal disperses in search of a location to establish a new home range. Obviously, an animal would not want to disperse any farther than needed or it is expending energy unnecessarily and reducing its chances of survival. On the other hand, it should not settle into a suboptimal home range if a better home range means spending only a bit more energy. How is it that animals find these potential home ranges and decide to settle in one?

Dispersal capabilities vary widely among species, being quite limited for some species (e.g., red tree voles; Hayes 1996) considered dispersal specialists and quite extensive for others considered dispersal generalists (e.g., Pacific jumping mice; Gannon 1988). Dispersal, especially natal dispersal, is a process that is critical to the growth and spread of populations, stability of metapopulations, recolonization of vacated habitat, and flow of genes and bodies among subpopulations. Body mass and trophic level seems to influence natal dispersal distances in mammals: large species disperse farther than small ones and carnivores disperse farther than herbivores and omnivores (Sutherland et al. 2000). Similar relationships have been documented for carnivorous bird species, but not for herbivorous or omnivorous birds (Sutherland et al. 2000). Bowman et al. (2002) found that dispersal distances are often associated with the home-range size of the animal more than with the body mass of the animal. Birds and mammals tend to have a dispersal pattern that follows a negative exponential

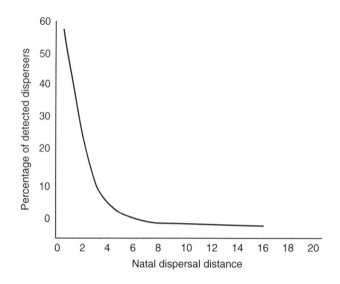

FIGURE 12.1   When dispersal distances are scaled to the median dispersal distance for the organism (in this case 29 species of mammals), a negative exponential pattern emerges. (Redrafted from Sutherland, G.D. et al. 2000. *Conserv. Ecol.* 4: 16. With permission.)

curve; that is, most animals disperse a small distance (relative to the median distance dispersed by the species) and few individuals disperse long distances (Figure 12.1). For many species, dispersal distances decline as the number of home-range diameters increases (Bowman et al. 2002). This is an intuitively appealing way to consider dispersal because animals cruise around their home ranges regularly. To move another home-range diameter would not seem to be too energetically costly, but as dispersal continues over multiple home-range diameters, the ability of the organism to expend that energy and still survive comes into question quickly. It may be more energy efficient for an animal to settle into a suboptimal patch and survive there than to continue to disperse and risk the chance of dying by dispersing through low-quality patches.

So what conditions are necessary for successful dispersal across a landscape to allow recolonization of a patch or introduction of new alleles into a population? The probability of successful dispersal across a landscape can be considered a function of the probability of three interacting probabilities: encounter, survival, and continuing. To examine the concepts behind successful dispersal, let us first consider a simple conceptual model, and then we will complicate it with reality.

UNDERSTANDING THE PROBABILITY OF SUCCESSFUL DISPERSAL

There are three dominant effects that determine if an individual dispersing from one patch is likely to reach another patch where it can find the resources to survive. Let us consider the patch from which it is leaving as the *source patch* and the patch to which it is going as the *target patch*. First, an animal must travel in a direction that causes it to encounter the target patch. Many animals do not seem to disperse in a given direction, but rather travel in random directions (Figure 12.2). Although the actual path that an animal takes may appear to be as that of a pinball in a pinball machine, when it settles, it will likely be in some random direction unless some totally inhospitable condition (i.e., the ocean) exists in one direction (Figure 12.3). Consequently, the probability that a dispersing animal will travel in a direction that would lead it to even encounter a target patch is a function of the angle formed from the center of the source patch to the outermost edges of the target patch (P_e, Figure 12.3). In our case, if the angle was 36°, then the probability of an animal encountering the

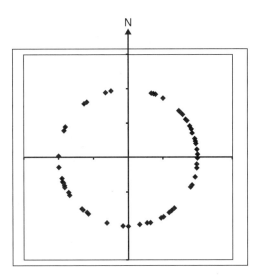

FIGURE 12.2 Dispersal directions of 61 white-tailed deer in Pennsylvania. Data and figure provided by Dr. Eric Long and used with permission (Long 2005).

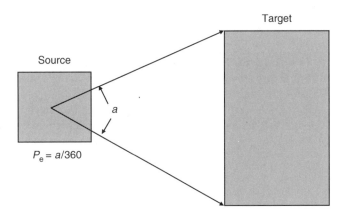

FIGURE 12.3 Conceptual diagram illustrating the probability of an animal encountering a target patch if dispersal is random. As the target increases in size or decreases in distance from the source, the probability of encounter increases.

target from the source would be .10 or a 10% chance. Another way of looking at this is that for every 100 animals dispersing at random from the source, approximately 10 would be expected to encounter the target. Now consider what happens to that angle if the target patch were closer to the source; the angle increases and the probability of encountering the target increases. Increase the size of the target and the probability of encounter increases. Increase the number of targets in various directions and the probability of encounter increases. So, from this very simple example we can see that if we want to maintain a metapopulation through dispersal, then more and larger target patches close to the source patch will increase the likelihood of maintaining the population. Dispersing animals not only have to encounter the target, they also have to survive the trip.

The probability that an animal will survive a dispersal event is dependent on its fitness as it moves from one patch type to another. Recall that fitness is related to the probability of surviving and reproducing in a patch. When dispersing, reproduction is usually not a concern, but survival is a requirement. Consequently, if we can understand what the probability of survival might be for a species in each intervening patch as it moves across a landscape, we can predict the likelihood that it will survive its dispersal trip. One way of measuring this survival is using a time-specific probability of survival, such as its daily probability of survival. Do you know what your daily probability of survival is? I assure you that the life insurance companies have estimated this for you! It is not 100%, unfortunately; you just never know . . . ; but it is high. Say your daily probability of survival was 99.9%, then you would have an annual probability of survival of $.999^{365} = .694$ or a 69.4% probability of living a year. Not great odds! Now consider how that might change if you were living in Antarctica for that year, or Camden, New Jersey (the most dangerous U.S. city in 2004) vs. living in Newton, Massachusetts (the safest U.S. city in 2004). Where you are matters, and for dispersing individuals with no prior knowledge of the resources available in intervening patches, the probability of survival can fluctuate considerably along the trip. But the overall probability that an animal would survive a dispersal event is not only related to the time-specific difference in survival probabilities among the patches along its route but also a function of the time it spends in each patch. Consequently, the movement rate through each intervening patch must be considered. So, the probability of survival in a patch is

$$PS_x = PSD^d,$$

where PS_x = probability of survival in a patch x; PSD = daily probability of survival in patch x, and d = the number of days spent in patch x (Figure 12.4). For a dispersing individual, d will be dependent on the rate of movement (distance per day) and the distance traveled in the patch. If the species dispersing is a salamander and patch x is a closed canopy forest that would protect it from dessication, then it could probably spend quite a long time in patch x and survive (assuming it can find food). But if patch x is an interstate highway, then the salamander can run as fast as it can and spend as little time on the highway as possible, and still face a very low probability of survival. But even in this case, that probability is not 0. Send enough salamanders across the highway and a few will make it (humans have also used this approach — this was the basis for a common military strategy for invasion during wars!). Nonetheless, these patches with an inherently very low probability of survival are considered *barriers*; those with high probabilities of survival can be considered *corridors*. But it is most useful to think of these intervening patches as having

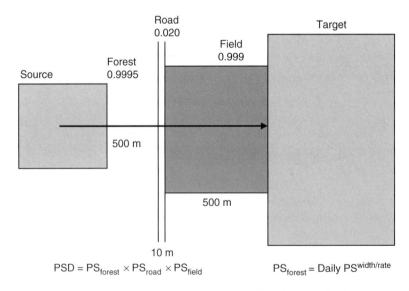

FIGURE 12.4 Conceptual framework for understanding the likelihood of an animal surviving a dispersal trip through various patch types, each with its associated own time-specific probability of survival. Barriers are those patch types with very low daily probabilities of survival.

differential permeability to moving organisms. And it is the product of the probabilities associated with all intervening patch types that yield an overall probability that the animal will survive the dispersal event (Figure 12.4). Species that are more mobile and have higher movement rates will realize higher survival rates in low-quality patches than less mobile species. Consequently, forest managers concerned about maintaining regional biodiversity tend to be concerned with providing habitat for low-mobility species such as salamanders, frogs, and small mammals (Spies et al. 2007).

But there is still one other factor that must be considered when understanding how likely it is that an animal might travel to the target from the source. Recall Figure 12.1. Most animals do not disperse more than one home-range diameter from their natal home range. Only a few go two home-range diameters, fewer go three home-range diameters. The farther the distance between the source and the target, the greater the likelihood that the animal will settle, and perhaps settle in low-quality habitat, perhaps even a sink (see chapter 2). Consequently, the probability of continuing the dispersal (1 — probability of settling) also influences the likelihood that an animal will make it to the target (Figure 12.5).

If you put these three factors together (the product of the probabilities) to understand the likelihood that a specific patch on the landscape might be colonized from individuals from another patch, the likelihood becomes overwhelmingly small as the distance between patches increases, as the size of the target patches decreases, and as the intervening habitat becomes more inhospitable (Figure 12.6). Some of the problems associated with low survival probabilities can be overcome by sheer numbers of dispersers. For example, consider that the number of successful dispersers among patches is a function of the cumulative probability of survival per individual among all patches crossed, animal density in the source patch, source patch size, and the percentage of the population that are dispersers:

$$NSD = PS \times AD \times AREA \times PROP$$

where NSD = number of successful dispersers, PS = cumulative probability of survival across the patch mosaic, AD = animal density per unit area, AREA = source habitat area, and PROP = the proportion of the population that are dispersers. For species with high reproductive rates, a low

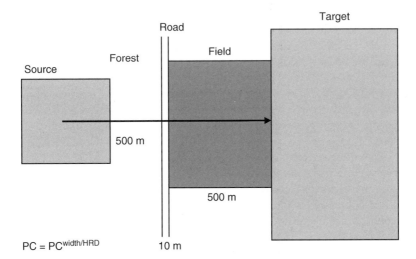

FIGURE 12.5 Conceptual framework for understanding the likelihood of an animal continuing a dispersal trip through various patch types, each with its associated own time-specific probability of survival. Settling into a patch with low fitness may be a reasonable survival strategy but may not allow a metapopulation structure to persist.

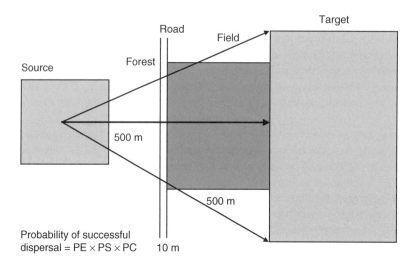

FIGURE 12.6 Conceptual framework that combines the probability of encounter, the probability of survival, and the probability of continuing during a dispersal event from a source to a target. The overall probability associated with an individual dispersing to the target is the product of these probabilities.

probability of successful dispersal may not be such a huge obstacle because of sheer numbers of dispersers — it is just that most dispersers die. Plants use this strategy quite often. But for species that have low reproductive rates, are rather immobile, occur at low densities, and may have daily survival probabilities that vary considerably among patch types, isolation can be a significant problem. These are often the species of most concern to conservation biologists wishing to conserve regional biodiversity. So what can be done about this problem? We can manage our forests to influence the connectivity among the habitat patches on a landscape for a species.

CONNECTIVITY AND GAP-CROSSING ABILITY

Connectivity refers to the degree to which the landscape facilitates or impedes movement among habitat patches and, therefore, the permeability of the landscape to dispersing individuals as measured by their survival. Connectivity is largely a function of the size and arrangement of disjunct patches (i.e., the area and isolation effects) the permeability of the intervening patches, or the physical connections among habitat patches via corridors.

Where a barrier occurs on a landscape impeding the movement of dispersing organisms in a particular direction, a corridor may facilitate movement across this barrier. Overpasses and underpasses built specifically for animals crossing highways have been quite successful in many areas (Jackson and Griffin 2000).

In theory, some of the negative effects of isolation can be mitigated by identifying and maintaining fixed or dynamic connections across the landscape. Connectivity can be managed as either (1) continuity — the physical connectivity of habitat or (2) connectedness — the functional connectivity of habitat. Corridors represent one type of connection that can be designed for those species with a poor ability to cross gaps in connectivity. Maintaining a more hospitable matrix condition between patches is an alternative to corridors. In either case, it may be helpful to think of the demographic basis for developing connections. Using the life history characteristics of species can allow us to understand which species are relatively more at risk from isolation effects compared to others. Maintaining connectivity for species that show adverse effects from isolation (e.g., California red-backed voles) may be a much higher management priority (Mills 1995) than species that disperse across complex landscapes more freely (e.g., deer mice).

What ultimately influences the connectivity of the landscape from the organism's perspective is the scale and pattern of movement relative to the scale and pattern of intervening patches (With 1999). The size, number, and distribution of habitat patches influence the physical connectivity of habitat across the landscape, and are the primary determinants of connectivity. So under what conditions does connectivity become an issue? Clearly, if most of a landscape is high-quality habitat for a species, then there should be little concern about connectivity. But if only 1% of a landscape is high-quality habitat in many tiny patches, then connectivity may be a huge concern. Determining the point at which a reduction in habitat area leads to disconnected habitat would seem to be an important piece of information when managing species in complex landscapes. Dr. Kimberly With (1999) developed a model to at least conceptually understand the relationship between habitat area and connectivity.

With (1999) developed a simple landscape structure in which black squares were habitat and white squares were not habitat for a species, and then used simple percolation model (testing connectedness from one side of the landscape to another) to assess at what points we would expect to see habitat connectivity disappear from a landscape. Habitat loss and fragmentation was simulated by removing habitat at random in one set of simulations and in clumps in another set of simulations (Figure 12.7a). As With (1999) decreased habitat area in the randomly fragmented landscapes, she observed a threshold at 50% habitat area where connectivity declined abruptly and fell to 0 connectivity when approximately 30% of the habitat area remained (Figure 12.7a). The probability of having a connected landscape in clumped fragmentation landscapes began to decline slowly when less than 65% of the habitat remained and fell to 0 at about 20% of the habitat remaining. The latter

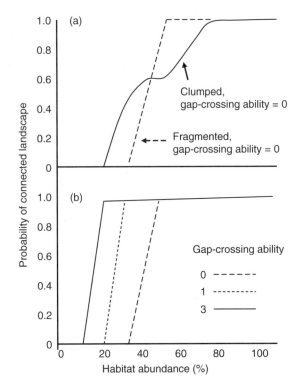

FIGURE 12.7 Simulated effects of landscape fragmentation in two patterns on the probability of maintaining a connection across a landscape (a) as affected by the species gap-crossing ability (b). (Redrafted from With, K.A. 1999. Is landscape connectivity necessary and sufficient for wildlife management? In Rochelle, J.P., L.A. Lehman, and J. Wisniewski (eds.), *Forest Fragmentation: Wildlife and Management Implications.* Brill Press, Netherlands. With permission.)

finding was due to the random chances of losing a connection vs. maintaining a connection being highly influenced by the pattern of clumps of habitat on the landscape, with clumps more likely to be either connected or disconnected compared to random fragmentation. Consequently, for species that require a connection across a landscape, connectivity would seem to become an issue when about 30 to 50% of the landscape remains as habitat; when habitat area drops below 30%, maintaining or restoring connectivity is an important consideration for the species.

But some species do not need a complete connection to move across a landscape. Some species are quite well adapted to crossing inhospitable areas that represent gaps in the connection across the landscape, while others are not (Bélisle and Desrochers 2003); and some species can cross wider gaps (e.g., goshawks) than others (e.g., clouded salamanders). So, With (1999) also examined the potential effect of gap-crossing ability on these thresholds. As you would expect, the better able a species is at crossing a gap in habitat across a landscape, the greater the level of fragmentation of its habitat it can tolerate and still be able to move across the landscape (Figure 12.7b). For species with the ability to cross three blocks of nonhabitat, the threshold for a disconnected landscape fell to 10 to 15%. From this work, several management implications emerge. First, although habitat area is the landscape feature most associated with animal abundance across complex landscapes, connectivity emerges as an important feature when habitat area has declined to 30 to 50%, or less, depending on the species' gap-crossing ability. Second, when habitat area falls to or below this threshold, planning will be needed to either provide corridors or a matrix condition that is more permeable to the species.

What is missing from the previous example of the inter-relationship of habitat area and connectivity is the concept of differential permeability of habitat patches across a landscape. The permeability of the matrix is not black or white, all or nothing; it is indeed shades of gray in its permeability to dispersing organisms. Permeability influences the degree to which the two patches are isolated or not.

MANAGEMENT APPROACHES TO CONNECTIVITY

So what is the best way to increase connectivity within a landscape? Unfortunately, little information is available on the gap-crossing abilities of most forest-associated species (Bélisle and Desrochers 2003). Further, gap-crossing abilities are species specific; and dispersal is often seasonal. Cold–wet winters — and hot–dry summers — may represent periods of movement facilitation or reduction for various species.

When the area of habitat for a species falls below 30 to 50% of the landscape (With 1999), managers should begin to think about identifying connections across the landscape to reduce risks associated with isolation if the intervening matrix is indeed isolating remaining patches of habitat. One approach is to purposefully manage a portion of the matrix to be very permeable for a species by developing a corridor between the patches. Corridors require an organism to follow a "path" through the otherwise inhospitable matrix if it is to have a high probability of moving from one patch to another. Connections may be in the form of static or dynamic corridors connecting specific patches on the landscape. *Static corridors* are those that are identified on a map and on the ground and purposefully managed over time to maintain connectivity in that location. *Dynamic corridors* "float" across the landscape over time such that different portions of the intervening matrix are managed to provide connectivity at all times — it is just that the specific stands that provide the connection change over time. This approach provides much more flexibility for the land manager to manage more stands for multiple values (Sessions et al. 1998).

Alternatively, the matrix condition as a whole could be managed to be made more permeable to dispersing organisms (*matrix management*). Such an approach does not require the animal to follow a path, but rather move at random through the matrix. For species that are able to cross gaps, placing small patches of good-quality habitat close to one another between the two larger high-quality patches creates a *stepping stone* approach to connectivity. All of these techniques allow otherwise isolated patches to be connected through planning or patch management. In managed forests, this is usually achieved through harvest planning and silvicultural decisions (Bettinger et al. 2001).

CASE STUDY: MATRIX MANAGEMENT FOR A
WIDE-RANGING SPECIES

The northern spotted owl is listed by the U.S. Fish and Wildlife Service as a threatened species under the Endangered Species Act. As such, it has received more attention than probably any other species in the United States for the past few decades in efforts to allow the species to recover. Generally northern spotted owls are accepted as a species that is associated with forests with certain structural features of old-growth forests (Carey et al. 1990) (although they also occur in younger forests when these features are present). The Northwest Forest Plan was developed to ensure habitat for this and many other species associated with old forest, and to allow the northern spotted owl populations to recover (FEMAT 1993). The habitat conservation strategy developed for the northern spotted owl identified large blocks of forest on federal lands that establish a metapopulation structure for the species (Thomas et al. 1990). These areas were termed late-successional reserves (LSRs) (Figure 12.8). The LSRs largely lie in the U.S. Forest Service lands because of the size and distribution of the remaining spotted owl habitat as well as habitat for marbled murrelets and other late-successional associates. Under the Northwest Forest Plan, areas between the LSRs are connected through a system of corridors falling largely along riparian management areas that extend 50 to 100 m from streamsides (to provide habitat for salmonid fish as well as connecting LSRs for terrestrial late-successional associated species). The plan also used a matrix-management approach, proposing management of the intervening matrix in a way that allows stands to develop with the structural complexity needed to support dispersing spotted owls as the stands mature. In addition, forests on Bureau of Land Management (BLM) lands, which fall largely on alternating sections (one section = 1 mi^2), provide a stepping stone connection among LSRs (Figure 12.8). Consequently, much of the basis for habitat

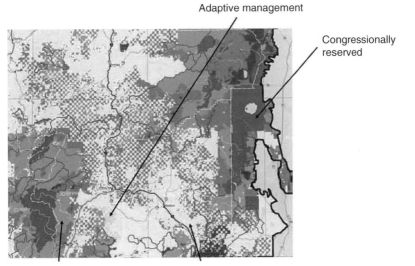

FIGURE 12.8   Land allocation pattern under the Northwest Forest Plan. Notice the checkerboard pattern of Bureau of Land Management lands providing stepping stone connections among LSRs as well as the matrix lands. Not shown are extensive riparian reserves designed to enhance connectivity across the landscape. (Map from FEMAT 1993).

management to recover this species is underpinned by providing landscape connections among LSRs. These connections utilize matrix management, stepping stones, and corridors to achieve landscape connectivity for northern spotted owls and other species across the northwest forested landscape.

SUMMARY

Although habitat area seems to be the most important feature affecting the occurrence and abundance of vertebrates on managed forest landscapes, at low levels of habitat availability connectivity can become important. It is at these low levels of habitat availability that landscape pattern would be expected to impact the abundance and distribution of vertebrates. Connectivity mitigates some of the adverse effects of low levels of habitat area, providing individuals within a population the habitat they need to disperse among patches. Connectivity provides opportunities for exchange of genetic information and the ability to repopulate otherwise isolated areas as habitat improves. Connectivity is effective if dispersal is adequate to meet these population goals. Successful dispersal is influenced by the likelihood that an animal will encounter a patch, survive the dispersal event, and continue dispersal before settling into a patch regardless of its quality. Providing static or dynamic corridors, stepping stones, or undertaking matrix management can facilitate the likelihood of successful dispersal.

REFERENCES

Bélisle, M., and A. Desrochers. 2003. Gap-crossing decisions by forest birds: an empirical basis for parameterizing spatially-explicit, individual-based models. *Landsc. Ecol.* 17: 219–231.

Bettinger, P., K. Boston, J. Sessions, and W.C. McComb. 2001. Integrating wildlife species habitat goals and quantitative land management planning processes. In Johnson, D.H., and T.A. O'Neill (managing editors), *Wildlife Habitat Relationships in Oregon and Washington.* Oregon State University Press, Corvallis, OR.

Bowman, J.J., A.G. Jaeger, and L. Fahrig. 2002. Dispersal distance of mammals is proportional to home range size. *Ecology* 83: 2049–2055.

Carey, A.B., J.A. Reid, and S.P. Horton. 1990. Spotted owl home range and habitat use in southern Oregon Coast Ranges. *J. Wildl. Manage.* 54: 11–17.

Forest Ecosystem Management Assessment Team (FEMAT). 1993. *Forest Ecosystem Management: An Ecological, Economic, and Social Assessment.* USDA For. Serv., USDC National Oceanic and Atmos. Admin., National Marine Fish. Serv., USDI Bur. Land Manage., Fish and Wildl. Serv., National Park Serv., and US Environ. Protect. Agency. Portland, OR.

Gannon, W.L. 1988. *Zapus trinotatus. Mammalian Species* 315: 1–5.

Hayes, J.P. 1996. *Arborimus longicaudus. Mammalian Species* 532: 1–5.

Jackson, S.D., and C.R. Griffin. 2000. A strategy for mitigating highway impacts on wildlife. In Messmer, T.A., and B. West (eds.), *Wildlife and Highways: Seeking Solutions to an Ecological and Socio-Economic Dilemma.* The Wildlife Society.

Long, E.S. 2005. Landscape and demographic influences on dispersal of white-tailed deer. Ph.D. Dissertation. The Pennsylvania State University, State College, PA, USA.

Martin, K.J., and W.C. McComb. 2003. Small mammals in a landscape mosaic: implications for conservation. In Anthony, R.G., and C. Zabel (eds.), *Mammal Community Dynamics in Pacific Northwest Forests.* Cambridge University Press, Cambridge, p. 732.

Mills, L.S. 1995. Edge effects and isolation: red-backed voles on forest remnants. *Conserv. Biol.* 2: 395–403.

Sessions, J., G. Reeves, K.N. Johnson, and K. Burnett. 1998. Implementing spatial planning in watersheds. In Kohm, K.A., and J.F. Franklin (eds.), *Creating a Forestry for the 21st Century: The Science of Ecosystem Management.* Island Press, Washington, D.C.

Spies, T.A., B.C. McComb, R. Kennedy, M. McGrath, K. Olsen, and R.J. Pabst. 2007. Habitat patterns and trends for focal species under current land policies in the Oregon Coast Range. *Ecol. Applic.* 17:48–65.

Sutherland, G.D., A.S. Harestad, K. Price, and K.P. Lertzman. 2000. Scaling of natal dispersal distances in terrestrial birds and mammals. *Conserv. Ecol.* 4: 16.

Thomas, J.W., E.D. Forsman, J.B. Lint, E.C. Meslow, B.R. Noon, and J. Verner. 1990. *A Conservation Strategy for the Northern Spotted Owl: Interagency Scientific Committee to Address the Conservation of the Northern Spotted Owl.* USDA For. Serv., USDI Bureau of Land Management, Fish and Wildlife Service, and National Park Service, Government Printing Office, Washington, D.C.

With, K.A. 1999. Is landscape connectivity necessary and sufficient for wildlife management? In Rochelle, J.P., L.A. Lehman, and J. Wisniewski (eds.), *Forest Fragmentation: Wildlife and Management Implications.* Brill Press, Netherlands.

13 Approaches to Biodiversity Conservation

So far we have focussed on habitat management for individual species. And for some forest wildlife goals, that is an appropriate approach. But oftentimes, especially on public lands, conservation of the full suite of living organisms present on a site, on an ownership, or in a watershed is an objective while also meeting other societal objectives such as drinking water, recreation, aesthetics, and timber production. So by now you must be asking, "How in the world can we possibly manage forests to conserve the hundreds if not thousands of species that occur within a forest with one owner, let alone multiple owners?" Using a species-by-species approach is clearly untenable. Consequently, biologists propose approaches for biodiversity conservation that consider a hierarchical set of steps that are designed to minimize risk of losing a species while taking into account uncertainty in our decision-making process.

WHAT IS BIODIVERSITY?

Scientists define *biodiversity* as the genes, organisms, populations, and species of an area, and the ecosystem processes supporting them (Figure 13.1). Key principles that are often included in the definition of biodiversity are those of structure, composition, and function occurring at various scales of space and time. Most nonscientists view biodiversity as the collage of species, and many equate biodiversity with those species that are rare and wild. Clearly, for scientists and managers to be effective in meeting the expectations that society has for conserving biodiversity, the collage of species must be addressed. Indeed, some of the most challenging aspects of biodiversity conservation are deciding how to understand complexity and uncertainty, protect both known and unknown species, and communicate these approaches to the public. There is a triad of biodiversity perceptions, biodiversity concepts, and biodiversity assessments (Figure 13.1), viewed by the public, scientists, and managers, respectively, that must be interconnected if we are to successfully address biodiversity issues.

Species are usually considered the primary currency of biodiversity conservation. But even conservation of species presents challenges. Rare, threatened, and endangered species garner much attention politically, and species that are hunted or are aesthetically appealing (e.g., songbirds and wildflowers) are often used as focal species or as special-interest species when making biodiversity decisions. But these are just examples of species that could or should be considered during management. Over 1.6 million species have been described on the earth and this is only a fraction of what occurs here — many have not yet been described. And patterns of species richness (the number of species in an area) for one taxonomic group do not reflect patterns of other groups very well (Flather et al. 1997), limiting our ability to use one group as a surrogate for another in planning. So there are clear challenges to ensuring that we do not lose biodiversity across the earth at a rate significantly different from what would be expected if technologically advanced humans did not have such a profound effect on the earth's resources.

Planners and managers usually assume that genes will be successfully conserved among individuals within a species if we can ensure the long-term viability of populations throughout each

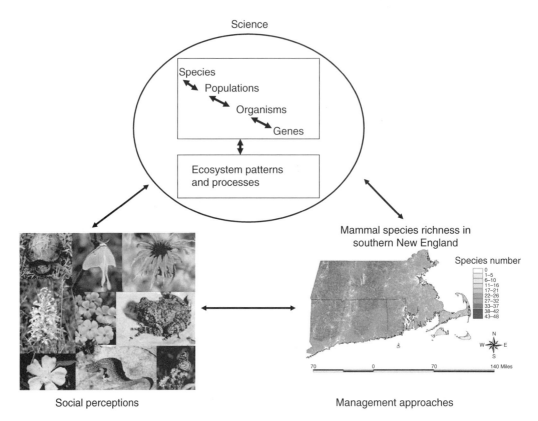

FIGURE 13.1   The scientific concept of biodiversity (top) is a set of processes and conditions that interact to reflect the breadth of life on the planet. Biological complexity is often perceived as a collage of life by nonscientists (bottom left; Photos by Dr. James Petranka, used with permission). Scientific concepts can be used to conserve the collage of life by developing maps of species richness for various groups of organisms (bottom right, from southern New England Gap Program, UMass–Amherst) or by using coarse-filter/fine-filter approaches (see text).

species geographic range. Policies or actions that eliminate a species from part of its geographic range is assumed to reduce the genetic diversity of the species and increase the risk that the species would be less able to tolerate perturbations to its habitat in the future, ultimately leading to extinction (Lomolino and Channell 1995). This rule of thumb, "keep the gene pool (and the species) intact by keeping the subpopulations especially along the periphery of a geographic range" is an important assumption that probably is true for many species. We certainly see geographic variation in phenotypes (what an individual looks like), diets, habitat selection, and home-range sizes within many species of vertebrates. A reasonable assumption is that these differences reflect some evolutionary advantage to the species in those places. Very rarely have these assumptions been tested (but see Lomolino and Channell 1995), and so the approach follows the precautionary principle.

SETTING BIODIVERSITY GOALS

Most biodiversity objectives reflect the paraphrased text of Aldo Leopold: ". . . the first rule of intelligent tinkering is to save all the pieces." Indeed, the pieces are the genes, organisms, populations, species, and supporting ecosystem processes (Figure 13.1). These are the very things that are implicitly part of the integrated filter approach to biodiversity conservation. The key word in this quote is "all" and the phrase begs the question, "How much of each?" The answer, obviously, is "Enough!" Saving all the pieces is a noble goal. Indeed, it is a rule of thumb for people who care about seeing the collage of life on this planet persist for future generations to enjoy. But these people are only part of society. Indeed, in some societies, cultures, and places, this group may well be in the minority. Or society may embrace the noble goal of saving all the pieces, but they may follow that by asking how much is enough? And at what price will it be provided? Take, for instance, the recovery of wild stocks of salmon in the Pacific northwest of the United States. Years of research indicate that there probably are some key factors all working together to cause wild salmon stocks to be at less than 10% of the historic levels. If society truly wants salmon to recover to historic levels, then: (1) remove some or all dams to improve passage, (2) do not mix wild genetic stocks with hatchery fish genes, (3) reduce or eliminate sport and commercial fishing, (4) restore freshwater conditions to be acceptable for spawning, and (5) allow all spawning fish to enter the stream and die to provide stream nutrients (Compton et al. 2006). Remove a source of hydropower? Increase electricity bills? Use coal or nuclear fuels for electricity? Do not allow salmon harvest? Will society agree to this? Not likely. And this is in a wealthy society. Consider the overgrazing situation in the dry tropical forests of South America that has led to desertification. Tell the campesino to stop grazing for a few years to allow the rangeland to recover (and it would), and he and his family will starve. Not likely. So setting biodiversity goals must consider the genetic resources, the species, the ecological processes, *and* the goals and objectives of the people affected by a decision. And because goals will change over time, plans designed to meet goals now must be adaptable to allow future goals to be met.

HOW DO WE CONCEPTUALIZE "BIODIVERSITY" TO BE ABLE TO CONSERVE IT?

Given the complexity associated with biodiversity and recognizing that it is a resource that society values, what is a scientist, manager, planner, or decision maker to do to ensure that biodiversity is conserved for future generations? How can we hope to understand and consider the needs for all species in a planning area? Generally, a tiered approach to decision making is used that considers the needs of some species explicitly but assumes that the needs of others will be met through a more generalized strategy of habitat protection and/or management. So scientists simplify the problem by taking a logical step-wise approach, albeit with significant assumptions. The *filter approach* is often used as a basis for reducing the risk of losing a species from an ecosystem (Hunter 1999; Figure 13.2). In this approach, three management strategies, termed "filters" are used. These filters are analogous to management filters designed to "catch" species in the management approaches and minimize the risk of losing species. The three filters are coarse, meso, and fine filters, each with a set of assumptions about how the combinations of these three types of filters can be employed to "capture" species in a management strategy.

COARSE-FILTER APPROACHES

The coarse filter is applied to the landscape by describing the distribution of biophysical classes (e.g., vegetation classes, slope classes, stream classes, etc.) that occur in an area of concern, and documenting the arrangement and connectivity of these biophysical classes across the landscape. These current conditions may then be projected into the future under various alternative management assumptions or compared with past conditions to see how much they have changed over time. The current and possible future conditions are often compared with some reference conditions. Recently, that comparison has quite often been with the historical range of variability (HRV) in one or more ecosystem indicators (Landres et al. 1999; Figure 13.3).

It is important to understand that when using the HRV as a reference condition, the objective is *not* to return to a condition that once occurred in the past, but rather consider the range of conditions that species likely encountered in the past and the process that led to those conditions. Biologists often assume that the species persisted within these ranges of conditions and processes. The more

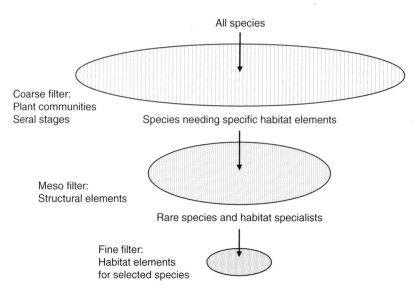

FIGURE 13.2   Coarse-filter goals are met using vegetative types and successional stages that are likely to meet the needs for many species in a planning area. Some species require specific habitat elements provided within a meso filter. For those that are not likely to be met using this approach, a fine-filter (single-species) analysis is conducted.

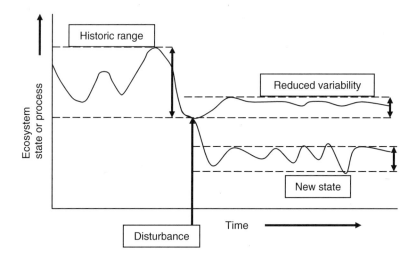

FIGURE 13.3   Use of the HRV allows managers to consider the implications of future conditions following a disturbance on reducing the variability in a system or creating an entirely new state of conditions in an ecosystem.

the current and likely future conditions depart from the HRV, the greater is the *risk* that genes or species may be lost from the system. For instance, consider the likely distribution of one ecosystem indicator, open and early-successional conditions, in New England (Figure 13.4). As European humans introduced new technology and approaches to land management, pre-European fluctuations in this indicator began to change following forest clearing and eventually farmland establishment and subsequent abandonment (DeGraaf and Yamasaki 2001). The departure from the HRV was significant with most forestland converted to open land in much of southern New England. What was the risk to biodiversity of this departure? Global extinction of a few species, most notably

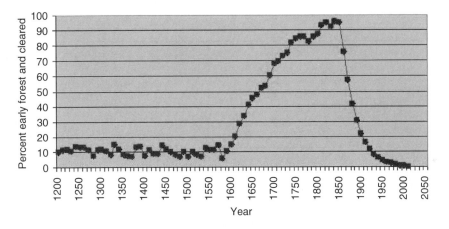

FIGURE 13.4 Generalized changes in open and early-successional forest conditions following European colonization in New England. While increases in this ecosystem state provided habitat for many species others were lost. Now, because of a much lower availability of this condition, species associated with this condition are at risk.

FIGURE 13.5 Black bears were extirpated from southern New England during the 1800s and did not return to the area for over 100 years. Now they are common and can cause damage to homes, property, and crops in the region. (Photo with permission by Karl J. Martin.)

passenger pigeons, the likely loss of species that we had not identified by the time they were lost, and regional extinction of forest-associated species such as fishers, moose, white-tailed deer, black bears (Figure 13.5), wild turkeys, and beavers. As the forest returned and the amount of open land declined, these latter species have occupied the area once again from forests to the north. But now the amount of open grassland and early-successional conditions is very low, probably lower than it was historically, and we now see species that are considered at risk because we have less of these conditions than we did historically. Bobolinks, eastern meadowlarks, and chestnut-sided warblers are all considered species associated with these open and early successional conditions, and are species of concern in the region (Vickery et al. 1992; Figure 13.6).

FIGURE 13.6 Early-successional forest is being created on MassWildlife lands and privately owned lands using landowner incentives program (LIP) funds in response to regional declines in species associated with shrublands and grasslands.

The organisms and values that currently occur in the forests of a region have persisted through centuries of natural- and human-induced disturbances. Some did not survive past deforestation. Some may only now be recovering. To assess risk to species in a desired future condition (DFC) for a land-scape, you should consider the plan within the context of the HRV for key ecosystem indicators. Consider the likely representation of early-, mid-, and late-seral forests during the early 1600s in the New England region. Proportions of these successional stages are inherently variable annually, dec-ade to decade, and century to century, as hurricanes, ice storms, floods, fires, and other disturbances occurred with varying levels of frequency and severity. But these proportions clearly now depart from historic ranges. Early-seral shrub stage and grassland conditions are now poorly represented in the northeastern landscape. Owing to past forest clearing, only a very small proportion of the land-scape is now in late-seral condition. Consequently, species associated with any of these conditions may be considered at greater risk than species associated with mid-successional conditions.

Further, it is important to think of the range of variability that once occurred not just as seral condition but also plant community representation. What should be a reasonable proportion of representation of vegetation types across the landscape to reflect the conditions that these species have evolved with? Because of recent land uses some plant communities are very rare. Prairie communities in the Great Plains and in the Willamette Valley of Oregon have been reduced to a tiny fraction of what they were historically. Old-growth forests in New England are less than 1% of historic conditions. Should we restore these systems to historic levels? Can we? Probably not. But we may wish to focus on recovery of a greater proportion of these types to minimize further losses of species from the region.

Finally, it is important to consider the spatial scales of the patch types that are created across the landscape. In many cases, most patches should be of small size, fewer of larger size, and very few very large patches (of course, small and large are relative to the sizes represented over time following disturbances and regrowth). Once you have considered what the conditions are with respect to these broad patterns of vegetation and successional stages using a coarse-filter approach, then you should be able to predict which plant communities and successional stages are likely underrepresented in the region. Of those that are underrepresented in the region, which could be represented on the landscape that you are managing? Given the contribution of your management actions to improving regional representation of underrepresented types, what should be the desired future representation of these types on your forest? Answering these questions will help to design a coarse-filter strategy for a landscape.

The departure of key ecosystem indicators, especially ecosystem processes, from historical conditions under which the species persisted can be useful in understanding if species are likely to be at risk of local or global extinction. But what indicators do you choose when making this coarse-filter assessment? Whitman and Hagan (2003) evaluated over 2000 biodiversity indicators and "simplified" this list to 137 indicator groups. Even this is an overwhelming number of indicators for managers to address, and so Whitman and Hagan (2003) proposed that decision makers and scientists should work together with managers to identify indicators relevant to the values identified by stakeholders. Once the group of indicators is identified, then considering historical conditions may be problematic, or not even particularly useful in some instances. If current human values differ markedly from the historical range of conditions and processes, then simply using the indicator without a historical reference may still provide a context for setting coarse-filter goals.

MESO-FILTER APPROACHES

Not all species may be caught in the coarse filter. Some require certain structural elements that must be present in plant communities and seral stages to ensure that they will likely persist in the management area. Hence, a meso-filter approach that considers the sizes, distribution, and abundance of structural elements such as snags, logs, hollow trees, and the other elements presented in Chapter 3 are distributed across the landscape at a range of spatial scales (Hunter 2004). These are often inventoried and managed at the stand level, but it is the distribution of these habitat elements among stands across a landscape that will influence habitat quality for many species, especially those having home ranges exceeding a stand in size. How many of these elements are needed? Again approximating a range of conditions that would be expected following historical disturbances and regrowth provides one context for estimating the numbers, especially where habitat relationships studies have not been conducted. Where data are available, then data-driven habitat relationships can influence decisions regarding how much, what size, and where to provide these elements. An excellent example is use of DecAID to guide management of dead wood across a landscape based on existing habitat relationships data (Mellen et al. 2005).

FINE-FILTER APPROACHES

But the combination of coarse- and meso-filter approaches may not provide suitable habitat for all of the species in a landscape. Some species are likely to simply be rare enough, have low reproductive rates, have large territories, or have been adversely affected by habitat loss (or other factors) so that their populations are low and they require special attention. Consequently, a "fine filter" is constructed that maintains the coarse-filter structure and the meso-filter elements, but takes special management actions to conserve the set of species identified for fine-filter consideration. Those species that may need to be considered more carefully to ensure that their needs are met in the coarse-filter approach might include those based on the following criteria (Figure 13.7).

Risk: Species that are rare or already at risk of declining in abundance so as to become locally or regionally extinct or are already designated as threatened or endangered through a regulatory status.

Narrow niche breadth: The species is restricted to specific successional stages, especially those that are or may become uncommon or disconnected in the future, or is sensitive to environmental gradients such as moisture gradients or elevational gradients.

FIGURE 13.7   Examples of species selected as foci for fine-filter analyses based on (a) risk (Florida panther photo from USDI USFWS digital library), (b) Narrow niche breadth (many species of neotropical migrant birds such as this hermit thrush are selective of vegetative structural stages, (c) Ecological function (raptors such as this red-tailed hawk play a key role in energy transfer among trophic levels), (d) umbrella species such as wild turkeys have large area requirements and use multiple vegetation conditions (photo by Michele Woodford and used with her permission), (E) species of high economic importance such as moose (photo by Mike Jones and used with his permission), (f) species for which we have limited data or knowledge such as many species of reptiles including this western fence lizard, and (g) species that have high public interest due to risks associated with them such as this prairie rattlesnake.

Ecological function: Keystone species whose effects on one or more critical ecological processes or on biological diversity are much greater than would be predicted from their abundance or biomass (Aubry and Raley 2002). Also, *link species* that play critical roles in the transfer of matter and energy across trophic levels or provide a critical link for energy transfer in complex food webs (e.g., insectivorous birds; Cohen and Briand 1984).

Management focal species: These are umbrella species that, because of their large area requirements or use of multiple habitats, encompass the habitat requirements of many other species, or species that are representative of certain conditions that are now or are likely to be uncommon in the future on the landscape (Lambeck 1997).

Economic importance: Game species from which local economies and stakeholder groups derive benefit from hunting, or species that inflict costs on forest owners and managers.

Cryptic Species: Those species for which we have limited data or knowledge and need to be explicitly considered during management, often using expert advice.

Public/regulatory interest: Species that society has expressed interest in because of media events (e.g., rattlesnakes), public policy (e.g., migratory birds) or human health concerns (e.g., carriers of west Nile virus).

These criteria can help managers narrow the list of species that must be explicitly considered in a fine-filter approach. Once identified, the specific habitat elements needed in appropriate arrangements, sizes and numbers can be provided in the forest to ensure that these species have their habitat requirements met.

Note that the preceding list does not mention guilds or indicator species. *Guilds* are groups of species that share common resources such as cavity-nesting birds, bark-foraging birds, or forest-floor insectivores. *Indicator species* are species that are assumed to be surrogates for other species having similar resource needs. In the early ecological literature, however, indicator species were used to indicate certain environmental conditions (e.g., water pollution). Indicator species in forest management are used quite differently. For instance, pileated woodpeckers are often used as a management indicator species for cavity-nesting birds (Landres et al. 1988). But use of both guilds and indicator species as convenient ways to manage for multiple species is fraught with problems. Because each species has its own set of habitat requirements, no species can ever be a reasonable surrogate for another. Indeed, tests of individual species responses to forest management indicate that, although several species may belong to the same guild, they each respond differently to a forest management treatment (Mannan et al. 1984). Consequently, it is important to consider management strategies that focus on species, habitat elements, or broad vegetative conditions, and not seek "short cuts" that may lead to misleading management strategies, and increase the level of uncertainty in meeting biodiversity goals.

CHALLENGES TO MANAGING BIODIVERSITY

The filter strategy is based on many assumptions. But several factors will likely influence the degree to which a protection of biodiversity will be effective using this technique. The spatial scale over which the decision is made, its context, and the level of spatial detail used in the decision all contribute to effective management. Similarly, the temporal framework within which the decision is made is critical. Will the decision meet the concerns of constituents now? Ten years from now? 100 years? What is the appropriate time frame? And all decisions are couched within a number of factors associated with the uncertainty of ecological and sociological processes. How do we effectively consider the uncertainties so that the decisions made are effective yet still reflect this uncertainty? Each of these factors must be considered in detail.

SPATIAL SCALE

Land ownership implies a certain level of commitment to part of the earth, and that commitment is expressed through the accumulation of individual landowner behaviors over space and time. It would seem obvious that one landowner making a decision to manage for cavity-nesting birds in a stand on her land is easy. Just leave a certain number of trees or snags of certain sizes, and the goal is reached. Or is it? How will the actions of her neighbors influence the likelihood that these and other biodiversity objectives will be met on her land? And how will her actions influence the achievement of her neighbor's goals to provide a corridor for migrating elk? Can she trust her federal neighbors to follow through on their commitments to follow their plans even as administrations change? Will her private neighbors sell their land? Subdivide it? Will the state impose restrictions on private land management that inhibit her ability to achieve her goals on her land and that of her neighbors? Will an NGO intervene to offer a conservation easement and purchase development rights? All of these questions, driven by social values, are played out on the patchwork quilt of the landscape occupied by landowners and their neighbors (Figure 13.8). Effective decisions must consider this spatial context for the property or properties being managed or reserved.

Landscape management goals often are formed based on a larger regional plan at a large spatial scale (e.g., Northwest Forest Plan) and are implemented through cumulative actions made at small

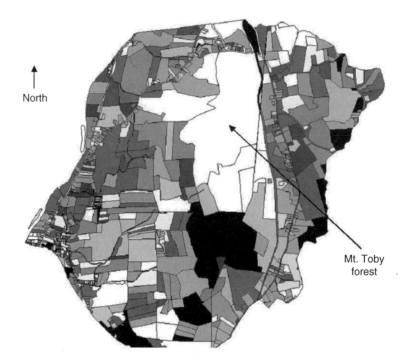

North

Mt. Toby
forest

FIGURE 13.8   Management of a 300 ha forest in western Massachusetts is heavily influenced by the goals, objectives, and actions of many adjacent landowners, illustrated here by all tax lots within a few kilometers of the forest.

spatial scales over time (e.g., stand prescriptions and forest plans). The policy guides the actions (e.g., how many wildlife trees to leave in a clearcut, and how wide a riparian buffer strip should be), but decisions must be made locally to determine where and how often these should occur.

Not only must the sociopolitical framework be considered when making biodiversity conservation decisions, but also the species and ecological processes must be considered. Large territory and home-range sizes of some species, combined with the need to ensure that an adequate number of individuals of each must be maintained, may dictate the appropriate spatial scale. How large an area do we need to consider in order to make effective decisions that will include habitat for marbled salamanders? Northern goshawks? Wolverines? Where you draw the line taxonomically in your assessments and decisions will influence the spatial scale associated with the decision-making process. Similarly, the dominant ecological processes that might influence the outcome of the decisions also drive spatial scale decisions. Wild fires, insects, disease, wind, ice, and climate change all have ranges of frequencies, sizes, and intensities associated with various locations on this earth. The spatial scales associated with these disturbances must also be considered. Some managers choose to manage spatial scales such that these natural disturbances are "captured" within the spatial extent of the landscape being managed (Poiani et al. 2000).

TIME

Stakeholders in the outcome of a landscape management plan often view effective time frames as days, weeks, maybe years, and sometimes decades. We all at least try to plan for our financial security throughout our lives, and so we are used to thinking in multiple decades. And most people want to leave a legacy of their values to the next generation. But we humans have a more difficult time thinking in terms of multiple lifetimes. Yet, some plans made to achieve biodiversity goals may not be apparent for many decades (e.g., the recovery of nesting habitat for spotted owls) or several centuries (e.g., returning the Oregon Coast Range to within the HRV) (Spies et al. 2007). Decisions to designate a part of the landscape that contains many square kilometers of young plantations as late-successional reserves, which may take three human lifetimes (200 years or more) to develop as intended, may leave many constituents unable to see how recent decisions are effectively leading to the intended goal. It is important not just to consider human lifetimes when considering the appropriate temporal scale for projecting likely effects of biodiversity decisions. It is equally as important to understand the effects of decisions relative to the multiple generations of key species affected by the decisions. Consider long-lived species such as box turtles and Puerto Rican parrots. These are such long-lived species (40 years or more) that, by the time declines in populations are detected, the options for recovery may be very limited. Similarly, recovery when it is possible may take multiple generations — hundreds of years. Consequently, the appropriate time frame for considering effectiveness of biodiversity management over large landscapes is driven by the interface of several key actions:

1. Human schedules for implementing the plan
2. The inherent rates of growth and disturbance affecting the vegetative (and occasionally the physical) components of the environment
3. The potential and realized rate of population growth for key species
4. The rate of movement and colonization of habitat for key species.

Consequently, there is no standard appropriate time frame associated with all biodiversity management plans. Consider the potential for documenting responses of organisms to management actions. The species in Table 13.1 are ranked by their potential longevity in years. This would be the maximum time needed for a complete turnover in a generation. The number of generations are then portrayed for 40-, 100- and 200-year rotations in a forest managed using even-aged systems. If a species can generate multiple generations on a site before it is harvested, then any single individual faces less chance of being displaced to other home ranges.

Clearly, natural disturbances have probabilities of occurring at a range of frequencies that also could lead to displacement for these species. As rotation lengths depart from the historical range of natural disturbance frequencies, the risk to long-lived species (which often have low reproductive rates) increases. In addition, because they are so long lived, changes in populations can be subtle, making it difficult to detect population declines. Conversely, documenting recovery of these species can also be difficult, requiring long periods of population monitoring.

TABLE 13.1

Approximate Longevity in Years and as Expressed in Number of Generations per Rotation under Even-Aged Management, for Six Species with Very Different Life Histories

Species	Longevity (years)	Rotation length (years)		
		40	**100**	**200**
Short-tailed shrew	3	13	33	67
Winter wren	4	10	25	50
Spruce grouse	5	8	20	40
Red-cockaded woodpecker	16	3	6	13
Great-horned owl	27	1	4	8
Box turtle	50	1	2	4

There are at least two dominant additional factors that must be considered when making biodiversity decisions over large multiowner areas. Land tenure can influence achievement of goals, particularly if the parcels being sold or inherited change owners having one set of core values to another. Rotation lengths also influence the time that a forest will be suitable for a set of species. Hence, an understanding of these transition probabilities can be particularly important in understanding the likely trajectory of landscape change over the planning period.

UNCERTAINTY

One of the greatest uncertainties facing conservation biologists and land planners is development of conservation and management strategies with incomplete information about the suite of species under consideration. Based on past research, we know enough about the habitat used by many species to at least develop reasonable management plans. However, for some species we know nearly nothing, and then there are all of the undescribed species that we have yet to discover.

Consequently, it is important that biodiversity planners recognize that a reasonable course of action is to follow the precautionary principle and err on the side of conservation rather than resource extraction where that is the primary trade-off faced by planners. This approach places the burden of proof on managers to demonstrate that there is minimal risk to these species of extracting resources. Then, once the plan is developed, we can monitor using techniques that will add to the information available to make decisions. Survey and manage, formal monitoring protocols, and moving the system toward the HRV while monitoring key resources are approaches that can improve knowledge and reduce uncertainty over time.

Planning requires those making decisions to define the spatial and temporal grain (finest level of information needed), extent (outer bounds of the planning problem), and context (surrounding landscape conditions). Once the spatial and temporal context for biodiversity decisions has been decided (perhaps one of the most critical first decisions), the decision makers suddenly find themselves faced with a number of uncertainties that influence the effectiveness associated with resulting decisions. One key uncertainty is the continued social commitment to biodiversity values. Societal values change and the decision-making process must be adaptable to that change (Figure 13.9). Values placed on deer have evolved from largely utilitarian to protection, recreation, nuisance, and finally to public health concerns. How will values change for spotted owls? Townsend's big-eared bats? Burying beetles? We tend to think of goals and objectives as being relatively stable. But they will change, and our decisions regarding biodiversity protection should reflect an ability to adapt to new values.

There also are a number of biophysical uncertainties: fires, floods, invasive plants and animals, disease, and global climate change, to name a few. For many of these factors we have information that can help us understand probabilities of occurrence over time at varying spatial scales. Hence, the uncertainty, or likelihood, can be quantified. In doing so we can also assign risks associated with these events. Many forest "health" issues are framed within this risk assessment paradigm. For instance, an "unhealthy" forest is often considered one with a higher than expected chance of wildfire, disease, or insect irruption, and the ensuing ecological and social effects can be predicted. Further, these risks can be expressed as a departure from HRV. But in many systems we cannot return many attributes of a system to fall within the HRV. For instance, in the northeastern United States there are no longer any passenger pigeons, American chestnuts, and wolves. Atlantic salmon are effectively gone, and society will not likely pay for restoration to historical levels. Although some aspects of the northeastern hardwood forests and associated streams have recovered, some aspects will never recover. Hence, the past may help inform decisions, but it is not always a good template. And even the ability for the past to inform decisions is weakened as we see systems develop in ways that they never have developed in the past. Hence, uncertainties regarding how systems might develop, and how aspects of biodiversity might respond, proliferate as we become increasingly unable to use history as a guide.

FIGURE 13.9   Prescribed burning was used to maintain grasslands and grass–shrub savannahs in pitch pine forests of Massachusetts to maintain conditions important to many species of vertebrates and invertebrates. Over time, public values may shift because of air quality, smoke, and risk of fire escapement, placing systems like these at greater risk.

Finally, there are political uncertainties. Although political decisions are (usually) an outcome of societal values, our political system and those of other countries can result in decisions being made that result in significant constraints (or freedom) to achieve biodiversity goals. Honoring the Kyoto agreement, 9/11, economic recovery, going to war — these are decisions made by a few individuals who were entrusted with making wise decisions on behalf of society. Once made, these decisions have significant effects on the certainty with which biodiversity decisions will truly achieve the intended objectives.

Further, changes in policy such as modifications to the Endangered Species Act, Clean Water Act, the National Forest Management Act, and others are not only likely they are inevitable given the changes in society and politicians that we can expect over the next 100 years. Even now we are seeing a society that is becoming more and more divided on many issues important to maintaining a balanced and functional society. As philosophical beliefs of our elected and appointed officials wax and wane, so will the degree to which policies provide the legal framework for biodiversity decision making.

SUMMARY

Biodiversity is represented by the genes, organisms, populations, and species of an area, and the ecosystem processes supporting them. Contemporary approaches to biodiversity conservation typically take a filter approach. Coarse-filter strategies use the proportional representation of plant communities, successional stages, and other classes of biophysical conditions as a guide to providing habitat for most species most of the time. Meso-filter strategies further consider the habitat elements present within these biophysical classes in a way that "captures" even more species. Fine-filter strategies consider the needs for a few individual species that are of high social importance. This approach is based on numerous assumptions and is influenced greatly by the choice of spatial and temporal scales over which it is employed. Further, uncertainty in the effectiveness of the strategies often leads managers to follow the precautionary principle.

REFERENCES

Aubry, K., and C.M. Raley. 2002. The pileated woodpecker as a keystone habitat modifier in the Pacific Northwest. In Laudenslayer Jr., W.F., P.J. Shea, B.E. Valentine et al. (eds.), *Proceedings of the Symposium on the Ecology and Management of Dead Wood in Western Forests*. USDA Forest Service General Technical Report PSW-GTR-181.

Cohen, J.E., and F. Briand. 1984. Trophic links in community food webs. *Proc. Natl. Acad. Sci. USA* 81: 4105–4109.

Compton, J.E., C.P. Andersen, D.L. Phillips et al. 2006. Ecological and water quality consequences of nutrient addition for salmon restoration in the Pacific Northwest. *Frontiers in Ecol. Environ.* 4: 18–26.

DeGraaf, R.M., and M. Yamasaki. 2001. *New England Wildlife: Habitat, Natural History and Distribution*. University Press of New England, Hanover, NH.

Flather, C.H., K.R. Wilson, D.J. Dean, and W.C. McComb. 1997. Mapping diversity to identify gaps in conservation networks of indicators and uncertainty in geographic-based analyses. *Ecol. Appl.* 7: 531–542.

Hunter Jr., M.L. (ed.). 1999. *Maintaining Biodiversity in Forest Ecosystems*. Cambridge University Press, Cambridge, UK.

Hunter Jr., M.L. 2004. A meso-filter conservation strategy to complement fine and coarse filters. *Conserv. Biol.* 19: 1025–1029.

Lambeck, R.L. 1997. Focal species: a multi-species umbrella for nature conservation. *Conserv. Biol.* 11: 849–856.

Landres, P.B., J. Verner, and J.W. Thomas. 1988. Ecological uses of vertebrate indicator species: a critique. *Conserv. Biol.* 2: 316–328.

Landres, P.B., P. Morgan, and F.J. Swanson. 1999. Overview of the use of natural variability concepts in managing ecological systems. *Ecol. Appl.* 9: 1179–1188.

Lomolino, M.V., and R. Channell. 1995. Splendid isolation: patterns of geographic range collapse in endangered mammals. *J. Mammal.* 76: 335–347.

Mannan, R.W., M.L. Morrison, and E.C. Meslow. 1984. Comment: the use of guilds in forest bird management. *Wildl. Soc. Bull.* 12: 426–430.

Mellen, K., B.G. Marcot, J.L. Ohmann, K. Waddell, S.A. Livingston, E.A. Willhite, B.B. Hostetler, C. Ogden, and T. Dreisbach. 2005. *DecAID, The Decayed Wood Advisor for Managing Snags, Partially Dead Trees, and Down Wood for Biodiversity in Forests of Washington and Oregon. Version 2.0.* USDA For. Serv. Pac. Northwest Res. Sta. and USDI Fish and Wildl. Serv., Oregon State Office, Portland, OR.

Poiani, K.A., B.D. Richter, M.G. Anderson, and H.E. Richter. 2000. Biodiversity conservation at multiple scales: functional sites, landscapes, and networks. *BioScience* 50: 133–146.

Spies, T.A., B.C. McComb, R. Kennedy, M. McGrath, K. Olsen, and R.J. Pabst. 2007. Habitat patterns and trends for focal species under current land policies in the Oregon Coast Range. *Ecol. Appl.* 17: 48–65.

Vickery, P.D., M.L. Hunter Jr., and S.M. Melvin. 1992. Effects of habitat area on the distribution of grassland birds in Maine. *Conserv. Biol.* 8: 1087–1097.

Whitman, A.A., and J.M. Hagan. 2003. *Final Report to the National Commission on Science for Sustainable Forestry. A8: Biodiversity Indicators for Sustainable Forestry*. Manomet Center for Conservation Sciences, Brunswick, ME.

14 Landscape Management Plans

Plans. You probably have some plans for the weekend, or the semester, or your life. But the more complicated the plans and the longer the time frame, the less certain you can be that you will realize your plans. There is uncertainty, but to move forward without a plan, to manage haphazardly, probably will not allow you to reach a goal or a set of goals either for your life or for biodiversity.

ESTABLISHING GOALS

In order for society to achieve biodiversity goals, the filter approach (Chapter 13) or some other comprehensive approach must be described in a plan that is then implemented and monitored to ensure that the risk of losing species from a region is minimized. Plans must have goals. Goals are developed by the landowner, land manager, and stakeholders, who have a vested interest in the future of the land. Goals are a reflection of the desires and actions of stakeholders such as affected publics and nongovernmental organizations (NGOs). These goals usually reflect the priorities for ecosystem structure and composition, plant communities, and individual species as defined in Chapter 13, and also by the economic, aesthetic, or cultural goals for the region. Although involving interested publics in plan development is imperative, under contentious circumstances, the initial investment in stakeholder involvement can be significant. But in the long run it is usually worth the effort.

Regulatory Goals: Although private forest landowners may have goals for their forest that are not driven primarily by biodiversity protection, they must still abide by federal and state laws (see Chapter 19). Some goals are prescribed in federal, state, and occasionally local policies. Probably the most powerful environmental protection law in the United States is the Endangered Species Act (ESA). The ESA is a species-by-species approach to addressing species at risk of being eliminated from all or a significant portion of its geographic range. Once listed, a Species Recovery Plan must be developed, implemented, and monitored to provide the basis for recovery and removal from the list. Habitat restoration to allow species to recover from emperilment represents one type of fine-filter goal that can be combined with other such goals. Coarse-, meso-, and fine-filter strategies combine to minimize adverse effects of human actions on biodiversity loss (see Chapter 13).

Landowners must consider the effects of their actions on protected species, especially those protected under the ESA. Private landowners, corporations, state or local governments, or other nonfederal landowners who wish to conduct activities on their land that might incidentally harm (or "take") wildlife species listed as endangered or threatened under ESA must first obtain an *incidental take permit* from the U.S. Fish and Wildlife Service. To obtain a permit, the applicant must develop a *habitat conservation plan* (HCP) which is designed to offset any harmful effects the proposed activity might have on the species. An approved HCP allows management to proceed while continuing to promote the conservation of the species of concern. This approach is allowed under the "no surprises" regulation in the ESA and provides assurances to landowners participating in these efforts that they will be allowed to proceed with management after a HCP has been approved without a "surprise" that new regulations or other restrictions will be imposed on them.

The incidental take permit allows managers to manage forests as described under a HCP that has been approved by the regulatory agency (usually the U.S. Fish and Wildlife Service). In this instance, take does not mean directly killing the animal but rather removal of *critical habitat* for the species, at least in the near term. In the past, court battles have resulted from the debate over the issue of removal of critical habitat constituting take, leading to judicial decisions that have broad implications (e.g., the Sweet Home Decisions). The outcome of those battles was the HCP approach that is consistent with the required Recovery Plan for the listed species. Specific decisions regarding what constitutes "take" of species through habitat modification often requires that site-specific decisions be made to address particular issues (e.g., timber sales). Species Recovery Plans and HCPs require an understanding of not only the effects of management actions on the target species over large multiownership areas but also often must consider the effects on other species, including people, both regionally and locally.

HCPs are often used to ensure that both the landowners and regulatory agencies are in agreement about the goals and objectives for the land, the resulting level of take, and the likely effects of take on the portion of the population covered by the plan. Landscape management plans may or may not be HCPs depending on the goals of the landowner and the presence or absence of federally listed species. The U.S. Forest Service has long used harvest-planning models, wildlife habitat relationships models, and other tools to assist with development of forest plans, required under the U.S. National Forest Management Act. States also address complex planning problems using landscape management plans. For instance, habitat issues on the Elliott State Forest in Oregon prompted the Oregon Department of Forestry to develop a landscape management plan that interfaces with a HCP to meet habitat requirements for northern spotted owls, marbled murrelets, and other species, while also considering economic effects on local communities (Oregon Department of Forestry 2006). Many states have natural heritage programs and lists of sensitive species at risk of being lost from the state or province. Some, such as Massachusetts, have their own state endangered-species lists requiring landowners to take particular actions to avoid habitat damage for state-listed species. Because land managers often are faced with more than one species that is of regulatory concern, multispecies management plans, and HCPs are becoming more common.

Nonregulatory Goals: Goals usually do not stop with regulated species. Game species, ecological keystone species, and others may be the focus of a particular landscape, and clearly human interests must also be accommodated. Protection of cultural sites, recreation areas, aesthetics, and economic income must all be balanced with the habitat management approaches chosen by the planners and managers. Without a formalized plan in place, the myriad of possible effects of management on all of these values can be overwhelming to those managing a landscape. And without a plan the risk of taking an action that has long-lasting adverse consequences is likely to increase.

Nongovernmental organizations also are using landscape management plans to aid in large-scale planning efforts. The Nature Conservancy uses models of ecoregion structure and composition in combination with principles of landscape ecology to identify areas of potential high priority for protection or recovery (Groves et al. 2002, Poiani et al. 2000). Even more species-specific groups such as the Ruffed Grouse Society and the Wild Turkey Federation may employ land-scape management plans at times to facilitate management on public and private lands (Ferguson et al. 2002, Yahner 1984). In so doing, these groups are ensuring, to the degree possible, that the needs of those species of most concern to their constituents are met, though they also freely recognize the need to consider many other species, ecosystem services, and social values as well.

Goals reflecting conservation of biodiversity may also be driven by economics. Forest industries now often seek green certification (see Chapter 18) as a way of assuring that forest practices are sustainable, including measures taken to conserve biodiversity.

Large-scale ecoregional assessments are often used as the context for landscape management plans that address biodiversity protection among other social values. Often these assessments rely on landscape management plans as the mechanism to implement the regional plan. Strategies for biodiversity protection are often established at regional scales that guide landscape plans, which guide stand prescriptions. Regional strategies are realized by implementing stand prescriptions over landscapes and implementing landscape plans over regions. Guidance comes from large spatial scales and implementation is cumulative over small spatial scales. Although one can argue the details of effectiveness and subsequent use of the information, efforts such as the Northwest Forest Plan (FEMAT 1993), the Columbia Basin Ecoregional Assessment (Wisdom et al. 1999), the Willamette Valley Alternative Futures analysis (Hulse et al. 2002), and the Sagebrush Ecoregional Assessment (Connelly et al. 2004, Knick and Rotenberry 2000) all contributed to a foundation or framework within which more local decisions could be made that could contribute to broader goals and objectives. Each of these assessments has relied on landscape management plans to achieve long-term goals.

CURRENT CONDITIONS

Classification of vegetation into plant communities and successional stages provides the basis for assessing coarse-filter goals, and subsequent identification of under- or over-represented communities on the landscape. This information and an estimate of conditions that are likely to occur on the ownerships outside the boundaries of the landscape (the context) being managed can help to guide the articulation of the desired future condition (DFC).

Stand maps for each ownership of the forest (cover type and age), in combination with information on roads, streams, underlying geology, known locations of sensitive species, and culturally important sites, can provide the basis for describing current conditions for many ecological conditions and social values. This information clearly should include an assessment of the distribution and connectedness of plant communities and seral stages, the levels of certain habitat elements associated with focal species, and availability of habitat for each focal species. Collectively, this information is used to describe the *current condition* of the area, and is the basis for development of a DFC for the landscape.

Once the coarse-filter goals have been established and the species of concern have been identified, the specific habitat elements needed by each species can be assessed over space and projected over time to understand how management alternatives might lead to changes in habitat quality for these species. The particular elements related to reproduction and foraging, including the spatial requirements, connectivity, and other attributes associated with habitat quality for each species should be identified and structured in a way that allows large-scale assessment of habitat for each species (McComb et al. 2002, Spies et al. 2007). In some cases, population viability analyses may need to be conducted, a topic that we will cover in detail in Chapter 16. With the use of geographic information systems (GISs), rapid assessments of current habitat availability for a suite of species is possible. In addition, these tools can help the planners to design a DFC for each species that are included in their fine-filter approach.

In addition, you should be able to develop a list of economic values and ecological services, as well as maps of habitat availability for selected species that could occur on the landscape now. This list should be developed in consultation with stakeholders.

DESIRED FUTURE CONDITIONS

Given the list of values generated by stakeholders, a number of questions should be addressed as you develop a description of the DFC:

1. Are there plant communities or seral stages that are under-represented on the landscape now as determined by comparison of current proportions to proportions represented in some reference condition (often the historical range of variability, if appropriate)?
2. Are levels of specific habitat elements within and among stands sufficient to minimize risk of losing species from the landscape?

3. How is the current pattern of habitat availability for each species contributing to the regional habitat availability for each species listed or protected under federal or state policies? How much latitude do you have in managing this habitat without risk to populations in and outside of the landscape?

4. Are there species or values that you would like to favor on the landscape in the future that currently may not be provided for now?

5. What species or species groups provide the greatest opportunity for contributing to local or regional populations?

6. Are there any species or values that are sensitive to land use that would likely be eliminated from the area in the future under current patterns of land use?

7. Could these species or values be accommodated wholly or partially on your landscape?

Once these questions are addressed, they become the basis for developing a DFC for each ownership that collectively achieves the DFC for the landscape. Just as no single stand condition will meet the needs of all wildlife species or values, neither will any single landscape management approach meet the needs of all species and values. There are always trade-offs to assess. Habitat for some species will decline over time, while that for others will increase. Careful planning of stand treatments over space and time will move a landscape toward a DFC in an attempt to meet multiple goals (Bettinger et al. 2001).

Often it is useful not to think of just one DFC but a set of DFCs in which one DFC leads to another over time to achieve a series of objectives over both space and time. This approach also allows the planner to think about how each DFC can be achieved while still providing a high likelihood that future DFCs can occur later in landscape development, even in the face of uncertain events such as natural disturbances or social change.

For each of the species that you identify as being of special interest, you will need to assess the habitat requirements that would need to be considered during implementation of the plan: stand management and harvest planning. Life history information, habitat suitability models, or habitat relationships models can help to ensure that these conditions are provided over sufficient areas with sufficient connectivity (see Chapter 12). Bettinger et al. (2001) demonstrated how harvest planning can be integrated with habitat goals for selected species to achieve multiple objectives in an economically efficient manner. Once these conditions have been described, then a DFC for the landscape can be articulated by describing in writing what you want, and by mapping these conditions to ensure that they are indeed feasible given other constraints and opportunities (Bettinger et al. 2001). These maps and descriptions of the DFCs should be developed in cooperation with stakeholders and vetted by all stakeholders affected by plan implementation.

PATHWAYS TO DFCs

Once you know what the current conditions are and have described the DFCs, it is then important to understand how you will implement management in a manner that will likely reach the DFC. It is important at the very least to develop a set of maps that clearly indicate how you will plan to move the forest from its current condition to the DFC through harvest planning, reserves, and other management approaches. There are several possible tools that you can use to project the future conditions of the forest. Interfacing stand growth models (e.g., Forest Vegetation Simulator; Teck et al. 1996) with GIS tools can allow simulation of landscape change over time (Figure 14.1). The Landscape Management System developed at the University of Washington and Yale University provides many opportunities for understanding changes in forest structure and composition over time and has been linked to habitat models for several species that allow an understanding of habitat changes over time (Marzluff et al. 2002). These kinds of tools help you to understand if there is or not a clear and achievable path from the current condition to each of the DFCs and allow comparisons with past conditions and likely achievement of goals (Figure 14.2).

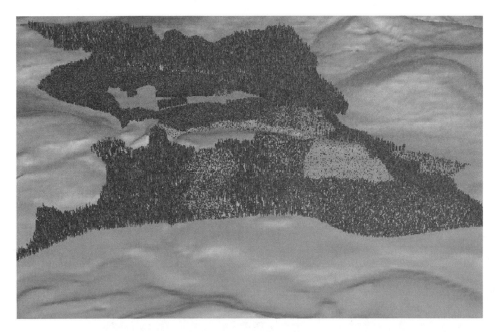

FIGURE 14.1 Example of one-time step in forest landscape pattern. (From McGaughey, R.J. 2001. Using data-driven visual simulations to support forest operations planning. In *Proceedings of the First International Precision Forestry Cooperative Symposium*. Precision Forestry, Seattle, Washington, pp. 173–179 and McCarter, J.M. et al. 1998; Landscape Management Systems.) Information on each stand and information among stands allows planners to understand if a DFC will likely be met.

DEVELOPING THE LANDSCAPE MANAGEMENT PLAN

With knowledge of current conditions, the conditions you would like to achieve, and how you might be able to change the landscape to achieve your goals, you are now ready to write a landscape management plan. You may write several plans, each feasible, but with different emphases in different plans (including a *no-action* plan) so that stakeholders, shareholders, and constituents can choose from them.

POLICY GUIDELINES FOR HCPS

When a landscape management plan also is a HCP as defined under Section 10 of the ESA, then certain guidelines apply when developing the plan (the following section is paraphrased from Oregon Department of Forestry 2006). Four tasks must be completed to determine the impacts that are likely to result from the proposed taking of a federally listed species:

1. Delineate the boundaries of the plan area
2. Collect and synthesize the biological data for all species covered by the HCP
3. Identify the proposed activities that are likely to result in incidental take
4. Quantify anticipated take levels.

Usually an impact assessment must be developed that meets requirements for an Environmental Assessment under the National Environmental Policy Act (NEPA) as well as Section 10 under the ESA. This assessment includes techniques that will be used to monitor, minimize, and mitigate impacts on the listed species; the funding available to implement the plan; and the procedures to deal with uncertainty. Section 10(a) of the ESA requires that permit issuance does not "appreciably

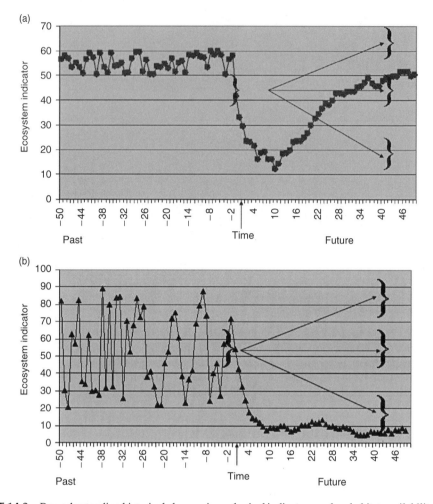

FIGURE 14.2 By understanding historical changes in ecological indicators such as habitat availability through present (time = 0) and likely future changes in the indicators, we can begin to see if future conditions are or are not likely to meet societal goals (brackets) for the indicator. Note that the trajectory (b) indicates a new indicator state in the future, while (a) indicates a recovery following a lag period.

reduce the likelihood of the survival and recovery of the species in the wild" (Oregon Department of Forestry 2006). The ESA does not require that HCPs result in the recovery of species covered under such plans. As most landowners manage only a portion of the geographic range of a species, they may contribute to population change but do not directly control it. Each HCP must contain measures that result in impacts that are consistent with the long-term survival of the listed species. Long-term survival includes the maintenance of genetically and demographically viable, well-distributed populations throughout the geographic range of each listed species. Actions that will meet the intent of long-term survival will vary from one species to another and must be approved by the regulatory agency.

In addition, Section 10(a) regulations require that a HCP describe the specific monitoring measures the applicants will conduct to ensure that the plan is being implemented as described and is effective in achieving its goals. Such an approach also helps both the land manager and the regulatory agency deal with unforeseen changes in habitat or populations. This approach also lends itself well to an *adaptive management* strategy where monitoring data may allow managers and regulators to agree on midcourse changes in the plan to better meet the goals of both parties (see Chapter 17).

The HCP should consider alternative management strategies, including at least one that would not result in take (usually a *no-action* alternative), and the reasons why that alternative is not selected (usually for economic reasons). The regulatory agency may also place additional requirements on the land managers including an implementation agreement, which is a legal contract that describes the responsibilities of all HCP participants.

GENERAL STRUCTURE OF THE LANDSCAPE MANAGEMENT PLAN

A landscape management plan can be structured in many ways similar to a stand management plan or prescription. It is important to have a written plan that provides a general schedule of predicted entries and treatments, because people charged with implementing the plan will likely change over time, especially on multiowner landscapes. Proposed actions and the schedule may need to be modified somewhat as new information comes available through monitoring activities and additional research, and so the plan should be viewed as a strategic plan (a general strategy for achieving goals) and not a tactical plan (a specific set of actions to be taken in certain places at certain times).

The plan should include:

Context: What is the historical, spatial, and temporal context for the forest? What opportunities and restrictions exist that influence development of the landscape plan? What policies drive current actions on the property? What are the physical, biological, and social resources that must be considered? This information can include the goals and purposes for having a plan and the process used to arrive at the goals (e.g., public meetings and stakeholder involvement).

Current Forest Condition: What are the current conditions with regard to areas of vegetative communities, seral stages, and stand sizes? What conditions exist for those species or resources that you would consider as part of your fine-filter approach? To comply with NEPA requirements, this section should include a description of the affected environment. This section should include at the very least:

1. The current state of vegetation (plant communities and seral stages).
2. Species of plants and animals likely to be affected by proposed treatments.
3. Habitat requirements of threatened or endangered species.
4. Location and extent of wetlands or other sensitive ecosystems.
5. The underlying geology and soils and the potential of those conditions to maintain the current species assemblages.
6. Current land-use patterns within the planning area and the influence of patterns outside of the area on conditions in the planning area.
7. Air and water quality issues apparent now, especially if they are contributing to impaired watersheds and airsheds under the Clean Water Act and Clean Air Act, respectively. If any impairment is noted, then remedial measures that are underway should be described.
8. Significant cultural resources should be described, especially those that are protected under antiquities acts. Any ongoing management designed to restore or maintain these resources should be described.

Desired Future Conditions: All of the factors listed above should be addressed in the DFC as components of the affected environment. What are the conditions that you would like to produce over time for each of several alternative plans? One of the alternatives should be the *no-action* alternative that can be compared with other alternatives. Often a *preferred alternative* is also described and is one that the developers of the plan find most desirable. Providing alternatives allows stakeholders the opportunity to comment on the alternatives and to compare the costs and benefits of various alternatives. This approach is required if NEPA is required. Specifically, the DFCs in each alternative should include:

1. *Coarse-Filter Goals and Objectives.* What are the general goals and objectives with regard to plant community representation, seral stages, and stand sizes? What is the rationale for these goals (based on the context above)?
2. *Mesofilter Goals and Objectives.* Standards and guidelines that provide the manager with the numbers, levels, and distributions of habitat elements that should be provided in the seral stages within each plant community. A rationale for these goals should be provided.
3. *Fine-Filter Goals and Objectives.* What species are considered explicitly in the management strategy? How do these goals complement coarse-filter goals? A rationale for selecting these species should be provided.

Management Actions to Achieve DFCs: What will you do to achieve your DFCs? How long do you think it will take to achieve them? How long will they last? How much will it cost? The answers to these questions can best be answered following projections of likely future conditions to ensure that there are clear system pathways to allow DFCs to be realized over space and time.

Monitoring Plans: What will you measure and how often will you measure to determine if your management plan was implemented correctly and if the actions were effective? How will you decide if you need to change your management plans? What thresholds must be reached before you make a change? See Chapter 18 for more guidance on developing monitoring plans.

Budget: What will implementation and monitoring of the plan cost? Where will the funds come from? In the event of a budget shortfall, what contingency plans are in place?

Schedule: What management actions are scheduled to occur during each decade for the next few decades? How will these actions be moving the forest into the DFC? Remember that a plan such as this should be revisited and revised periodically, often every 5 to 10 years, to allow incorporation of information gained through monitoring and through publications in the scientific literature.

References: Use references from *refereed journals* as much as possible to support any assumptions that were made in articulating the DFC. Contemporary approaches to finding information often involves searching the web using a readily available search engine, but information from unknown sources should be viewed with caution — it is easy to get what you pay for. Nothing! Or worse yet, incorrect information. Search engines that search the refereed literature can be quite helpful but be sure to check the original literature before citing it.

CONSIDERING ALTERNATIVE PLANS

When considering alternatives to large complex plans, such as the Northwest forest plan (NWFP), the landscape dynamics and resource outputs from a large area must be analyzed. The NWFP was developed to address biodiversity concerns on public lands over the geographic range of the northern spotted owl. Species that fell into a high-risk fine-filter group that needed particular attention to ensure that the plan met their needs included spotted owls, marbled murrelets, red tree voles, and many others. Indeed, over 1000 species were assessed to understand the risks of implementing 1 of 11 land management options under the NWFP. The species-by-species assessments, constituted the fine-filter assessment designed to ensure that those species that may not be captured by the coarse-filter planning strategy (moving more forest into later successional stages), would still be addressed and protected. But additional questions arise with a planning process like this. As older forests are allowed to develop, young forests decline in abundance and so do the species associated with younger forests. In particular, young forests that are structurally or compositionally complex (as would occur after a natural disturbance) may decline to an even greater degree as timber production is intensified on the remaining land base. Hence, focal species that are known to be associated with particular conditions

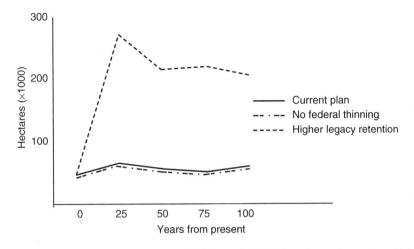

FIGURE 14.3 Projected response of western bluebirds to the NWFP and two alternative policies, Oregon Coast Range. (Redrafted from Spies, T.A. et al. 2007. *Ecol. Applic.* 17: 48–65. With permission from the Ecological Society of America.) Note that two alternatives had the same likely influence on habitat availability over time and produced identical projections.

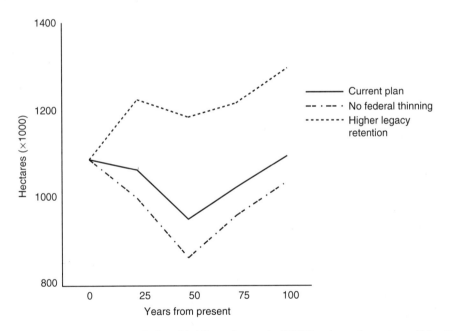

FIGURE 14.4 Projected response of olive-sided flycatchers to the NWFP and two alternative policies, Oregon Coast Range. (Redrafted from Spies, T.A. et al. 2007. *Ecol. Applic.* 17: 48–65. With permission from the Ecological Society of America.)

(snags in young forests, hardwoods, shrubs, etc.) in early-seral stages are selected as examples of how a species might respond to changes in forest conditions. By projecting habitat availability for these species forward in time, we can see how their habitat might change under current and alternative policies. Examples for early-successional snag associates (western bluebirds) and high-contrast forest edge associates (olive-sided flycatchers) are provided in Figures 14.3 and 14.4. In these cases, we can see that current policies either maintain or increase habitat for the species over time, but that some policies are likely to be better than others in providing these conditions. Hence, inferences can be made to other species that also might be associated with these conditions. It is important to remember though that these are not indicator species. Because each species has its own habitat requirements, the responses of one species will never accurately reflect the responses of the other.

FINDING SOLUTIONS TO LAND MANAGEMENT PLANNING PROBLEMS

With any forest consisting of multiple stands, there are a number of acceptable plans (plans that will meet goals at varying costs) and a number of unacceptable plans (plans that will not meet goals or will only succeed at unacceptable costs). Usually, there is more than one way of managing a landscape to achieve the DFCs. If ten managers each designed a management plan to achieve one set of DFCs, the ten plans probably would all be different in some ways. So given these multiple alternative plans, how can we find the best plan? Well, the best plan depends on who is doing the judging. About all we can usually hope for is a plan that is mutually acceptable to a group of stakeholders.

One way of finding not only an acceptable solution but also a good one is to develop a number of plans that are all considered acceptable. Each solution can then be ranked from highest to lowest for any number of values that the manager (or society) places on the landscape. Those plans with the lowest cumulative ranks are those that are likely to be better than those with higher total ranks (Table 14.1).

I provide a very simple example considering five plans developed to meet a set of DFCs for four resources in Table 14.1. Of the five plans, all of which are acceptable, plan D would seem to be the most desirable option from the standpoint of the four resources being considered. We could have any number of resources represented and we could use different rules on which to base our selection of the preferred alternative (no resource must rank below 3 for instance). These rules will be based on the values assigned by stakeholders in an iterative fashion until those involved can agree that the selected alternative is the one most likely to address public values over multiple ownerships now and into the future. Indeed, the most difficult (and most informative) part of the planning process can be the social debate necessary to ensure that the selected alternative is most likely to meet social goals.

Comparisons among alternative plans should be done jointly with affected constituents and planners, and scientists should be available to assist with interpretation of assumptions. It is entirely possible and indeed quite likely that arguments might ensue among the constituents, planners, and scientists once a set of options have been assessed. The arguments could result in a determination that none of the alternative plans are socially acceptable and that the planning process should start anew.

TABLE 14.1
Example of a Process of Prioritizing Values Associated with Four Resource Values in Each of Five Alternative Plans

Plan	Turkey	Timber	Aesthetics	Trout	Sum
A	3	2	3	3	11
B	4	4	1	1	10
C	5	1	4	5	15
D	1	3	2	2	8
E	2	5	5	4	16

If the stakeholders included turkey hunters, forest products industry, hikers, and fly-fishers, and each group ranked each alternative (1 = best, 5 = worst), then the sum of the ranks can provide an idea of that plan which represents the best compromise (lowest sum = plan D). Note that weights are not assigned to alternatives in this example.

A similarly likely outcome, especially given the number of assumptions on which future projections are based, is that the affected parties reach agreement on a preferred alternative and implement it, but based on monitoring data the observed responses are not what was expected. Should important societal values not be sustained due to these departures from predictions, there may yet again be a reason to start the planning process anew with a greater understanding of the assumptions that must be changed. But even if the system responds as anticipated and current social values are sustained, social values, including biodiversity goals, are not static. As societal expectations evolve, so do biodiversity goals. Stuff happens. Evolution of cultural mores in addition to unanticipated events (September 11, 2001, tsunamis, wild fires, or disease) can drastically alter the perceived values and importance of biodiversity in our cultures. It should be accepted that assessments will need to be revisited as new social issues emerge due to the dynamic nature of cultural values.

Plan Effectiveness

Plans abound. But do they achieve the intended goals? Recently, there have been several assessments of the effectiveness of Species Recovery Plans and HCPs in preventing further declines in populations of threatened species. Shilling (1997) noted that the number of threatened and endangered species in the United States is increasing monthly and *critical habitat* (as defined in the ESA is habitat required for continued existence of the species) is constantly being destroyed. Shilling (1997) contends that the application of HCPs may be making things worse, not better, for these species. Bingham and Noon (1997) also questioned the logic of allowing incidental take by having an approved HCP. They found that mitigation solutions are often arbitrary and lacked a basis in the habitat requirements for the species. They proposed that the concept of "core area" (that portion of an animal's home range that receives disproportionate use) be used as the basis for mitigating habitat loss within a HCP. Clearly, habitat requirements need to be addressed at an appropriate scale when mitigation measures are involved.

But plan effectiveness is always in doubt. All of the uncertainties described in Chapter 13 can rear their ugly heads in the face of the best-made plans. Wilhere (2002) made the case that HCPs entail a compromise between regulatory certainty and scientific uncertainty, and previous authors indicated that many HCPs do not adequately address scientific uncertainty. Monitoring the implementation and effectiveness of the plan is one way of acquiring and applying information to allow continual improvement of the plan (Wilhere 2002). Adaptive management has been promoted as a means of managing in the face of uncertainty, but few HCPs incorporate genuine adaptive management (see Chapter 17). Developing and implementing HCPs and other landscape management plans are costly both in monitoring the effectiveness of the plan and in the economic value of resources forgone to meet habitat needs. Wilhere (2002) proposed that economic incentives might encourage implementation of more effective plans because it would enable adaptive management. Incentives might include direct payments or tax deductions for reliable information that benefits a species. Indeed, it is quite possible that incentives are more likely to induce creativity in developing and implementing effective plans than regulations; regulations tend to homogenize systems, while incentives can encourage creativity.

SUMMARY

Goals for a landscape are usually set by stakeholders representing different ecosystem services and economic values associated with the landscape. Because landscapes are often large and complex with many competing values, landscape management plans are essential to reducing the risk of not achieving goals during management. Some goals are established by law; HCPs are one such regulation that allows "take" of critical habitat for federally listed species in the United States. The ability of a land manager to meet legal requirements and achieve biodiversity goals is dependent on developing plans, identifying a good acceptable plan, implementing it, and monitoring its effectiveness.

REFERENCES

Bettinger, P., K. Boston, J. Sessions, and W.C. McComb. 2001. Integrating wildlife species habitat goals and quantitative land management planning processes. In Johnson, D.H., and T.A. O'Neill (managing editors), *Wildlife Habitat Relationships in Oregon and Washington*. OSU Press, Corvallis, OR.

Bingham, B.B., and B.R. Noon. 1997. Mitigation of habitat "take": application to habitat conservation planning. *Conserv. Biol.* 11: 127–139.

Connelly, J.W., S.T. Knick, M.A. Schroeder, and S.J. Stiver. 2004. Conservation assessment of greater sage-grouse and sagebrush habitats. Western Association of Fish and Wildlife Agencies. Unpublished report. Cheyenne, WY.

Ferguson, R., A. Huckabee, and C. McKinney. 2002. *South Carolina's Forest Stewardship Management Plan*. National Wild Turkey Federation, Edgefield, SC.

Forest Ecosystem Management Assessment Team (FEMAT). 1993. *Forest Ecosystem Management: An Ecological, Economic, and Social Assessment*. USDA For. Serv., USDC Nat. Oceanic Atmos. Admin., National Marine Fish. Serv., USDI Bur. Land manage, Fish and Wildl. Serv., National Park Serv. and Environmental Protection Agency. Portland, OR.

Groves, C.R., D.B. Jensen, L.L. Valutis, K.H. Redford, M.L. Shaffer, J.M. Scott, J.V. Baumgartner, J.V. Higgins, M.W. Beck, and M.G. Anderson. 2002. Planning for biodiversity conservation: putting conservation science into practice. *BioScience* 52: 499–512.

Hulse, D., S. Gregory, and J. Baker. 2002. *Willamette River Basin Planning Atlas: Trajectories of Environmental and Ecological Change*. Oregon State University Press, Corvallis, OR.

Knick, S.T., and J.T. Rotenberry. 2000. Ghosts of habitats past: contribution of landscape change to current habitats used by shrubland birds. *Ecology* 81: 220–227.

Marzluff, J.M., K.R. Millspaugh, K.R. Ceder, C.D. Oliver, J. Withey, J.B. McCarter, C.L. Mason, and J. Comnick. 2002. Modeling changes in wildlife habitat and timber revenues in response to forest management. *For. Sci.* 48: 191–202.

McCarter, J.M., J.S. Wilson, P.J. Baker, J.L. Moffett, and C.D. Oliver. 1998. Landscape management through integration of existing tools and emerging technologies. *J. For.* 96: 17–23.

McComb, W.C., M. McGrath, T.A. Spies, and D. Vesely. 2002. Models for mapping potential habitat at landscape scales: an example using northern spotted owls. *For. Sci.* 48: 203–216.

McGaughey, R.J. 2001. Using data-driven visual simulations to support forest operations planning. In *Proceedings of the First International Precision Forestry Cooperative Symposium*. Precision Forestry, Seattle, Washington, pp. 173–179.

Oregon Department of Forestry. 2006. Final *2006 Elliott State Forest Management Plan (FMP)*. Oregon Department of Forestry, Salem, Oregon.

Poiani, K.A., B.D. Richter, M.G. Anderson, and H.E. Richter. 2000. Biodiversity conservation at multiple scales: functional sites, landscapes, and networks. *BioScience* 50: 133–146.

Shilling, F. 1997. Do habitat conservation plans protect endangered species? *Science* 276: 1662–1664.

Spies, T.A., B.C. McComb, R. Kennedy, M. McGrath, K. Olsen, and R.J. Pabst. 2007. Habitat patterns and trends for focal species under current land policies in the Oregon Coast Range. *Ecol. Applic.* 17: 48–65.

Teck, R., M. Moeur, and B. Eav. 1996. Forecasting ecosystems with the forest vegetation simulator. *J. For.* 94: 7–10.

Wilhere, G.F. 2002. Adaptive management in habitat conservation plans. *Conserv. Biol.* 16: 20–29.

Wisdom, M.J., R.S. Holthausen, B.C. Wales, D.C. Lee, C.D. Hargis, V.A. Saab, W.J. Haan, T.D. Rich, M.M. Rowland, W.J. Murphy, and M.R. Earnes. 1999. *Source Habitats for Terrestrial Vertebrates of Focus in the Interior Columbia Basin: Broad-Scale Trends and Management Implications*. USDA For. Serv. PNW Res. Sta., Portland, OR.

Yahner, R.H. 1984. Effects of habitat patchiness created by a ruffed grouse management plan on breeding bird communities. *Am. Midl. Nat.* 111: 409–413.

15 Ecoregional Assessments and Prioritization

Time is money. Time, money, and commitment are what make habitat management possible in stands, landscapes, and regions, but those resources are limited. Every forest manager has a budget and personnel limitations. Consequently, a manager will need to know where to invest those resources to have the greatest impact on the resources of interest. Getting the "biggest bang for the buck" is the approach that most managers want to take. For instance, consider a forest manager in Alabama with three primary goals: bobwhite quail, white-tailed deer, and timber. Patterns of food patches interfaced with cover are important to deer and quail (albeit at different spatial scales), but making a profit is important as well, and so the problem becomes one of optimizing habitat quality for the two game species while ensuring profitable timber production. One way to approach this problem is to view habitat for the two species as constraints on the timber production or alternatively view timber production as a constraint on habitat for the two species. In either case, the resulting decision is one where one group of resources is given more value than the other, and the decision resulting from the analysis can be implemented over space and time to achieve the desired goals (assuming some natural disturbance does not come along and change everything).

Now consider the problems likely to occur over much larger areas of space and time. How would you decide where to provide habitat for rare species throughout their geographic range in order to minimize risk of extinction? Or decide which parcels to buy before they are turned into housing developments? Or decide which nuclei of forests to protect from invasive species before they are overrun? Or decide how to coordinate management actions among landowners over a region to achieve biodiversity goals? Just as landscapes provide the context for stand prescriptions and regions provide the context for landscape management plans, global patterns of biodiversity provide the context for regional conservation strategies. But global patterns of biodiversity will only be conserved if the strategies are implemented among stands over landscapes and among landscapes over regions. Strategies are developed from the top down and implemented from the bottom up. Think globally, act locally. But within this context it is often difficult to know where to invest the time, money, and commitment to achieve these regional goals. Regional assessments can provide the context and prioritization analyses can provide the guidance for investments.

ECOREGIONAL ASSESSMENTS

Ecoregions are often used as the basis for assessments and habitat strategies are developed at this scale, which in turn guide landscape management plans. Ecoregions are areas of similar climate, topography, soils, and other factors influencing patterns of vegetation and the animals and the processes that support them occur and recur predictably (Table 15.1).

In Figure 15.1, ecoregions of the United States are displayed as generalized areas of climatically associated patterns of vegetation. This map is one of several attempts that have been made at mapping ecoregions, each with differences as influenced by the goals of the organization funding the work. Some systems of delineating ecoregions have greater detail than others (Omernick 1995). But ecoregions are not discrete entities. One will grade into another and no two places within any one are the same. The devil is in the detail. Clearly as you zoom in on any ecoregion, there is variability in patterns of soil, topography, climate, and vegetation that occurs locally, hence the need

TABLE 15.1
U.S. National Hierarchy of Ecological Units

Planning scale	Utility
Ecoregion	
Global	Broad applicability for modeling and sampling
Continental	Strategic planning and assessment
Regional	International planning
Subregion	Strategic, multiforest, statewide, and multiagency analysis and assessment
Landscape	Forest or area-wide planning and watershed analysis
Land unit	Project and management area planning and analysis

Source: From Bailey, R.G. 1980. *Descriptions of the Ecoregions of the United States.* Washington, D.C.: U.S. Department of Agriculture, Forest Service. Misc. Pub. 1391. With permission.

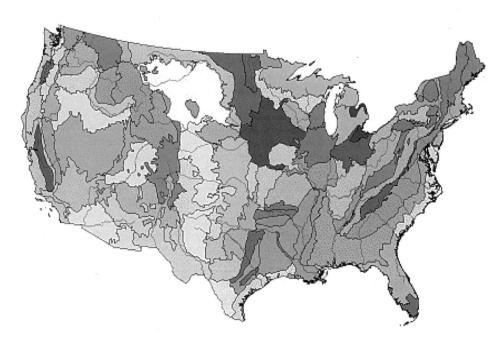

FIGURE 15.1 Ecoregions of the United States. See Bailey (1980) for codes. (Reprinted from Bailey, R.G. 1980. *Descriptions of the Ecoregions of the United States.* Washington, D.C.: U.S. Department of Agriculture, Forest Service. Misc. Pub. 1391. With permission.)

for a hierarchy of ecological units (Keys et al. 1995, McMahon et al. 2001, Nesser et al. 1994; Table 15.1). Hierarchical patterns of ecological units are useful because they do not necessarily follow political boundaries and can be useful in responding to issues that cross administrative and jurisdictional boundaries (Probst and Crow 1991). Using ecoregional units as the basis for assessments and development of coarse-filter conservation strategies is also intuitively appealing because disturbance forces and recovery patterns are often grossly consistent throughout a region. Hence, although local modifications are often needed during landscape planning, regional patterns provide a broader context for assigning goals and objectives that are related to the ranges of variability in ecosystem indicators (historic or future) seen in the ecoregion and can provide a logical link to population viability modeling efforts (Anderson and Bernstein 2003).

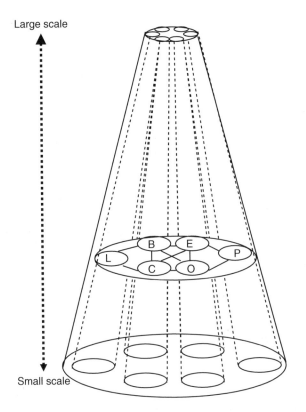

FIGURE 15.2 Organisms (O), populations (P), communities (C), landscapes (L), ecosystems (E), and biomes (B) all interact over a range of spatial scales, with many levels of interaction possible at small spatial scales than at large scales. (Redrafted from Allen, T.F.H., and T.W. Hoekstra. 1990. *J. Veg. Sci.* 1: 5–12. With permission from Opulus Press.)

Although ecoregions often have climatically or topographically defined boundaries, ecoregions are not a spatial scale per se. Some are large and some are small. In fact, the interaction of various ecological states and process can all occur over a range of spatial scales. Allen and Hoekstra (1990) and Hoekstra et al. (1991) made an excellent case that ecological functional units such as genotypes, organisms, populations, communities, landscapes, ecoregions, and biomes all interact over a range of spatial scales and there are more potential interactions among these units that can occur at small spatial scales than at large spatial scales (because the planet is only so big; Figure 15.2). So, although we can use ecoregional units as a basis for planning, ecoregional units may have, and often do have, genotypes, organisms, populations, communities, and landscapes that extend beyond ecoregional boundaries because ecoregions are developed in a hierarchical manner. This hierarchy is useful when designing conservation strategies for various species (Bailey 1987). Consider a community of large carnivores in the Rocky Mountains containing wolves, lynx, wolverines, and cougars. This community represents a collection of interacting species that clearly transcend some ecoregional units in their genotypes, individual organisms, and populations. Hence, using ecoregional units may be useful, but the appropriate level in the classification hierarchy must be interfaced with the spatial domains represented within each of these species if species conservation is a goal.

Indeed, it is the spatial scaling properties of the region and the species using the region that interface to provide information about the potential risk of losing species due to changes in patch areas, edges, and other factors that describe the spatial complexity of a region. Weins (1989) defined "domains of scale," or spatial patterns of patches that emerge as you perceive increasingly large areas of a region. For instance, given two regions, each with different land use or disturbance histories,

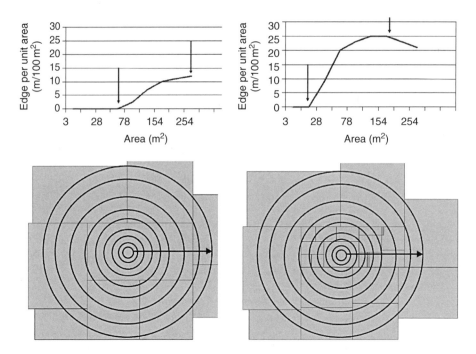

FIGURE 15.3 Simplified example of identifying domains of scales in portions of two landscapes. In a land-scape where patches are uniform and evenly distributed (left), a threshold in edge density emerges at 50 m^2 and approaches an asymptote at 254 m^2. In a more heterogeneous landscape (right) two thresholds emerge at 12 m^2 and at 202 m^2 where the patch heterogeneity is expressed in two domains of scale. Depending on the domains of scale associated with a species life history, one landscape may provide higher quality habitat than another.

the patch size and configuration might differ (Figure 15.3). Taking a transect from any point on the region and estimating an ecologically important metric over increasingly large areas produces a trend that should increase to an asymptote in a landscape in which all patches were uniformly distributed. But in many regions, patches are not uniformly distributed and the trend in landscape metrics is not asymptotic. Instead, there are thresholds that emerge from these analyses that define domains of scale. These domains can be used as information in regional assessments in two ways. First, the domains identified in current landscapes can be compared to a reference condition(s) such as the conditions that might be seen under the historical range of variability. An implicit assumption is that as the domains depart from the reference condition that there is increasing risk of losing species that are not well adapted to using these new domains of scale. In a more explicit analysis, the home range, territory, or metapopulation sizes for a species can be compared to the domains of scale associated with the species' habitat over a region. The greater the disparity between the domains of scale needed by a species and those found in a region, the greater is the risk to the species of not having adequate patches or connectivity (Figure 11.12).

In addition, it is often the edges between ecoregional units that can be of considerable interest during conservation planning and management prioritization. On a map these edges between eco-regions are sharp lines due to the need to classify units, but in reality they are blurred boundaries (Bailey 1980), oftentimes with rich plant and animal communities occurring at climatic ecotones. Hence, simply viewing the ecoregional unit as the basis for planning without recognizing the poten-tial importance of the ecoregional hierarchy and ecoregional ecotones, you may miss important drivers of species richness, particularly in the face of relatively rapid climate change where species shifts may be first noticed.

EXAMPLES OF ECOREGIONAL ASSESSMENTS

Federal, state, and nongovernment organization (NGO) groups have been involved in ecoregional assessments across the United States and Canada. The scope of the assessment represents a broad spectrum of space, processes, and political entities. Some, such as the Forest Ecosystem Management Assessment Team (FEMAT) limited their assessment to federal lands in the Pacific Northwest and focused largely on late successional species (FEMAT 1993). Others, such as Coastal Landscape Analysis and Modeling Systems (CLAMS) project considered all landowners, long time frames, and a multitude of processes and species (Spies et al. 2007a). In the CLAMS approach, both past and likely future conditions are considered in the face of current and alternative future policies. The Interior Columbia Basin Ecosystem Management Planning (ICBEMP) assessment considered a huge multistate area and resulted in an assessment of forest-related ecosystem process and species over the region (Wisdom et al. 2000; Figure 15.4). States have conducted much smaller ecoregional analyses, such as the Berkshire Ecoregion Assessment in Massachusetts (Fleming 2006). Oftentimes when states or federal agencies are involved in assessment, it stops at political boundaries although the ecoregion extends across boundaries. This was the case in the Berkshire assessment. The issue is further

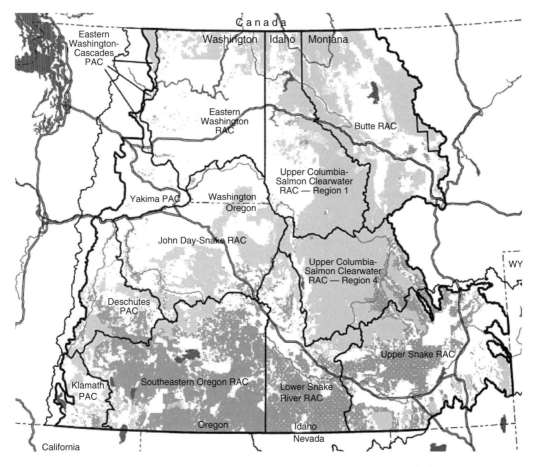

FIGURE 15.4   U.S. Forest Service and Bureau of Land Management lands covered by the ICBEMP. (Reprinted from the USDA Forest Service Interior Columbia Basin Ecosystem Management Plan. From Wisdom et al. 2000. *Source Habitats for Terrestrial Vertebrates of Focus in the Interior Columbia Basin: Broad-Scale Trends and Management Implications.* USDA Forest Service General Technical Report PNW-GTR-485.)

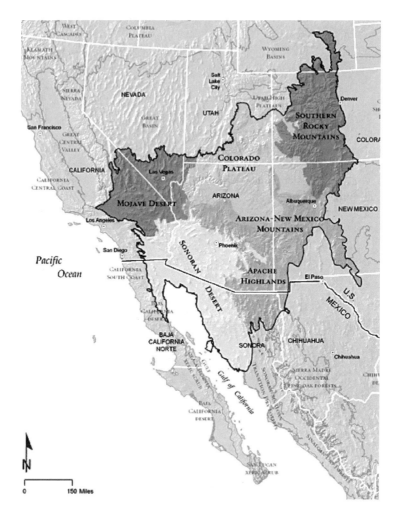

FIGURE 15.5   Example of a cross-border ecoregional assessment coordinated by TNC. Six ecoregions are included in this assessment. (From Marshall, R.M. et al. 2004. *An Ecological Analysis of Conservation Priorities in the Apache Highlands Ecoregion.* The Nature Conservancy of Arizona, Instituto del Medio Ambiente y el Desarrollo Sustentable del Estado de Sonora, agency and institutional partners. With permission.)

confounded when ecoregional analyses include international borders. FEMAT and ICBEMP largely stopped at the Canadian border, although the contributions of resources from Canada were considered as part of the context for the assessment. NGOs, especially The Nature Conservancy (TNC), also have used ecoregional assessments in their planning and prioritization work. NGOs are not restricted to political boundaries to the degree that states, provinces, and countries might be. For instance, TNC (2001) completed a multistate assessment for the Appalachian forests of Maryland, Virginia, Pennsylvania, and West Virginia and facilitated a cross-border assessment in the southwestern United States and Mexico that included six ecoregions (Marshall et al. 2004; Figure 15.5).

CONDUCTING AN ECOREGIONAL ANALYSIS

Will you ever be involved in conducting an ecoregional assessment? Well, maybe, but you will quite likely be developing landscape management plans or working within management plans that are tiered to an ecoregional assessment. In order to provide a useful context for landscape plans, ecoregional assessments should be effective in identifying likely risks to species, their habitats, and

---

TABLE 15.2
A Seven-Step Conservation Planning Framework

Step 1: Identify conservation targets
- Abiotic (physically or environmentally derived targets)
- Communities and ecosystems
- Species: imperiled or endangered, endemic, focal, keystone

Step 2: Collect information and identify information gaps
- Use a variety of sources
- Rapid ecological assessments, rapid assessment programs
- Biological inventories
- Workshops with species' experts

Step 3: Establish conservation goals
- Address both representation and quality
- Distribute targets across environmental gradients
- Set a range of realistic goals

Step 4: Assess existing conservation areas
- Gap Analysis (or CAPS or CLAMS type of assessment, see section on "Prioritizing management and assessing policies")

Step 5: Evaluate ability of conservation targets to persist
- Use criteria of size, condition, and landscape context
- Use GIS-based "suitability indices" to assess current and future conditions

Step 6: Assemble a portfolio of conservation areas
- Use site- or area-selection methods and algorithms as a tool
- Design networks of conservation areas employing biogeographic principles

Step 7: Identify priority conservation areas
- Use the criteria of existing protection, conservation value, threat, feasibility, and leverage to prioritize areas

Source: From Groves, C.R. et al. 2002. *BioScience* 52: 499–512.

---

plant communities over large areas. Recent systematic ecoregional assessments and associated conservation planning approaches probably have been more effective at conserving biological diversity than approaches of the past (Margules and Pressey 2000). Past approaches often resulted in a biased distribution of lands identified for protection or management specifically for biodiversity goals, with many areas occurring on lands not useful for other purposes such as high elevations and steep slopes (Scott et al. 2001). Groves et al. (2002) proposed a seven-step process to identify conservation areas (which may or may not require management) across ecoregions that is both efficient and cost effective and could be used as the basis for an ecoregional analysis (Table 15.2). Whereas large ecoregional assessments of the past have cost millions of dollars and taken 5 or more years to complete, this process has a median cost of $234,000 per plan (in 2002 U.S. dollars) and an average completion time of just under 2 years.

ASSESSING PATTERNS OF HABITAT AVAILABILITY AND QUALITY

Much of the data used to set goals, develop assessments, and establish priorities over regions come from remotely sensed sources such as satellite imagery (e.g., LANDSAT), light detection and ranging (LIDAR) data, aerial photography (orthophotos), and the resulting geographic information systems (GISs) data that are derived from these techniques. Consequently, the data are usually restricted to some minimum grain (perhaps a 30 × 30 m pixel for LANDSAT) and logical extent (simply from the standpoint of managing too many pixels of data). Using satellite data, reflectance values in various spectral bands are usually classified into conditions on the ground that are recognizable to humans as land-cover classes. These classes become the basis for further associations as habitat for various species. But each species has its own habitat requirements and so one classification system is unlikely to work well for all species. Nonetheless, these classified data are often related to the likely occurrence of various species and processes across the area of interest. Wildlife Habitat Relationships (WHR) models relate classified land types (seral stage and plant communities) to the occurrence of each of the vertebrate species that occur in an area (DeGraaf and Yamasaki 2001, Johnson and O'Neil 2001). For many species, the WHR models produce reasonable estimates of the availability of habitat for many species across a region (Block et al. 1994), but not all. Ideally classification would have to be tailored to each species to more accurately reflect the collection of habitat elements important to the species. Further, the grain of the assessment (say 30 × 30 m) may be too large to reliably capture the information needed to assess habitat for some species, both because one classification system does not apply to all species and some species use habitat at scales smaller than the grain size. Consequently, WHR models may be a useful guide to patterns and changes in habitat area and configuration for species over an area, but the reliability of the information varies considerably from species to species. Hence, these approaches only provide generalized estimates of habitat availability, indicating where the species could occur and not necessarily where it would find habitat of better or worse quality (as it might influence animal fitness). Indeed, knowledge of site-specific sizes, abundance, and distribution of habitat elements would be needed to understand specifically how habitat quality might change from place to place or over time in a region (McComb et al. 2002, Spies et al. 2007b). Some of this information could be extracted from aerial photos and from LIDAR data. These techniques can provide some information on individual trees and even smaller structures that can be seen from the air. Additional information can be derived from generally available GIS layers that include topography, soils, hydrography, and climate. These classified images and supplementary remotely sensed data can provide the information needed to assess habitat availability for many species. But habitat element information for some species would better be assessed from the ground. For instance, the amount of cobble in a stream important to torrent salamanders or the presence of hollow trees for swifts will not be reasonably reflected in remotely sensed data. Ground-plot information is needed.

Ground-plot data are systematically collected from multiresources inventories such as Forest Inventory and Analysis data on private (and many public) forest lands in the United States and these efforts can be used to infer patterns of habitat availability for species over space and time (Ohmann et al. 1994). In addition, many industrial land managers have continuous forest inventory plots distributed across their properties to monitor tree growth and death, and these plots can be adapted to allow collection of site-specific data on habitat elements as well. But, of course, these ground inventories are samples and not inventories, and so they have been of limited value when representing habitat availability in situations where both fine-scale habitat elements and patch sizes and patterns might be important.

Only recently have ground-plot data been interfaced with remotely sensed data to allow representation of habitat elements across complex regions (Ohmann and Gregory 2002, Spies et al. 2007b). Using this approach, the ground-plot data are georeferenced to the physical location, topography, climate, reflectance values, and many other features on GIS layers to create a subset of "informed" pixels (pixels with a ground plot within them). For all pixels that do not have associated ground data ("uninformed pixels"), the same descriptive characteristics are also estimated, but of course there is no corresponding ground-plot data associated with the uninformed pixels. To provide a "seamless" representation of ground-plot information, characteristics of informed pixels are used to "inform" those pixels without ground-plot data that are most similar in these descriptive characteristics. Hence, fine-scaled ground-plot data can be imputed to all pixels in the extent of the assessment (Ohmann and Gregory 2002). Once ground-plot data have been assigned to the uninformed pixels, then creating new categories of habitat based on the ground-plot data can be used to create habitat quality maps across the planning area (McComb et al. 2002, Spies et al. 2007b). These maps can then be used as the basis for assessing net gains and losses of habitat over space and time as well as population viability analysis (see Chapter 16) for species of high risk of being lost from the area in the future.

PRIORITIZING MANAGEMENT AND ASSESSING POLICIES

How would you decide which tools to use and approaches to follow to ensure that your biodiversity conservation goals will be effective? The tools available to assist in decision making for biodiversity protection have exploded in number and complexity over the past decade. Gordon et al. (2004) identified over 50 decision support tools that could be used to assist in biodiversity conservation. Choosing which to use, if any, is an overwhelming task and is highly dependent on the specific questions, goals, and objectives of the assessment (Anderson and Bernstein 2003, Johnson et al. 2007). The following are examples of a few commonly used and powerful approaches to assessments and prioritization.

COARSE-FILTER APPROACH

There are numerous examples of how estimates of habitat patterns, availability, and quality have been used to provide a means of prioritizing management decisions. These same techniques often can be used to assess alternative management plans or policies across the area of assessment. Because time and money are usually limited when making decisions regarding management to conserve biodiversity, prioritization of the areas to manage or protect become paramount.

One such approach is the Conservation Assessment and Prioritization System (CAPS), which uses a coarse-filter approach to parcel prioritization (Gordon et al. 2004). The CAPS approach was developed by Dr. Kevin McGarigal and uses potential biodiversity valuation that applies "biodiversity screens" to each patch in the landscape. These screens are applied to a map of predicted natural communities modeled from remotely sensed and GIS data. Biodiversity screens are models that reflect the content, context, spatial character, or condition of a patch to arrive at an index of potential biodiversity value. Stakeholders are involved in deciding how various parameters of the screens guide the identification of high-priority patches in the region. Parameters such as the size of a natural community patch, edge contrast, edge density, its proximity to water, the soil type, or road density (among many others) can be identified and weighted to help identify priority patches for management or protection. The result of applying a set of screens is a biodiversity value ranging from 0 (low value for biodiversity conservation) to 1 (high value) for each patch on the landscape, which then can be used to highlight those patches of highest value (Figure 15.6). The resulting high-priority patches represent areas that may receive special management practices or could be placed in reserves, conservation easements, or purchased from private landowners to protect species associated with the priority patch characteristics. The species receiving protection include species that are known to be associated with the priority patches as well as those represented in the "hidden diversity" or those species assumed to be associated with these patch conditions but which have not yet been identified. Further, the approach has been used not only in ecoregional assessments (e.g., the Berkshires), but also in mitigation to replace areas gobbled up by roads and development with patches of appropriate sizes and conditions.

INTEGRATED COARSE- AND FINE-FILTER APPROACHES

Many assessments use a combination of coarse-, meso-, and fine-filter approaches to understand current conditions across complex ecoregions. Some, such as the Willamette Alternative Futures

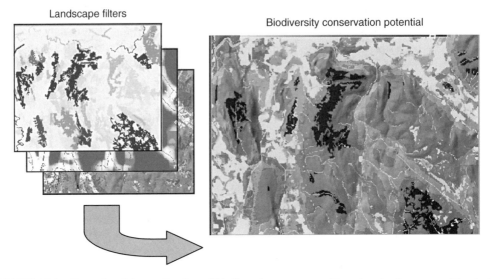

FIGURE 15.6 Example of the application of biodiversity screens or filters to a landscape resulting in the identification of high-priority blocks for management or protection. (Figure provided by Dr. Kevin McGarigal and used with his permission.)

approach (Hulse et al. 2002) used the likely changes in abundance and distribution of vertebrates across the ecoregion as a primary assessment of current and future effects of alternative future landscapes. They also considered the areal extent and distribution of various plant communities across the planning area, but did not assess the landscape metrics associated with the patches in a patch prioritization manner such as used in CAPS. Nonetheless, the results of this effort have been widely used to inform land-use planning decisions in the region so that planners can consider the effects on forest land and potential impacts on biodiversity of land use decisions. Another recent assessment, the CLAMS Project (Spies et al. 2007a) is an ecoregional assessment approach designed to analyze the ecological and socioeconomic consequences of various forest policies across multiple ownerships. The process includes complex set of interacting models that consider the disturbance and regrowth of forests as guided by forest management policies and the resulting patterns of plant communities and habitat quality for focal species across the region (Figure 15.7). The results can not only be used to assess alternative forest policies but also identify locations in the region that might be particularly important as core patches or linkages across complex multiownership landscapes. The projected changes in plant communities (as a coarse-filter index to protecting hidden diversity) as well as changes in habitat quality and distribution for focal species (those selected to represent certain ecological associations) are used to compare policy alternatives (Figure 15.8).

Fine-Filter Approaches

Other approaches to ecoregion assessments take a species-by-species approach to identifying areas for particular management or protection. Because most of the species assessments rely on WHR models to develop maps of occurrence of species, the underlying maps can also be used as a coarse-filter assessment as well. One such approach is a nationwide effort called Gap Analysis (Scott et al. 1993). The goal of Gap Analysis is to "keep common species common" by identifying those species and plant communities that are not adequately represented on existing conservation lands. By identifying habitat for all vertebrates in a region, Gap Analysis provides information that can be used to make decisions regarding vertebrate species conservation and management.

Gap Analysis consists of three main data layers, a land-cover layer, a layer showing the predicted distributions of vertebrate species, and a stewardship layer (Figure 15.9). These layers are used in a

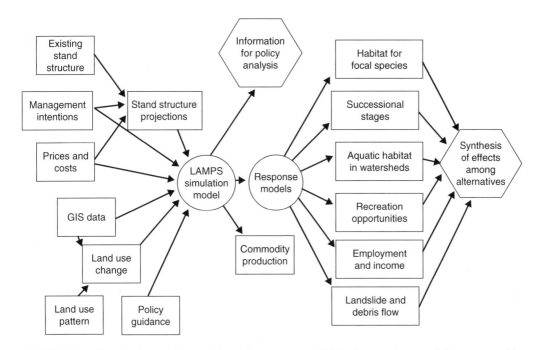

FIGURE 15.7 Complex interacting models produce current and likely future estimates of plant communities, habitat for focal species, and economic goods and services for the Oregon Coast Range under alternative forest policies. (From Spies, T.A. et al. 2002. In J. Liu and W.W. Taylor, eds. integrating landscape ecology into natural resource management. *Cambridge Univ. Press.* Cambridge, U.K. Pages 179–207. With permission.)

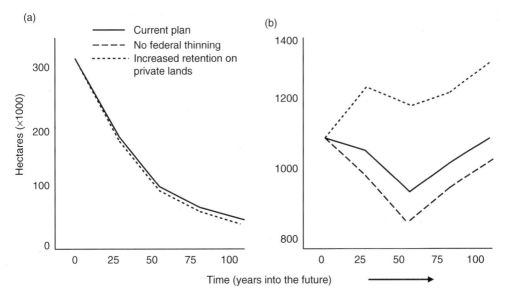

FIGURE 15.8 Projections of habitat availability for an example focal species (olive-sided flycatchers, b) and a plant community type (hardwoods, a) in the Oregon Coast Range under three policy alternatives: current policies (———), no thinning allowed on federal lands (— — —), and green tree retention on private lands (· · · · ·). (From Spies, T.A. et al. 2007b. *Ecol. Applic.* (in press). With permission from the Ecological Society of America.)

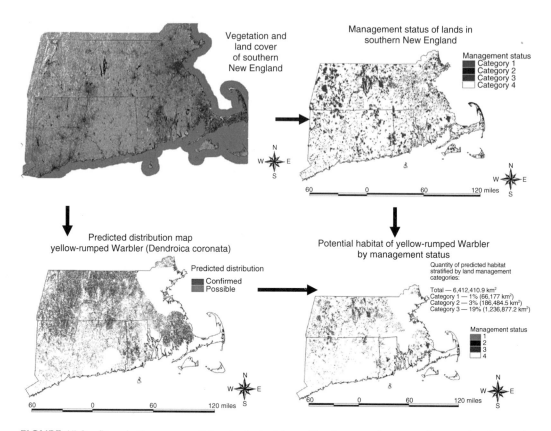

FIGURE 15.9 Steps in the process of Gap Analysis. Maps of land-cover using remotely sensed and ground-plot data (top left), species distribution based on geographic ranges and habitat relationships (top right), and management status (bottom left) are overlain to identify gaps in the protection network for each species (bottom right). (From Zuckerberg, B. et al. 2004. *A Gap Analysis of Southern New England: An Analysis of Biodiversity for Massachusetts, Connecticut, and Rhode Island.* Final contract report to USGS Biol. Resour. Div., Gap Analysis Program, Moscow, Idaho. With permission.)

Gap Analysis consisting of three primary steps. The first step is to map plant communities to develop a land-cover layer. Land cover is mapped using satellite data as well as other supporting information from existing GIS layers, areal photos, and ground-plot data.

The second step is to map predicted distributions of vertebrate species known to breed or use habitat in the region. Known, probable, and possible occurrences are used to define the geographic range of each species. Then a WHR model is developed for each species that relates the land-cover data to the likely occurrence of the species across the region. The process does not usually include any assessment of habitat quality or viability.

The third step of a gap analysis is to assign a land stewardship rank between 1 and 4 to each patch on the assessment area. Status 1 lands have the highest degree of management for conservation and status 4 lands have the lowest. Stewardship ranks are based on the long-term intent of the managing entity (owner or steward). Ranks are based on the following (Scott et al. 1993):

1. Permanence of protection from conversion of "natural" land cover to "unnatural" (human-induced barren, arrested succession, and cultivated exotic-dominated).
2. Amount of the tract protected, with 5% allowance for intensive human use.
3. Inclusiveness of the protection, that is, single feature such as wetland vs. all biota and habitat.
4. Type of management program and degree that it is mandated or institutionalized.

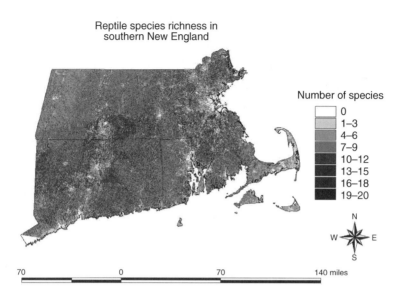

Reptile species richness in
southern New England

Number of species

- [ ] 0
- 1–3
- 4–6
- 7–9
- 10–12
- 13–15
- 16–18
- 19–20

70   0   70   140 miles

FIGURE 15.10 Example of a composite map of species richness for reptiles in southern New England. Protection or management of darker areas is more likely to capture more species than lighter areas. (From Zuckerberg, B. et al. 2004. *A Gap Analysis of Southern New England: An Analysis of Biodiversity for Massachusetts, Connecticut, and Rhode Island.* Final contract report to USGS Biol. Resour. Div., Gap Analysis Program, Moscow, Idaho. With permission.)

The fourth step is to analyze the representation of each species (or plant community) in areas managed for the long-term maintenance of biodiversity. To accomplish this, maps showing animal and plant community distributions are intersected with stewardship maps to identify areas where the species could occur, but that are not receiving protection based on management status or appropriate management. Identification of high-priority areas for protection or management can be based on individual species (Figure 15.9) or on species richness patterns (Figure 15.10). Gap Analysis has been completed for nearly every state in the United States and composite assessments across state lines now allow ecoregional analyses.

UTILITY AND EFFECTIVENESS OF ECOREGIONAL ASSESSMENTS

Of course, the ultimate test of effectiveness is in conserving species and preventing listing of species in the future through hierarchical planning, implementation, and monitoring. Few regional assessments have monitored effectiveness sufficient to provide evidence for having "saved" species that otherwise would have been regionally or globally eliminated. And measuring effectiveness requires that measurable goals are set to quantify success or failure (Tear et al. 2005). Only through monitoring over long-time periods can effectiveness be assessed and even then how can we know what we did not lose? Much of the effectiveness of assessment and planning is based on assumption and faith that to plan objectively and monitor the quantitative objectives is less risky than following unstructured approaches to species conservation.

SUMMARY

Ecoregional assessments are often used as a context for development and implementation of landscape management plans that provide direction for development and implementation of stand prescriptions. Ecoregions are hierarchically defined classes of vegetation, based largely on climatic and other physical features that seem to drive patterns of plant communities. The utility of ecoregional assessments in providing useful contexts for other efforts are highly dependent on matching the appropriate spatial hierarchy with the spatial and temporal scales associated with the species and communities of concern. Assessments can use coarse-, fine-, and meso-filter approaches, but most use some combination of these approaches. Usually, these approaches are designed to prioritize areas for protection or management. One example is the Gap Analysis Project, which estimates species occurrences in large areas to identify gaps in areas providing protection for species. The degree to which these efforts have been effective in protecting species is largely unknown and will likely come to light only after years of monitoring over large areas.

REFERENCES

Allen, T.F.H., and T.W. Hoekstra. 1990. The confusion between scale-defined levels and conventional levels of organization in ecology. *J. Veg. Sci.* 1: 5–12.

Anderson, M.G., and S.L. Bernstein (eds.). 2003. *Planning Methods for Ecoregional Targets: Matrix Forming Ecosystems.* The Nature Conservancy, conservation science support, northeast & Caribbean division, Boston, MA.

Bailey, R.G. 1980. *Descriptions of the Ecoregions of the United States.* Washington, D.C.: U.S. Department of Agriculture, Forest Service. Misc. Pub. 1391.

Bailey, R.G. 1987. Suggested hierarchy of criteria for multi-scale ecosystem mapping. *Landsc. Urban Plan.* 14: 313–319.

Block, W.M., M.L. Morrison, J. Verner, and P.N. Manley. 1994. Assessing wildlife–habitat–relationships models: a case study with California oak woodlands. *Wildl. Soc. Bull.* 22: 549–561.

DeGraaf, R.M., and M.Yamasaki. 2001. *New England Wildlife: Habitat, Natural History and Distribution.* University Press of New England, Hanover, NH.

Fleming, M. 2006. *Landscape Assessment and Forest Management Framework: Berkshire Ecoregions in Massachusetts.* Executive Office of Environmental Affairs, Boston, MA.

Forest Ecosystem Management Assessment Team (FEMAT). 1993. *Forest Ecosystem Management: An Ecological, Economic, and Social Assessment.* U.S. Government Printing Office: 1993-793-071.

Gordon, S.N., K.N. Johnson, K.M. Reynolds, P. Crist, and N. Brown. 2004. *Decision Support Systems for Forest Biodiversity: Evaluation of Current Systems and Future Needs.* Final Report Project A 10. National Commission on Science and Sustainable Forestry.

Groves, C.R., D.B. Jensen, L.L. Valutis, K.H. Redford, M.L. Shaffer, J.M. Scott, J.V. Baumgartner, J.V. Higgins, M.W. Beck, and M.G. Anderson. 2002. Planning for biodiversity conservation: putting conservation science into practice. *BioScience* 52: 499–512.

Hoekstra, T.W., T.F.H. Allen, and C.H. Flather. 1991. Implicit scaling in ecological research. *BioScience* 41: 148–154.

Hulse, D.W., S.V. Gregory, and J.P. Baker (eds.). 2002. *Willamette River Basin Planning Atlas: Trajectories of Environmental and Ecological Change.* Oregon State University Press, Corvallis, OR.

Johnson, D.H., and T.A. O'Neil (managing directors). 2001. *Wildlife–Habitat Relationships in Oregon and Washington.* Oregon State University Press, Corvallis, OR.

Johnson, K.N., S. Gordon, S. Duncan, D. Lach, B. McComb, and K. Reynolds. 2007. *Conserving Creatures of the Forest: A Guide to Decision Making and Decision Models for Forest Biodiversity.* National Commission on Science for Sustainable Forestry Final report. NCSSF, Washington, D.C.

Keys Jr., J.E., C.A. Carpenter, S.L. Hooks, F.G. Koenig, W.H. McNab, W.E. Russell, and M.L. Smith. 1995. *Ecological Units of the Eastern United States: First Approximation.* Geometronics, Southern Region, USDA forest service, Atlanta, GA.

Margules, C.R., and R.L. Pressey. 2000. Systematic conservation planning. *Nature* 405: 243–253.

Marshall, R.M., D. Turner, A. Gondor, D. Gori, C. Enquist, G. Luna, R. Paredes Aguilar, S. Anderson, S. Schwartz, C. Watts, E. Lopez, and P. Comer. 2004. *An Ecological Analysis of Conservation Priorities in the Apache Highlands Ecoregion.* Prepared by The Nature Conservancy of Arizona, Instituto del Medio Ambiente y el Desarrollo Sustentable del Estado de Sonora, agency and institutional partners.

McComb, W.C., M. McGrath, T.A. Spies, and D. Vesely. 2002. Models for mapping potential habitat at landscape scales: an example using northern spotted owls. *For. Sci.* 48: 203–216.

McMahon, G., S.M. Gregonis, S.W. Waltman, J.M. Omernik, T.D. Thorson, J.A. Freeouf, A.H. Rorick, and J.E. Keys. 2001. Developing a spatial framework of common ecological regions for the conterminous United States. *Environ. Manage.* 28: 293–316.

Nesser, J., L. Freehouf, W. Robbie, T.M. Collins, C.B. Goudey, R. Meurisse, W.H. McNab, J.E. Keys, Jr., W.E. Russell, T. Brock, and G. Nowacki. 1994. Ecoregions and subregions of the United States. In Bailey, R.G., P.E. Avers, T. King, and W.H. McNab (eds.). USDA forest service. Map published by US Geological Survey.

Ohmann, J.L., W.C. McComb, and A.A. Zumrawi. 1994. Snag abundance for primary cavity-nesting birds on nonfederal forest lands in Oregon and Washington. *Wildl. Soc. Bull.* 22: 607–619.

Ohmann, J.L., and M.J. Gregory. 2002. Predictive mapping of forest composition and structure with direct gradient analysis and nearest neighbor imputation in coastal Oregon, USA. *Can. J. For. Res.* 32: 725–741.

Omernick, J.M. 1995. Ecoregions: a framework for managing ecosystems. *George Wright For.* 12: 35–51.

Probst, J.R., and T.R. Crow. 1991. Integrating biological diversity and resource management. *J. For.* 89: 12–17.

Scott, J.M., F. Davis, B. Csuti, R. Noss, B. Butterfield, C. Groves, H. Anderson, S. Caicco, F. D'Erchia, T.C. Edwards Jr., J. Ulliman, and G. Wright. 1993. Gap analysis: a geographic approach to protection of biological diversity. *Wildl. Monogr.* 123.

Scott, J.M., F.W. Davis, R.C. McGhie, R.G. Wright, C. Groves, and J. Estes. 2001. Nature reserves: do they capture the full range of America's biological diversity? *Ecol. Applic.* 11: 999–1007.

Spies, T.A., G.H. Reeves, K.M. Burnett, W.C. McComb, K.N. Johnson, G.E. Grant, J.L. Ohimann, S.L. Garman and P. Bettinger. 2002. Assessing the ecological consequences of forest policies in a multi-ownership province in oregon. In J. Liu and W.W. Tayler, eds. Integrating Landscape Ecology into Natural Resource Management. *Cambridge Univ. Press*, Cambridge, U.K. Pages 179–207.

Spies, T.A., K.N. Johnson, K.M. Burnett, J.L. Ohmann, B.C. McComb, G.H. Reeves, P. Bettinger, D.J. Miller, J.D. Kline, and B. Garber-Yonts. 2007a. Assessing forest policies in the Coastal Province of Oregon: and overview of biophysical and socioeconomic responses. *Ecol. Applic.* (in press).

Spies, T.A., B.C. McComb, R. Kennedy, M. McGrath, K. Olsen, and R.J. Pabst. 2007b. Habitat patterns and trends for focal species under current land policies in the Oregon Coast Range. *Ecol. Applic.* 17: 48–65.

Tear, T.H., P. Kareiva, P.L. Angermeier, P. Comer, B. Czech, R. Kautz, L. Landon, D. Mehlman, K. Murphy, M. Ruckelshaus, J.M. Scott, and G. Wilhere. 2005. How much is enough? The recurrent problem of setting measurable objectives in conservation. *BioScience* 55: 835–849.

TNC (The Nature Conservancy). 2001. *Ecoregions of North America. The Nature Conservancy.* Western Conservation Science Center, Boulder, CO.

Weins, J. 1989. Spatial scaling in ecology. *Funct. Ecol.* 3: 385–397.

Wisdom, M.J., R.S. Holthausen, B.C. Wales, C.D. Hargis, V.A. Saab, D.C. Lee, W.J. Hann, T.D. Rich, M.M. Rowland, W.J. Murphy, and M.R. Eames. 2000. *Source Habitats for Terrestrial Vertebrates of Focus in the Interior Columbia Basin: Broad-Scale Trends and Management Implications.* USDA Forest Service General Technical Report PNW-GTR-485.

Zuckerberg, B., C.R. Griffin, and J.T. Finn. 2004. *A Gap Analysis of Southern New England: An Analysis of Biodiversity for Massachusetts, Connecticut, and Rhode Island.* Final contract report to USGS Biol. Resour. Div., Gap Analysis Program, Moscow, Idaho.

16 Viable Populations in Dynamic Forests

I have heard some foresters and biologists say, "We manage habitat, not populations." The *Field of Dreams* approach to wildlife management: build it and they will come. But will they? Can they given their mobility and intervening conditions now and 50 years from now? And if they do come to the new habitat, how long will they last there? And how does their presence contribute to ensuring that the greater population is not going extinct? All of these questions relate to understanding population viability.

Population viability analysis (PVA) is a structured approach to examining population performance, to link population performance with the quantity, quality, and distribution patterns of habitat, and to predict population extinction or persistence. The approach is the marriage of the concepts of demography, population dynamics, habitat selection, and landscape dynamics. Many of the concepts covered in this book to this point are brought to bear on assessment of population viability in forested landscapes. The current techniques have been developed in response to the concern for the persistence of small or isolated populations, a response to law (The Endangered Species Act and The National Forest Management Act), and large-scale regional assessments. These techniques are both elegant and uncertain. PVA is a modeling exercise and, as some have said, "All models are wrong, but some are useful." The results from a PVA can be useful in guiding management alternatives if uncertainty is understood and included in the decision-making process.

EXTINCTION RISKS

It's a hard world out there. As far as we know, all populations will eventually go extinct (including ours!). Large populations will likely last for many generations; small populations are much more vulnerable to extinction. The crux of the population change issue is encapsulated in this formula:

$$N_{(t+1)} = \lambda N_{(t)},$$

which states that the population size at time $t + 1$ (the future) is equal to a change coefficient λ multiplied by the current population $N_{(t)}$. Should λ be a negative number over a long time (long relative to the generation length for the species), then concerns arise regarding the potential for the species to become extinct. At least four factors likely influence the risk of extinction, especially in small populations.

First, vital rates of populations vary from year to year and place to place simply by chance alone. Birth rates, death rates, and reproductive rates within an age class typically have a mean value over time in a stable population, but there is also variability associated with that mean value. In some species, that variability is small (species with low but predictable reproduction such as bears and humans), and for others it is quite large (species with irruptive populations such as voles and quail). These levels of variability in vital rates apply irrespective of whether the population is large or small, but in small populations the effect can be quite dramatic. A failure to reproduce for a year (or more, depending on the life span) in a population of 1000 voles may mean a temporary crash followed by recovery, but a similar event in a population of 10 animals may lead to a population decline from which it cannot recover due to random shifts in sex ratios and the difficulty in finding mates.

This *demographic stochasticity* is an inherit property of populations and the variability represented in these vital rates is likely an evolutionary response to unpredictable events of the past.

Second, there are also exogenous events that interact with habitat elements to influence the likelihood that an animal will die, reproduce, or move. Climatic changes from year to year, for instance, can have a significant effect on population change over time (Olson et al. 2004). For instance, an exceptionally rainy year can cause nest failure for some bird species. There is some probability that a very wet year will occur, but whether it happens or not is a chance event. There is a probability, but the timing of the actual event is uncertain. These uncertain events, unrelated directly to habitat structure and composition, represent *environmental stochasticity*. Random events such as storms, droughts, and epizootic diseases that do not affect habitat structure and function cause vital rates to fluctuate considerably from year to year and place to place. In small populations, these effects can be magnified. An event that increases mortality by 50% in a population of 1000 animals for 1 year may simply result in population recovery from the remaining 500 individuals over the next few years. But, a similar event in a population of 10 individuals might result in an extremely skewed sex ratio by chance alone, causing the population to go extinct.

Third, *natural catastrophes* are extreme cases of environmental uncertainty such as hurricanes, fires, and epizootics that can cause massive changes in vital rates unrelated to habitat structure and composition. When Hurricane Hugo blasted the Francis Marion National Forest, the majority of nesting cavities for red-cockaded woodpeckers were destroyed (Hooper et al. 1990). These sorts of events cause fluctuations in vital rates that far exceed the expected year-to-year variability seen in most populations, and can have devastating consequences for small populations. Dennis et al. (1991) predicted a reasonably high probability of population persistence for Puerto Rican parrots when not accounting for hurricanes, but Hurricane Hugo nearly decimated the population, potentially changing the probability of recovery for this species.

Fourth, in small populations, chance has a huge effect on otherwise subtle changes in genetic materials present in the population. In many instances, the variety of gene expressions, or *alleles*, represents the potential adaptability of the species to environmental uncertainties. In large populations in dynamic environments, we would expect to see some reasonably high level of heterozygosity (genetic variability) in the population. As populations decline, by chance alone, some alleles may dominate in a population, leading to a preponderance of individuals being homozygous (genetically more uniform) for some traits. This phenomenon is called *genetic drift*. These changes may make many of the individuals less adaptable to environmental uncertainty associated with those traits. In small populations this genetic shift can result from increased levels of inbreeding by closely related individuals. In cases where inbreeding leads to reduced fitness in the populations, *inbreeding depression* occurs. In extreme cases, alleles may be entirely eliminated from populations as they decline in abundance. Regaining that genetic variability would occur either through immigration from surrounding populations (if there are any) or from mutations (most of which are not beneficial). Restrictions in genetic expressions can also be seen as animals disperse to unoccupied patches of habitat and establish a new population. As a population beginning from just a few individuals is more likely to have a narrow range of alleles in the newly establishing population, the genetic variability in the population can remain narrow as the population grows, a process termed the *founder effect*. A similar process can occur when a population goes through a *genetic bottleneck* or a dramatic decline in the population to low levels where alleles are lost in the process. During population recovery following the decline, a narrower range of allelic expression may be seen. Founder effects and genetic bottlenecks can lead to populations that are larger (i.e., recovered) but more vulnerable to environmental uncertainties because the alleles needed to cope with those uncertainties had been lost.

Because these four factors (demographic stochasticity, environmental stochasticity, natural catastrophes, and loss of genetic variability) are exacerbated in small populations, conservation biologists and forest planners spend considerable effort ensuring recovery of small populations. Population viability analyses are a set of tools available to planners to address these issues when investigating the potential effects of alternative management strategies.

GOALS OF PVAs

There are two major goals usually associated with a PVA: (1) to predict short- or long-term rates of change and (2) to predict the likelihood of extinction (Beissinger and Westphal 1998). At the very least a manager would like to know what the likely trend in a population might be under various management scenarios or forest plans. Knowing that a population is decreasing or increasing can influence the direction of future management efforts if the species is listed as threatened or a potential pest. But if the decline is so severe that questions are raised regarding the potential for a population to go extinct, then a new goal emerges for the analysis.

Population viability analysis is an approach often used to predict the probability that a population will go extinct in a given number of years, for instance, to estimate a 95% probability of extinction within 1000, 100, or 10 years. The assumption behind these analyses is that, the shorter the potential time to extinction, the greater is the risk of losing the species, simply because the actual time to extinction can never be predicted given all the uncertainties associated with an extinction event.

Most PVA models are demographic in nature, employing age-specific fecundity and survival rates, but genetic implications have been considered in some models as well, especially when managing endangered species. Genetic PVA models incorporate estimates of effective population size, which is the number of breeding individuals effectively contributing to allele frequencies in a population. The effective population size is usually smaller than the absolute population size unless breeding is random and the chance of inbreeding is negligible, two assumptions likely to be violated in small populations. From the standpoint of potentially losing alleles in a population, especially in small populations, effective population sizes must be considered.

PVA MODELS

There are five predominant types of demographic models that have been used in analyses of population trends and extinction analyses (Beissinger and Westphal 1998). First, analytical models have been used to examine behavior of a system (or assumptions behind other models) and not usually used to make population predictions (Beissenger and Westphal 1998). Analytical approaches also have been used to simply relate populations to current and likely future conditions using statistical techniques such as regression, logistic regression, or classification and regression tree analyses. These approaches may be used to examine how a population might react to a change in abundance of habitat or connectivity or other factors that do not explicitly take into consideration the demographics of the population or its movement capabilities. Analytical approaches such as these may in fact miss changes in populations that could be caused by unrelated changes in birth or death rates. Developing associations does not prove a cause-and-effect relationship, but only that several things are related to one another in some way. For instance, using a regression relationship between number of wood-peckers and number of snags would be useful for some planning processes, unless some other factor, such as West Nile virus, caused the population to decline independent of snag density. Knowledge of birth, death, and survival rates can help to consider both density-dependent and density-independent causes of population change.

A second type of analysis that could be used to assess population trends is a deterministic single-population model. These models are generally based on a Leslie Matrix or a matrix of survival rates and reproduction rates in each of several age classes to predict change in populations over years (Leslie 1945). In this approach, estimates of birth and death rates by age class are developed from field data such as banding returns or radio telemetry data. These estimates are used to calculate survival rates in conjunction with information on reproduction in each age class to calculate population changes from one time step to another. This approach is among the simplest models requiring the least amount of data, but it assumes that demographic rates are constant (Caswell 2001). This kind of model has been used to assess changes in marbled murrelet populations for example (Beissinger 1995).

Stochastic single-population models overcome the assumption of constant demographic rates and include variability in estimated demographic rates in the calculations. Interestingly, this approach was adapted from Leslie models and first used in development of forest management models (Usher 1969). Estimates of variability derived from field data are incorporated into birth and death rates (and hence survival probabilities) and reproduction rates to allow multiple population projections. Because these variances represent the stochastic properties of population change, each projection produces a unique trajectory and ending population size. These can then be averaged or summarized to develop confidence intervals around population trends and probabilities associated with extinctions. This is a quite commonly used approach because it attempts to introduce reality into projections and has been used for species such as brushtail possums in managed forests of Australia (Lindenmayer et al. 1993). Both deterministic and stochastic population models are typically applied to single populations. When a population is segregated spatially into interacting subpopulations, a different approach is usually taken.

Metapopulation models are used to assess the interacting dynamics among subpopulations representing a metapopulation structure. The metapopulation dynamics are incorporated into the demographic model by using patch-specific demographic rates and dispersal probabilities between the patches (Beissinger and Westphal 1998). Dispersal rules are developed based on patch size and distances between patches. Patch quality, usually indexed to a carrying capacity, can be assumed to vary among subpopulations as well. This approach has been broadly used with many species that are assumed to have a metapopulation structure. For instance, Beier (1996) used a metapopulation model to assess cougar population trends in the western United States. When the condition between the patches becomes important to population trends and the configuration and dynamics of the patch conditions is important, spatially explicit approaches become more useful in assessing risks to species.

Spatially explicit models are designed to consider population dynamics on complex landscapes with varying matrix conditions between patches. These approaches also lend themselves well to assessing the interacting dynamics of populations and underlying vegetation conditions. The approach incorporates a spatial distribution of resources related to habitat quality as well as movement rules for dispersing animals to assess responses to land-use changes or management policies (Wilhere and Schumaker 2001). Because of the spatial detail needed and the dynamics associated with underlying resource layers, these sorts of models typically require enormous amounts of data, and so are most often used with species representing a very high risk of loss and associated very high economic importance. Most recently these approaches have been used for northern spotted owls (Lamberson et al. 1994), but also in a broader capacity for a suite of species in a large ecoregional assessment (Schumaker et al. 2004).

CONDUCTING A PVA FOR A FOREST-ASSOCIATED SPECIES

Because forests are inherently dynamic due to disturbance and regrowth, typically a dynamic, spatially explicit PVA model is used. The Program to Assist in Tracking of Critical Habitat (PATCH) is an example of the type of model typically used for these sorts of analyses (Wilhere and Schumaker 2001). There are eight primary steps to conducting such an analysis using PATCH (other models require similar, but not identical, steps). First, maps of habitat quality for the species over the analysis extent must be developed. This usually entails application of a wildlife habitat relationships model (WHR; Johnson and O'Neil 2001) but may entail a more detailed habitat quality estimate (McComb et al. 2002). The maps must be developed both now and for each time step into the future over a period of time deemed adequate to assess population trends. These maps should be based on likely changes in forest structure and composition caused by management and natural disturbances expected to occur over the projection period. If you wish to compare population trends among alternative

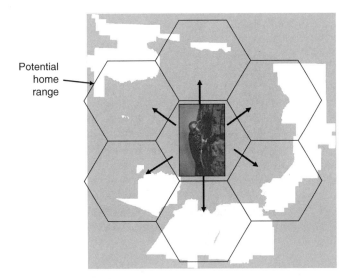

FIGURE 16.1   Simulated home ranges (hexagons) are overlain on maps of habitat availability to estimate the distribution of potentially occupiable home ranges over the extent of the analysis.

plans or policies, then you will need a different series of maps for each alternative management strategy being assessed.

Second, home ranges or territories are assigned to each map. These are usually represented by cells (squares or hexagons) of a size typical of the home range for the species, and overlain on the habitat maps (Figure 16.1). The habitat quality is then aggregated through rules representing the amount of habitat of a certain quality that would be needed by a species in its home range to likely achieve a certain level of reproduction or survival. This is done for each in the series of maps for each management alternative.

Next, demographic information for the species must be estimated for each age and sex class (although analyses are often restricted to females, and then a sex ratio is estimated to extrapolate to a total population). Age classes may be aggregated into stages if the parameters do not change appreciably from one year class to another. These parameters include both the estimated average and associated variance for birth rates, reproduction rates, movement rates, and movement direction (if it is not random). These data are typically extracted from published studies, although for high-priority species field data collection may be needed to ensure more accurate estimates of these parameters. At the very least, experts on the species are consulted to provide reasonable estimates.

These demographic data are then explicitly related to the habitat quality estimates assigned to the home range cells such that lower birth and survival rates occur in lower-quality patches and vice versa (Figure 16.2). This must be done for each cell on each map in the time series for each management alternative. The actual assignment of a value (e.g., survival) to a cell is typically conducted in a randomized manner. For instance, the survival values assigned to a cell in a given year will reflect the range of values associated with that parameter in that habitat quality class so that demographic stochasticity is represented. But these assignments are made so that the average value among all cells in a habitat quality class is equal to the mean of the published estimates for that habitat quality class.

At this point, individuals or pairs of animals are assigned to each cell on the landscape at time $t = 0$ (current conditions) such that they represent the known or estimated distribution of organisms over the area of assessment. If the distribution is not known, then a random assignment is made. The landscape is now "seeded" with individuals of various age classes and sexes in home ranges consisting of varying habitat quality. The survival, reproduction, and movement rates associated with each cell are then applied to each individual on the landscape and projected forward one time step, much the way a Leslie Matrix projects a population forward in time. At the same time, the underlying map of

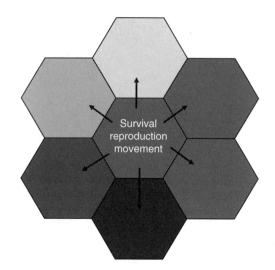

FIGURE 16.2 Demographic parameters are assigned to each home range (hexagon) based on its likely habitat quality. Should an individual arrive at a home range, then a random draw of parameters from the range of those associated with the home range quality are assigned to that individual in that time step.

habitat quality is changed to reflect forest disturbance and regrowth. This process is repeated for the number of time steps in the projection period. But because assignments of demographic parameters represent a stochastic process, many projections must be made to understand probabilities associated with population trends and extinctions. Once many projections have been made, then averages, probabilities, and confidence intervals can be assigned at each time step (Figure 16.3). The resulting estimate might show a declining trend and an estimated time to extinction of 150 years. Should you believe this estimate? Perhaps, in a relative sense, by comparing this estimate to estimates produced for alternative management plans (Figure 16.4). There are so many assumptions and uncertainties represented in each estimate that it is most useful to compare estimates at each time step between alternative management approaches (e.g., following NEPA, a no action and a preferred alternative). Indeed the best use of PVAs is in making projections to compare alternatives, selecting a preferred alternative, implementing it, monitoring the responses, and rerunning the PVA with additional data, allowing continual adaptation and improvements to population estimates. Using this process, we also learn about the habitat and demographic characteristics of the species that seem to most influence population change, and can address those factors during planning and management activities.

EXAMPLES OF PVA ANALYSES

GRIZZLY BEAR

One of the earliest attempts at conducting a PVA was for grizzly bears in Yellowstone National Park (Shaffer 1983). Subsequent approaches proposed by Dennis et al. (1991) and summarized by Morris et al. (1999) provide an estimate of persistence of this large predatory mammal in the lower 48 states of the United States.

The approach taken by Dennis et al. (1991) is somewhat unique in that the data on which the analysis is based includes exhaustive counts of individuals in the population including age and sex cohorts. Despite such exhaustive data, the approach is still based on a number of assumptions using multiple stochastic single-population modeling approaches (Morris et al. 1999):

1. The year-to-year variation in the counts reflects the true magnitude of environmentally driven variation.

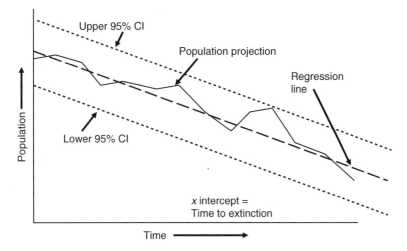

FIGURE 16.3   The results from multiple model runs are used to create an average and confidence intervals (CI) in populations over time. With declining populations, calculating the x intercept can provide an estimate of the time to extinction.

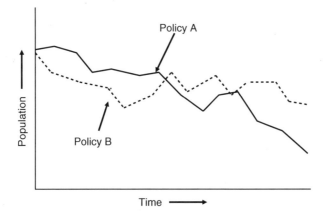

FIGURE 16.4   Trend lines are compared between alternative management strategies or policies. Those policies likely to provide the most options for the future (longest time to extinction) are assumed to represent less risk to the species.

2. Interannual environmentally driven variation is not extreme.
3. The population growth rate is not density dependent.

Dennis et al. (1991) calculated the extinction time for the population of grizzly bears in the Greater Yellowstone ecosystem. They then extrapolated some of the detailed data from the Yellowstone population to other areas including the Selkirk Range in British Columbia (BC), Northern Divide in Washington and BC, and the Yaak River Valley in Montana to understand the interacting probabilities of population extinction among these known subpopulations of grizzly bears. The analysis assumes that there is no current movement among the subpopulations, and hence single-population analyses were conducted rather than a metapopulation analysis. They found that for the 500-year projection, there was a 9.6% chance of extinction of all populations (Morris et al. 1999). Considering the populations individually, protecting the Yellowstone population provided the greatest opportunity for reducing extinction risk, but adding the Northern Divide population decreased the probability of extinction from 0.342 (for the Northern Divide alone) to 0.134 (Figure 16.5; Morris et al. 1999).

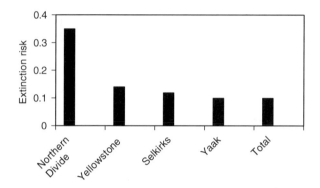

FIGURE 16.5   Extinction risk for cumulative grizzly bear populations among a collection of protected areas. (Redrafted from Morris, W. et al. 1999. *A Practical Handbook for Population Viability Analysis*. The Nature Conservancy, Arlington, VA. With permission from The Nature Conservancy.)

By adding additional subpopulations, the benefits to reductions in extinction decrease only modestly from that point on (Figure 16.5). So if we wished to be sure that grizzly bears were to remain a component of U.S. ecosystems in the lower 48 states, then maintaining all of these populations would be a reasonable approach, despite concerns over human safety and property damage. But in this example, the estimate of likely extinction (9.6%) is lower than what we might actually expect because those populations that are closest to one another are likely to be affected by similar factors influencing their populations (e.g., weather, disease, fire, etc.). But even these simple analyses can be used to understand the potential risks of extinctions if certain local populations are lost.

Marbled Murrelet

McShane et al. (2004) described a stochastic metapopulation model used to assess marbled murrelet persistence within six different zones in the Pacific Northwest. They developed a "Zone Model" in which they estimated population projections for each of six zones within the species geographic range in California, Oregon, and Washington for 100 years into the future. They assumed that there would be no change in population vital rates over 40 years, and they did not incorporate habitat changes into the model. Similarly, they did not incorporate possible effects of oceanic regime shifts (which could affect foraging efficiency and hence survival and reproduction) into the model. The model is a female-only, multi-aged, discrete-time stochastic Leslie Matrix model (Caswell 1989). They found that all zone populations are in decline (over 40 years) with declines of 2.1 to 6.2% per decade (Figure 16.6). Further, they predicted an extinction within 40 years in two zones and within 100 years in three zones (only one zone population was predicted to extend beyond 100 years) and that the probability of extinction over 100 years is 16%. By modifying their model parameters, they found a reduced rate of decline in two zones if oil spills and gill nets were eliminated as a source of mortality, pointing out the need to consider not just habitat-mediated effects on population changes.

Northern Spotted Owl

Noon and McKelvey (1996) summarized a spatially explicit dynamic landscape model used to compare population viability for northern spotted owls among 10 policy alternatives during the Forest Ecosystem Management and Assessment Team's (FEMAT 1993) planning exercise. Of the 10 policy options, Option 7 had the smallest reserve area and the least stream protection; Option 1 was the most restrictive, with the preferred option (Option 9) fitting approximately midway between 1 and 7. These initial analyses assumed no regrowth of suitable habitat over the projection horizon (Figure 16.7a through 16.7d). Projections of Option 9 revealed a likely smaller and more disjunct owl population for Oregon than under Option 1 (Figure 16.7b), but Option 9 retained large blocks of habitat within

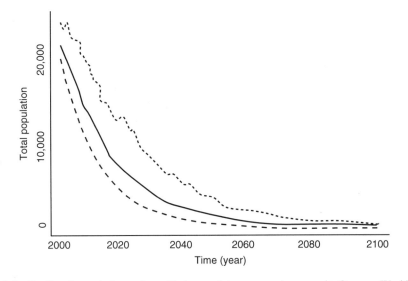

FIGURE 16.6 Predicted populations of marbled murrelets among six zones in Oregon, Washington, and California. Solid line is the mean, dashed lines indicate confidence intervals. (From McShane, C. et al. 2004. Evaluation report for the 5-year status review of the marbled murrelet in Washington, Oregon, and California. Unpublished report. EDAW, Inc. Seattle, Washington, D.C. Prepared for the U.S. Fish and Wildlife Service, Region 1. Portland, OR.)

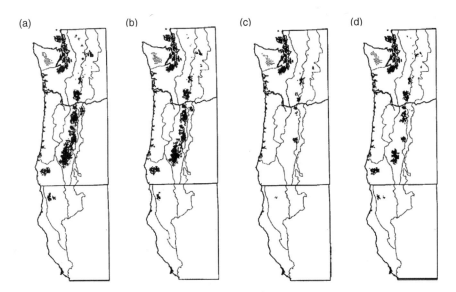

FIGURE 16.7 Modeled pair occupancy (dots represent locations occupied at a rate >70% during simulation runs of 100 years) if all current habitat were retained (4a), and assuming implementation of FEMAT Options 1, 7, and 9 (b, c, and d, respectively). (From Noon, B.R., and K.S. McKelvey. 1996. *Ann. Rev. Ecol. Syst.* 27: 135–162. With permission from Annual Reviews.)

the Cascades, and more stable populations were expected in these blocks (Figure 16.7d). Option 7 (Figure 16.7c) was predicted to support much lower populations except in the Olympic Peninsula (where virtually all federal land is reserved under all options). Hence, Option 9 was selected as the preferred alternative to provide a reasonable protection for owls while allowing a modest level of timber extraction.

MODEL ERRORS AND UNCERTAINTIES

Models are abstractions of the real world developed to aid in decision making and to allow managers and scientists to understand complex interacting systems. All model results are incorrect, and we often do not know how incorrect they really are. Independent data are needed to verify model predictions, but independent data (data not used to parameterize the model) showing adverse responses of populations of rare species to management actions are not only unlikely but they are often illegal! So those using the results from models such as PVAs should proceed with caution and be fully aware of the potential errors in the models and results. There are four dominant cases of errors that result in uncertainty in PVA model results (Beissinger and Westphal 1998).

POOR DATA

Population viability analysis models are often described as "data hungry." They require a broad suite of demographic information on a variety of habitat types ideally collected over many years to adequately represent inherent demographic and environmental stochasticity. Hence, data should be broadly representative, robust, and unbiased. Such data can be very costly, and data collection requires strict protocols and large sample sizes. The northern spotted owl is often used as the example of the cost associated with gaining adequate information over its geographic range to enable credible PVA projections. Millions of dollars have been spent on data collection for this species and millions of dollars are involved in the product values associated with decisions driven in part by PVA analyses. Rarely do we have the amount and quality of data for other species that we have for spotted owls. For many species the vital rates needed to parameterize a PVA are gleaned from the literature and usually from studies conducted not specifically designed to provide these parameter estimates. For the vast majority of species, PVA projections are better thought of as hypotheses than of estimates. If the estimates are used in decision making, then a set of assumptions and caveats regarding data quality should accompany the population projections.

DIFFICULTIES IN PARAMETER ESTIMATION

Consider a species' life history from birth to death. Throughout life there are events that influence individuals and populations. One critical part time in an individual's life that has an influence on survival is the period immediately after it leaves a nest or den. It is a juvenile, naïve to the world, and vulnerable to many factors. Survival during this juvenile stage is a parameter that often emerges as one that greatly influences population dynamics, but it is also one of the most difficult parameters to estimate (Wiens et al. 2006), especially for species that are too small for radio transmitters to be attached to them or species that are migratory or disperse widely. In these cases, the technical aspects of PVA modeling may far outweigh our technical capabilities of collecting the data needed to parameterize the models.

WEAK ABILITY TO VALIDATE MODELS

Although there are independent data that can be used to validate various components of a PVA model (e.g., occupancy rates, survival in some age classes, and sex ratios), there are few and often no data available to validate the projections for models. Indeed, due to lag times associated with population responses to management alternatives, some of which may be novel, it will likely take decades or centuries before monitoring data can provide independent tests of model performance. Fortunately, when monitoring data are collected, the resulting data can be used to continually improve model performance and accuracy. Oftentimes, though, funding available to develop recovery plans or management plans may not continue as needed over time to acquire the necessary monitoring information.

EFFECTS OF ALTERNATIVE MODEL STRUCTURES

There are several widely used PVA models available now and there will likely to be more in the future, each attempting to include more reality into the model abstractions (Gordon et al. 2004). Each model structure has its own set of strengths and weaknesses with regard to dealing with potential effects of different types of forest management, stochastic events, animal scaling properties and movements, competitors, predators, diseases, and parasites, among other factors influencing population dynamics (Beissinger and Westphal 1998). Consequently, projections resulting from one model structure may differ considerably from another. Indeed, projections of future conditions from a range of model structures conducted by different scientists may be one approach to understanding the uncertainties associated with population projections. If multiple projections using alternative model structures are somewhat consistent in their predicted extinction rates or population trends, then perhaps more faith can be placed in the results, given the caveats presented in the previous three limitations of all models.

INTERPRETING RESULTS FROM PVA PROJECTIONS

Given the caveats regarding uncertainty and potential errors described in the previous section, it is important to consider how best to use the results of population viability analyses. First and foremost, do not believe the predictions, at least not with the precision that is often implied by the projections. Predicting the future is easy, predicting it reliably is impossible. Think of weather predictions for instance. Predicting the weather a week or more into the future (*The Farmer's Almanac* not withstanding) is much less reliable than predicting it tomorrow, or in the next hour or next minute. The farther into the future the predictions are made, the less reliable the predictions are likely to be. Long-term predictions may be useful to understand factors such as lag effects and stochastic events, but should be viewed with considerable caution. Hence, PVA projections are best used over relatively short time periods when making decisions, while using the long-term projections as a context for near-term decisions.

The medieval philosopher William of Occam once stated, "One should not increase, beyond what is necessary, the number of entities required to explain anything" (known as Occam's razor). "Keep it simple stupid" (the Kiss principle) is the saying that my high school math teacher espoused. Do not use complex models when simpler, easier-to-comprehend models will provide you with estimates that are of value in decision making. Of course, there is always a trade-off, because populations and ecosystems are inherently complex. The natural tendency of decision makers is to use the models that best approach reality (most complex) even when they may not be necessary to reach an informed decision.

Model results, if used cautiously, can be used to understand relationships among the various factors influencing population change and to diagnose causes of decline and potential for recovery. These *heuristic* aspects of modeling (using modeling to teach us something about the system) may represent the most useful approach to PVA modeling in that it can help generate hypotheses regarding the factors that seem to be most likely to lead a population to extinction. These hypotheses can then be tested in field and controlled settings to allow us to hone in on key factors that can lead to a more efficient population recovery.

Trends are more important than numerical predictions because numerical predictions will change as parameters are improved. So it is best to use the projection results in a relative rather than absolute sense (Beissinger and Westphal 1998). Comparing extinction probabilities, population trends, or times to extinction among management options is more appropriate than saying that the population will go extinct in 40 to 65 years, for example. These relative comparisons become even more valuable when independent scientists using a variety of model structures all produce estimates that support (or contradict) one another. Decisions based on PVAs are usually made based on the credibility of the results, and the credibility is usually much greater in a relative than absolute sense.

Population viability analysis results are often used within a process called *risk analysis*. Risk analysis is a structured way of analyzing decisions and the potential effects of those decisions when the outcomes are uncertain. The process involves identifying options, quantifying or assigning probabilities, and evaluating and selecting management options. This approach is often used when there are multiple interest groups with different and conflicting objectives, the outcomes of management alternatives are uncertain, and any decision may have serious consequences. Risk analysis attempts to structure and quantify management options to help the decision maker understand the consequences of action or inaction and choose a decision path.

SUMMARY

Fine-filter analyses often entail an assessment of population trends and risks of extinction of species from all or a significant part of its range. These assessments are integral to the development of recovery plans for threatened species and have been conducted for a wide range of species during ecoregional assessments. There is a range of model structures used to assess future population trends representing a range of ecosystem complexities that are included in the model structures. Those that consider a dynamic landscape, particularly relative to the generation length of the species under consideration, are often selected for use with forest-associated species. Because of the complexities of the model structures, however, the results of these projections are best interpreted in a relative rather than an absolute sense to allow comparisons among management alternatives. Further, trends predicted into the near future are usually more reliable than long-term projections. Issues such as data quality, inadequate validation, and environmental uncertainties all influence the utility of these projections.

REFERENCES

Beier, P. 1996. Metapopulation modeling, tenacious tracking, and cougar conservation. In McCullough, D.R. (ed.), *Metapopulations and Wildlife Management*. Island Press, pp. 293–323.

Beissenger, S.R. 1995. Population trends of the marbled murrelet projected from demographic analyses. In Ralph, C.J., G.L. Hunt Jr., M.G. Raphael, and J.F. Piatt (eds.), *Ecology and Conservation of the Marbled Murrelet*. USDA For. Serv. Gen. Tech. Rep. PSW-GTR-152.

Beissenger, S.R., and M.I. Westphal. 1998. On the use of demographic models of population viability in endangered species management. *J. Wildl. Manage.* 62: 821–841.

Caswell, H. 1989. *Matrix Population Models: Construction, Analysis, and Interpretation*. Sinauer Associates, Sunderland, MA.

Caswell, H. 2001. *Matrix Population Models: Construction, Analysis, and Interpretation*. 2nd edn. Sinauer Associates, Sunderland, MA.

Dennis, B., P.L. Munholland, and J.M. Scott. 1991. Estimation of growth and extinction parameters for endangered species. *Ecol. Monogr.* 61: 115–143.

Forest Ecosystem Management Assessment Team (FEMAT). 1993. *Forest Ecosystem Management: An Ecological, Economic, and Social Assessment*. U.S. Government Printing Office: 1993-793-071.

Gordon, S.N., K.N. Johnson, K.M. Reynolds, P. Crist, and N. Brown. 2004. Decision support systems for forest biodiversity: evaluation of current systems and future needs. Final Report Project A 10. National Commission on Science and Sustainable Forestry.

Hooper, R.G., J. Watson, and E.F. Escano. 1990. Hurricane Hugo's initial effects on red-cockaded woodpeckers in the Francis Marion National Forest. Trans. No. Amer. Wildl. Natur. Resour. Confer., Wildlife Manage. Inst., Washington, D.C. 55: 220–224.

Johnson, D.H., and T.A. O'Neil, eds. 2001. *Wildlife–Habitat Relationships in Oregon and Washington*. Oregon State University Press, Corvallis, OR.

Lamberson, R.H., B.R. Noon, and K.S. McKelvey. 1994. Reserve design for territorial species — the effects of patch size and spacing on the viability of the Northern spotted owl. *Conserv. Biol.* 8: 185–195.

Leslie, P.H. 1945. On the use of matrices in certain population mathematics. *Biometrika* 33: 183–212.

Lindenmayer, D.B., R.C. Lacy, V.C. Thomas, and T.W. Clark. 1993. Predictions of the impacts of changes in population size and of environmental variability on Leadbeater's Possum, *Gymnobelideus leadbeateri* McCoy (Marsupialia: Petauridae) using population viability analysis: an application of the computer program VORTEX. *Wildl. Res.* 20: 67–86.

McComb, W.C., M. McGrath, T.A. Spies, and D. Vesely. 2002. Models for mapping potential habitat at landscape scales: an example using northern spotted owls. *For. Sci.* 48: 203–216.

McShane, C., T. Hamer, H. Carter et al. 2004. Evaluation report for the 5-year status review of the marbled murrelet in Washington, Oregon, and California. Unpublished report. EDAW, Inc. Seattle, Washington, D.C. Prepared for the U.S. Fish and Wildlife Service, Region 1. Portland, OR.

Morris, W., D. Doak, M. Groom, P. Kareiva, J. Fieberg, L. Gerber, P. Murphy, and D. Thomson. 1999. *A Practical Handbook for Population Viability Analysis.* The Nature Conservancy, Arlington, VA.

Noon, B.R., and K.S. McKelvey. 1996. Management of the spotted owl: a case history in conservation biology. *Ann. Rev. Ecol. System.* 27: 135–162.

Olson, G.S., E. Glenn, R.G. Anthony, E.D. Forsman, J.A. Reid, P.J. Loschl, and W.J. Ripple. 2004. Modeling demographic performance of northern spotted owls relative to forest habitat in Oregon. *J. Wildl. Manage.* 68: 1039–1053.

Schumaker, N., T. Ernst, D. White, J. Baker, and P. Haggerty. 2004. Projecting wildlife responses to alternative future landscapes in Oregon's Willamette Basin. *Ecol. Appl.* 14: 381–401.

Shaffer, M.L. 1983. Determining minimum viable population sizes for the grizzly bear. *Int. Conf. Bear Res. Manage.* 5: 133–139.

Usher, M.B. 1969. A matrix model for forest management. *Biometrics* 25: 309–315.

Wiens, J.D., B.R. Noon, and R.T. Reynolds. 2006. Post-fledgling survival of northern goshawks: the importance of prey abundance, weather and dispersal. *Ecol. Appl.* 16: 406–418.

Wilhere, G., and N.H. Schumaker. 2001. A spatially realistic population model for informing forest management decisions. In Johnson, D.H., and T.A. O'Neil (eds.), *Wildlife–Habitat Relationships in Oregon and Washington.* Oregon State University Press, Corvallis, OR.

17 Monitoring Habitat Elements and Populations

Implementation of any stand prescription or forest management plan is done with some uncertainty that the actions will achieve the desired results. Nothing in life is certain (except death!). Managers should expect to change plans following implementation based on measurements taken to see if the implemented plan is meeting their needs. If not, then midcourse corrections will be necessary. Many natural resource management organizations in North America use some form of adaptive management as a way of anticipating changes to plans and continually improving plans (Walters 1986).

ADAPTIVE MANAGEMENT

Adaptive management is a process to find better ways of meeting natural resource management goals by treating management as a hypothesis. The results of the process also identify gaps in our understanding of ecosystem responses to management activities. The process incorporates learning into the management planning process and the data collected during monitoring provides feedback about the effectiveness of preferred or alternative management practices. The information gained from the process can help to reduce the uncertainty associated with ecosystem and human system responses to management.

Adaptive management has been classified as both active and passive (Walters and Holling 1990). *Passive adaptive management* is a process in which the "best" management option and associated actions are identified, implemented, and monitored. The monitoring may or may not include unmanaged reference areas as points of comparison to the managed areas. The changes observed over time in the managed and reference areas are documented and the information is used to alter future plans. Hence, the manager learns by managing and monitoring, but the information that is gained from the process is limited, especially if reference areas are not used. Without reference areas we do not know if changes over time are due to management or some other exogenous factors.

Active adaptive management treats the process of management much more like a scientific experiment than passive adaptive management. Under active adaptive management, management approaches are treated as hypotheses to be tested. The hypotheses are developed specifically to identify knowledge gaps and management actions are designed to fill those gaps. Typically, the hypotheses are developed following modeling of the system responses (e.g., using forest growth models or landscape dynamics models) to understand how the system might respond and then use management to see if it responds as intended. Reference areas are used as controls to test responses of ecosystems and human systems to management. By collecting monitoring data in a more structured hypothesis-testing framework, responses can be quantified and used to identify probabilities associated with achieving desired outcomes in the future. Whereas passive adaptive management is somewhat reactive in approach (reacting to monitoring data), active adaptive management is proactive and follows a formal experimental design.

Adaptive management generally consists of six major steps (Figure 17.1). First the problem is assessed both inside and outside the organization. Public involvement in the process from the very beginning is the key to identification of points of concern and uncertainty. With information in hand

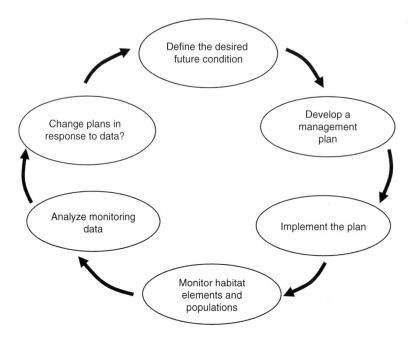

FIGURE 17.1 The adaptive management cycle is typically used to improve information used to make better management decisions.

from a series of listening sessions, a plan is designed that is considered to be the preferred or best plan among several alternative plans. Typically, reference areas are also identified to use as points of comparison. The plan is then implemented and various aspects of the plan are monitored to learn from the management actions.

Implementation monitoring is conducted to see if the plan is being implemented on the ground as it is described in the plan. Are standards and guidelines being followed? Are the appropriate number of snags, logs, and trees being left after harvest? Are harvest boundaries being respected? Are designated skid trails used? Answering these sorts of questions is important, because future stand or landscape conditions are often dependent on correct implementation of the plan. If the plan is not implemented correctly, then there is far greater likelihood that the goals of the plan will not be realized.

Effectiveness monitoring follows implementation monitoring and is designed to determine if habitat elements, populations, or processes are responding as expected and effectively achieving your management goals. Are trees growing as anticipated? Are trees and shrubs producing mast? Are focal species populations persisting and growing as anticipated? Answering these questions allows you to know if the management is effective and, if not, the results provide evidence for making changes to the plan.

Plans are almost always based on some assumptions. Monitoring of key processes such as tree and shrub growth rates, population changes, or changes in animal fitness can produce information that can be used to test assumptions. *Validation monitoring* provides the basis for reducing uncertainty associated with assumptions and provides a framework for understanding and interpreting the results of effectiveness monitoring.

With data in hand from implementation, effectiveness, and validation monitoring, managers can periodically evaluate the responses of the system to the management actions, understand better why they are seeing certain responses, and make adjustments to their management actions in a way that increases the likelihood that they will be more effective in the future. Once these adjustments have been made it is important to reassess the problem among involved publics to ensure that new concerns and opportunities are identified as a result of making the proposed adjustments. Then the entire process begins anew and, at least theoretically, allows managers to continually improve their ability to meet desired goals.

We may monitor habitat types, habitat elements, or populations depending on the goals of the monitoring program. Clearly, it is important to know if the habitat structure, function, and patterns are developing as expected under the plan. Managers may also want to know how populations respond to management. Monitoring populations can be critical in assessing and identifying the potential or actual impacts of management on the persistence of a species over some or all of its geographic range. Ideally, the monitoring should allow managers to evaluate (either directly or indirectly) the effects of management actions on factors associated with animal fitness such as fertility, recruitment rate, survivorship, and mortality. These types of data are very expensive to acquire, and so more often monitoring of populations is based on occurrence or abundance.

A properly designed monitoring effort allows managers and biologists to understand the long-term dependency of selected species on various habitat components. Resource availability is dynamic for all or most species. Consequently, the challenge when developing a monitoring plan is to assess whether changes in occurrence, abundance, or fitness in a population is independent of or related to changes in habitat availability and quality (Cody 1985). Managers will ideally want to identify and incorporate the interactions between habitat and population change in order to make informed management decisions through an adaptive management process (Barrett and Salwasser 1982). The ability to understand why populations are changing in certain directions will depend on the species' habitat requirements and coincident monitoring of the population and the habitat. The ability to correlate animal and habitat data allows decision makers to better predict the effects of management on populations. However, monitoring plans must address two critical considerations to allow for an accurate and sound comparison of population and habitat data (Jones 1986). First, both animal and habitat data must be collected on the same site. This allows monitoring programs to document the fluctuations in population density and distribution with respect to changes in the physical and spatial arrangement of habitat elements. Second, the level of detail identified for sampling for both species and their habitats must be determined before associations can be made. This level of sampling will ultimately depend on the objective of the monitoring program. Most monitoring programs deal with presence/absence detected data for many species of rare plants and animals, and generally more detailed data are collected on habitat elements in an effort to describe habitat conditions. However, if the objective of the monitoring protocol is assessing habitat quality and its influences on a species' demography, then data must be collected on a species' ability to survive and reproduce (e.g., mortality, survivorship, predation, parasitism; Cody 1985).

DESIGNING MONITORING PLANS

The first step in the process of developing a monitoring plan is to identify, clearly and concisely, the questions to be answered by the monitoring data. The questions must be focussed. Once each question has been articulated, then the following steps should be taken.

SELECTION OF RESPONSE VARIABLES

Considering your management plan as a hypothesis, what are the *response variables*? Provide the rationale for selecting specific indicators or attributes. What is it that will be measured? Why was this indicator selected over others? What are the benefits associated with this indicator? What are the limitations? What are the key habitat elements and population responses that are described in your desired future conditions (DFC)? Desirable characteristics of indicators include (Vesely et al. 2006):

- Those with dynamics that are consistent with the element or population of interest
- Are sensitive enough to provide an early warning of change
- Have low natural variability

- Provide continuous assessment over a wide range of environmental conditions
- Have dynamics that can be easily attributed to either natural cycles or anthropogenic stressors
- Are distributed over a wide geographical area and are very numerous
- Are harvested, endemic, alien, species of special interest, or have protected status
- Can be accurately and precisely estimated
- Have costs of measurement that are not prohibitive
- Have monitoring results that can be interpreted and explained
- Are low impact to measure
- Have measurable results that are repeatable with different personnel.

DESCRIBE THE SCOPE OF INFERENCE

Data must be collected in a manner that provides an unbiased estimate of your response variable from the planning area. Samples should be allocated in a randomized or stratified random manner with points selected from a pool representing the entire *scope of inference*. The scope of inference represents the space and time over which your data can be used to assess changes in the response variable with some known level of certainty. Extrapolating data beyond the scope of inference is done with increasing uncertainty as one departs more and more from the conditions sampled from within the scope of inference. Indeed, *broadcasting* from the monitoring data (extrapolating to other units of space outside of the scope of inference) and *forecasting* (predicting trends into the future from existing trends) must be done with great care because the confidence limits on the projections increase exponentially beyond the bounds of the data.

Oftentimes managers wish to sample large areas so that the results of the monitoring effort can be used more efficiently, but they quickly face a trade-off. That trade-off is to monitor over a large spatial extent so that results are broadly applicable vs. sampling over a small area with less variability to increase the precision of the data (and more likely detect trends). The variability in the indicator likely will increase as the spatial extent of the study increases. As the variance of the indicator increases, the probability of detecting a difference between treatments or of detecting a trend over time will decrease. Funding for the monitoring program often dictates what represents a reasonable level of sampling intensity. Generally, though, smaller replicated sites from throughout a larger scope of inference can provide information that is more broadly applicable, but yet would have sufficient statistical power to detect changes.

DESCRIBE THE EXPERIMENTAL DESIGN

The experimental design will depend on the goals for the monitoring program. Consider how the analysis will be used. Will the analysis be used to assess occurrence, trends, patterns, or effects? Estimating occurrence may entail some estimate of probability of occurrence at a site with an associated estimate of confidence. Estimating trends often involves a time-series regression with confidence intervals to understand both the slope of the trend and the uncertainty associated with the data. Estimating patterns may involve use of an analysis of variance (ANOVA), *t*-test, or multivariate analysis (e.g., principal components analysis) to understand if the response variable differs between or among areas having different management actions. Estimating effects of management on a response variable typically requires a before–after control-impact (BACI) approach so that we can understand the causes (management action) and effects (relative level of response) associated with our implemented plan.

Because monitoring data often are collected over time to detect trends or effects, the data often are not independent from one time period to the next. The data collected at one time are related to the conditions when data were collected at a previous time, and this violates a basic assumption when using standard statistical techniques such as ANOVA or regression. This issue can lead to the conclusion that trends exist when they really do not simply due to this temporal dependence, because the estimate of variance that is not accurate (too small), leading to a false conclusion (Hurlburt 1984). Repeated-measures analyses are often necessary to ensure that estimates of variance between or among treatments reflect this lack of independence (Foster 2001).

SAMPLING INTENSITY, FREQUENCY, AND DURATION

How many samples do I need? That is the question that is most often asked when planning a monitoring program. Use of existing data or conducting a pilot study can help address sample size questions. Oftentimes a stabilization approach is used to assess sample size. For instance, if data on animal density were collected from 20 sites extending out from some central location and the variance represented in the data is plotted over number of samples, then the variance should stabilize at some number of samples. Once that asymptote in variance has been reached, then adding additional samples is not likely to influence your estimate of the inherent variability in the response variable, at least not under current conditions. It is also useful to consider how your estimate of variance might change as the plan is implemented so that your sample intensity in the future is also adequate to address your monitoring question. A similar approach can be taken to identify the number of samples that might be needed to establish the probability of occurrence of a species at a site (Figure 17.2).

How often should I collect data? Some habitat elements and populations change very slowly (e.g., snag fall rates) over time, while others change quickly (browse biomass following a disturbance). The rates of change in the element should dictate the frequency with which monitoring data are collected. Sampling snag fall every year for 100 years is both inefficient and unnecessary given the likely changes that would be seen from year to year either in snags or the species that use them. Sampling every 5 or 10 years would provide useful information at a fraction of the cost.

How long should I monitor before I can stop? Sampling duration depends on the time that you think it would take to achieve the DFC and the time that you feel you should monitor the DFC to ensure that it is likely to persist once it is reached. By using an adaptive management approach, with continual improvements in management approaches, monitoring may continue indefinitely with some response variables being dropped, others added, and some retained as the process continues.

MONITORING HABITAT ELEMENTS

To assess the changes in habitat availability or quality following management, monitoring must measure and document appropriate habitat elements important to the key species, communities, or ecosystems of concern to the managers and associated publics (see Chapter 3 for a documentation of habitat elements). These data are collected over various scales; however, the relevant scale at which to collect and interpret these data will be defined by the characteristics of the organisms

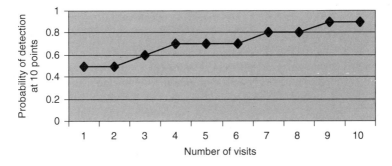

FIGURE 17.2 Hypothetical change in probability of detecting a species with increasing number of visits or samples. Note that an asymptote is reach at 9 or 10 samples. Sampling beyond this point will not likely improve the precision of your estimate.

of concern. The data should be related to one of the four levels of habitat selection described in Chapter 2: geographic range, home range, resource patch selection, or ultimate resources for each species identified in the monitoring plan.

Measurement of many habitat elements has already been incorporated into existing monitoring efforts associated with ground-plot measurements of vegetation changes (e.g., USDA Forest Service Forest Inventory and Analysis efforts) and may represent existing data that can be used in a monitoring program. A number of techniques exist for measuring habitat composition and structural conditions using ground plots. If these data are collected at permanently marked points, then it is possible to document changes in habitat components over time. This is especially important, as the carrying capacity of habitat for a species is dynamic and depends on a number of resources that can fluctuate over time and space due to natural (e.g., floods, hurricanes) and anthropogenic (e.g., silviculture) disturbances.

Remotely sensed data are being used more and more frequently in association with ground-plot information to assess patterns of change over large areas. Maps allow users to better define their sampling framework and are often used as the basis for stratifying their sampling design. Data for maps can be based on a number of sources including aerial photographs, satellite imagery, orthophotos, or topographic maps (Vesely et al. 2006).

Aerial photographs capture the reflectance of items on the earth's surface on a photographic film. Aerial photos are often used as the first step in assessing the pattern and distribution of habitat elements over large areas, but photos have distortion. *Orthophoto maps* are corrected aerial photos with topographic information superimposed. If available, these types of data can be very useful. Aerial photography is usually available in black and white, true color, and color-infrared, and each has its own advantages. Infrared photographs are particularly useful when classifying vegetation types (Vesely et al. 2006) because healthy green vegetation is a very strong reflector of infrared radiation and appears bright red on color-infrared photographs. When sampling over large areas for the purpose of assessing patterns of vegetation, dominant trees, water, and other items that can be seen from above, aerial photography is a much less expensive and more efficient technique than ground plots. But because some habitat elements are not seen from above, ground plots stratified by vegetation class are often used in conjunction with aerial photos. The utility of the photo comes with correct classification of habitat elements, and so skilled photo-interpretation is the key to successful use of aerial photos for monitoring.

Satellite imagery has become a popular method for a more crude classification of vegetation and other ground features than is possible using aerial photographs. The LANDSAT program placed satellites into orbit around the earth to collect environmental data about the earth's surface (Richards et al. 1999). As with aerial photos, the reflectance of various wavelengths of light is captured, but unlike aerial photos, satellite data are simply reflectance values (numbers) associated with discrete places (pixels) on the earth. The values are specific to a particular place at a particular time and a spectrum of reflectance values (brightness) that are assigned to each pixel (Richards et al. 1999). The area that each pixel covers on the ground represents the resolution or grain of the information and as technology continues to advance the pixel size is getting smaller and smaller (LANDSAT pixels are typically 30 m×30 m). As with aerial photos, satellite data must be classified to be of use. Once classified, resulting maps can provide a record of change in classified elements over large areas over time. Land-use change analyses are commonly conducted using satellite data; however, recently scientists have begun to integrate ground plot data with satellite data to provide more detailed estimates of change in fine-scale habitat elements (Ohmann and Gregory 2002). The accuracy of aerial-photo-interpretation and classified satellite imagery depends on ground-truthing (visiting spots on the ground to see if the class is accurately represented) and subsequent accuracy assessments.

The utility of aerial photos and satellite data is realized when the classification scheme is designed to assess habitat for the species of interest. Biologists often use a hierarchal approach to stratify vegetation types based on dominant and subdominant vegetation, landform, soil composition, or other factors that are deemed pertinent by the managers and biologists (Kerr 1986). Once a criterion is determined for classification, the landscape can be separated intro discrete units so that any additional ground samples can be stratified among vegetation types. The Resource Inventory Committee of British Columbia has outlined a good approach to vegetation stratification (Resource Inventory Branch 1998):

1. Delineate the project area boundary.
2. Conduct a literature review of the habitat requirements of the focal species. If there is enough available and accurate information on the habitat requirements of the species, then it may be possible to identify those vegetation components that relate to habitat quality. However, caution should be used when relying on habitat associations from previous studies since many studies may not be applicable to the region of study or the species of concern.
3. Develop a system of habitat stratification that you expect will coincide with species habitat requirements.
4. Use maps, aerial photographs, or satellite imagery to review and select sample units that are reflective of the study area.
5. Evaluate the availability of each habitat strata within the study area.

It is important to keep in mind that one classification scheme will not meet the needs of identifying habitat for all species. If we (humans) classify vegetation as we see structure and composition, then those patterns may or may not relate well to the way that various species respond to patches of vegetation. A classification system that is designed for each species is most likely to reasonably allow an understanding of how habitat is changing for each species over space and time.

MONITORING FOR SPECIES OCCURRENCE

In an inventory design generally you wish to be "$x\%$" confident that you have detected the species if it is really there. Assume that you are concerned that management actions will impact a habitat element or a species that could be present in an area. How sure do you want to be that the species occurs on the proposed management area? Do you want to know with 100% confidence if a species occurs in an area proposed for management? Or can you be 95% sure? 90%? The answer to that question will dictate both the sampling design and the level of intensity with which you inventory the site to estimate presence and absence. The more rare or cryptic the species, the more the samples that will be needed to assess presence, and if the species is rare enough, then the sampling intensity can become logistically prohibitive. In that case, other indicators of occurrence may need to be considered.

Consider the following possibilities when identifying indicators of occurrence in an area (Vesely et al. 2006):

1. Direct observation of a reproducing individual (female with young)
2. Direct observation of an individual, reproductive status unknown
3. Direct observation of an active nest site
4. Observation of an active resting site or other cover
5. Observation of evidence of occurrence such as tracks, seeds, and pollen
6. Identification of habitat characteristics that are associated with the species.

Any of these indicators could provide evidence of occurrence and hence potential vulnerability to management, but the confidence placed in the results will decrease from number 1 to 6 for most species based on the likelihood that the fitness of individuals could be affected by the management action.

Opportunistic observations of individuals, nest sites, or habitat elements can be of some value to managers, but often are of not much use in a monitoring framework except to provide preliminary or additional information. For instance, GPS locations of a species observed incidentally over a 3-year period could be plotted on a map and some information can be derived from the map (known locations). The problem with using these data points in a formal monitoring protocol is that they are not collected within an experimental design. There undoubtedly are biases associated with where people are or are not likely to spend time; species detectability among vegetative, hydrologic, or topographic conditions; and varying detectability among age or sex cohorts. Consequently, this information should be maintained, but rarely would it be used as the basis for a formal monitoring design.

MONITORING TRENDS

Long-term monitoring of populations to establish trends is often used within monitoring programs. Such monitoring programs provide information on changes in populations or habitat availability, but they do not necessarily indicate why populations are changing. For instance, consider the changes in woodcock populations over a 27-year period (Figure 17.3). Clearly, the number of singing male woodcock has declined markedly over this time period. This information is very important in that it indicates that additional study is needed to understand why the changes have occurred. Are singing males simply less detectable in 1995 than they were in 1968? Are populations actually declining? If so, are the declines due to changes in habitat on the nesting grounds? Wintering grounds? Migratory flyways? Is the population being overhunted? Are there disease, parasitism, or predator effects that are causing these declines? Are these declines uniform over the range of the species or are there regional patterns of decline? Analysis of regional patterns indicates that the declines may not be uniform (Figure 17.4). Indeed, declines are apparent in the northeastern United States, but not uniformly throughout the Lake States. So it would seem that causes for declines are probably driven by effects that are regional. Woodcock provide a good example of the need to consider the scope of inference in design of a trends monitoring protocol. The Breeding Bird Survey data indicate that there are areas where declines have been significant (Sauer et al. 2005), and the work by Bruggink and Kendall (1995) indicate that the magnitude of the declines in some areas are perhaps even greater than might be indicated by the regional averages. Approaches to understanding the potential causes for change in abundance of a species can be much more informative when considering changes to a management plan than simply examining trends. The causes then would be addressed at the more local scale in a manipulative manner that would allow assessment of cause and effect relationship.

The design of a trend monitoring program should carefully consider the scope of inference, and if the scope of inference is large (geographic range), monitoring may necessitate coordination over large areas among multiple stakeholders. Site-specific trend analyses will probably be of limited value

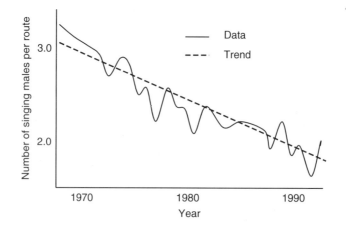

FIGURE 17.3 Trends in woodcock populations over time. Note that the annual data are variable but that the slope is sufficiently steep to allow detection of a trend. (From Bruggink, J.G., and W.L. Kendall. 1995. *American Woodcock Harvest and Breeding Population Status, 1995*. U.S. Fish and Wildlife Service, Laurel, Maryland.)

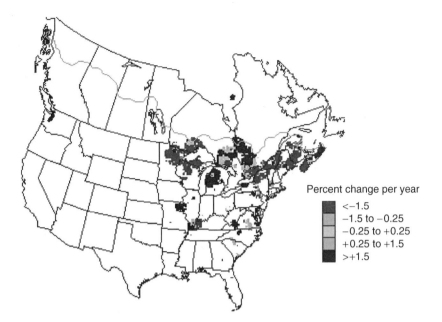

FIGURE 17.4 Geographic distribution in woodcock population changes. Note that population declines are not uniformly distributed throughout the range of the species. (From Sauer, J.R. et al. 2005. *The North American Breeding Bird Survey, Results and Analysis, 1966–2004*. Version 2005.2. USGS Patuxent Wildlife Research Center, Laurel, MD.)

in many instances because the fact that species x is declining at site y is probably not as important as knowing why the species is declining at site y. Using trend monitoring to detect increases or declines in abundance is perhaps best applied to high-priority species, or to allow development of associations with regional patterns of habitat availability. These associations then allow the opportunity for a more informed development of hypotheses that can then be tested in manipulative experiments to identify causes for changes.

One aspect of trend monitoring that must be considered carefully is the sampling intensity needed to detect a change in slope over time. Consider Figure 17.3. Annual data were quite variable over the 27-year period, but a trend is still detectable because the slope was so steep. Annual variability caused by population fluctuations and sampling variance often can prevent detection of a statistically significant change in slope. This problem is exacerbated when the slope is not so dramatic. Hence, there are several factors that must be considered carefully in the design of a trend-monitoring plan:

1. What is the spatial scale over which you wish to understand if habitat elements or populations are declining or increasing? If it is not the geographic range of a species or subspecies, then what portion will be monitored and how will the information be used?
2. Is the indicator that is selected as unbiased as possible and not likely to vary among time periods except those caused by fluctuating populations?
3. Given the inherent variability in the indicator that is being used, how many samples will be needed each time period to allow detection of a slope of at least $x\%$ per year over time?
4. Given the inherent variability in the indicator that is being used, at what point in the trend is action taken to recover the species or reverse the trend? Note that this must be well before the population reaches an undesirably low level, because the manager will first have to understand why the population is declining before action can be taken, and there may be a lag time in population response to proposed changes in management.

5. Will the data be used to forecast results into the future? If so, recognize that the confidence intervals placed on trend lines diverge dramatically from the line beyond the bounds of the data. Forecasting even a brief period into the future is usually done with little confidence unless the underlying causes are understood.

Finally, it is important to recognize that autocorrelation among data points is not only likely but also should be expected under traditional designs where data are analyzed using time-series regression.

CAUSE-AND-EFFECT MONITORING DESIGNS

If the monitoring plan being developed is designed to understand the short- or long-term effects of some management action on a population, then the most compelling monitoring design would take advantage of an approach that would assess responses to those actions. Monitoring conducted over large landscapes or multiple sites may use a comparative mensurative approach to assess patterns and infer effects (e.g., Martin and McComb 2002, McGarigal and McComb 1995). This approach allows comparisons between areas that have received management actions and those that have not and often is analyzed using an ANOVA approach. Alternatively, the BACI allows monitoring to occur on treated and untreated sites both before and after management has occurred (e.g., Chambers et al. 1999).

Although the BACI design is usually considered superior to retrospective analyses, BACI designs often are not logistically feasible. On the other hand, retrospective designs that compare treated sites to untreated sites raise questions about how representative the untreated sites are of the treated sites prior to treatment. In a retrospective design, the investigator is substituting space (treated vs. untreated sites) for time (pre vs. posttreatment populations). The assumption behind this approach is that the untreated sites are representative of the treated sites before they were treated. With adequate replication of randomly selected sites, this assumption can be justified, but often large-scale monitoring efforts are costly and logistics may preclude both sufficient replication and random selection of sites. Hence, doubt may persist regarding the actual ability to detect a cause-and-effect relationship using a retrospective approach, especially if the statistical power of the test is low.

BACI designs are more powerful and can establish cause and effect relationships, but they can often suffer from nonrandom assignment of treatments to sites simply due to the logistics involved in harvest planning. Often the location and timing of management actions do not lend themselves to strict experimental protocols. Lack of random selection may limit the scope of inference only to the sites sampled. Nonetheless, with some care in matching control sites to sites that will be treated in the future, there is more that can be learned about the effects of a treatment on a population using this approach than retrospective designs.

ARE DATA ALREADY AVAILABLE AND SUFFICIENT?

Before embarking on collection of new data to monitor management implementation or effectiveness, or validation of assumptions, it is always prudent to ask if data already exist to address the monitoring questions. But existing data may not be better than no data at all if the data are of poor quality or have inherent biases. Consider the following questions when evaluating the adequacy of existing data to address a monitoring question:

1. *Are samples independent?* Are observations in the data set representing units to which a treatment has been applied? Taking 10 samples from 1 harvest unit is not the same as taking 1 sample from 10 harvest units. In the former example, the samples are subsamples of one treatment area, and in the latter example, there is one sample in each of ten replicate units (Hurlbert 1984). Further, if the species under consideration has a home-range size smaller

than the average harvest unit size, then sampling the species in harvest units probably represent reasonably independent samples. If the species under consideration has a home range that spans numerous harvest units, then the selection of harvest units to sample should be based on ensuring to the degree possible that one animal is unlikely to use more than one harvest unit.

2. *How were the data collected?* What sources of variability in the data may be caused by the sampling methodology (e.g., observer bias, inconsistencies in methods, etc.). If sample variability is too high because of sampling error, then the ability to detect differences or trends will decrease.

3. *Were sites selected randomly?* If not, then there may be (likely is) bias introduced into the data that should raise doubts with regard to the accuracy of the resulting relationships or differences.

4. *What effect size is reasonable?* An *effect size* is the difference (or slope) that you could detect given your sample size, sampling error, and the probability of making an error (as indicated by an *alpha level*) when rejecting a null hypothesis (that there is no difference between treatments or no trend over time). Even a well-designed study may simply not have the sample size adequate to detect a difference or relationship that is real simply because the study was constrained by resources, rare responses, or other factors that increase the sample variance and decrease the effect size. Again, how this is dealt with depends on the question being asked. Which is more important, to detect a relationship that is real or to say that there is no relationship when there is really not? In many instances, where monitoring is designed to detect an effect of a management action, the former is more important (especially using the precautionary principle). In that case, the alpha level may be increased (from say 0.05 to 0.10 or more), but in doing so you will be proportionally more likely to say a relationship is real when it is really not.

5. *What is the scope of inference?* From what area were samples selected? Over what time period? Are the results of the work likely to be applicable to your area? The more different the conditions under which the data were collected from your area, the less confidence that you should place in the results.

Given the cautions indicated here, it is reasonable and correct to use data that are already available to inform and focus the questions to be asked by a monitoring protocol. For instance, results from the Breeding Bird Survey (Sauer et al. 2005) include a credibility index that flags imprecise, small sample size, or otherwise questionable results. For instance, yellow-billed cuckoos have shown a significant decline in southern New England over the past 34 years (Figure 17.5), but the data are deficient when considering regional changes in abundance due to low detection levels (Sauer et al. 2005). Further, an examination of the data would indicate that the one estimate in 1966 may be an outlier and may have an overriding effect on the results. In this example, it would be useful to delete the 1966 data and rerun the analyses and determine if the declining relationship still holds.

Use of existing data and an understanding of data quality can be of value in identifying areas of a management plan that are based on weak data or assumptions. Those factors that are based on assumptions or weak data and which seem quite likely to be influencing the ability to understand management effects should become the focus of questions to be answered by the monitoring plan.

MAKING DECISIONS WITH DATA

Once you have collected data, you need to decide what to do with them. Say that your monitoring data of population change over time under current management practices produced a chart similar to that in Figure 17.3. At what point along the *x*-axis do you decide that it is time to change your management approach? 1975? 1985? 1995? Do you wait and collect more data and make a decision in 2025? At what time is a decision to change management soon enough to reduce a declining trend

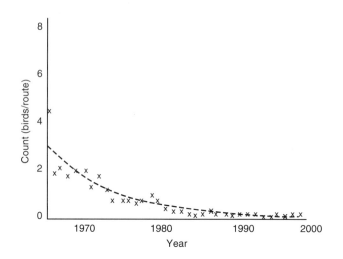

FIGURE 17.5 Changes in detections of yellow-billed cuckoos in New England. Limited detections and what appears to be an outlier in the early years of sampling limit our ability to say with certainty that there is a negative trend. But using the precautionary principle we might conclude that there is a negative trend and change management plans accordingly. (From Sauer, J.R. et al. 2005. *The North American Breeding Bird Survey, Results and Analysis, 1966–2004*. Version 2005.2. USGS Patuxent Wildlife Research Center, Laurel, MD.)

but not too late to make a change that may be moot? Those decisions should be clearly articulated in a management plan. We know that we cannot meet the needs of all species in the same stand or small forest at the same time; there will always be species that are increasing in abundance and others that are declining. When defining your DFC, the expected changes in area of habitat, populations, or frequency of occurrence of all species of concern can be described, and then monitoring can be conducted to see if trends are progressing as expected. Deciding when to make management changes can be based on when the rates of change depart from expected to a degree specified in the plan. It is important to consider these decision thresholds (sometimes referred to as "trigger points") prior to implementing a management plan rather than waiting until data have been collected and analyzed. Nonetheless, unanticipated results may arise during monitoring and this should trigger a reinitiation of the adaptive management process.

EXAMPLES OF APPROACHES TO MONITORING

Species respond to habitat availability and quality at multiple scales and management occurs over a range of spatial and temporal scales. Consequently, a monitoring plan usually takes either a management-centered approach or an organism-centered approach. Regardless of the approach taken, the scope of inference from the monitoring data will be influenced by the interaction between these two approaches and their inherent scales of space and time. The scale of the management actions relative to the scale associated with populations should help identify a set of questions that can be addressed by different data types. For instance, consider the following examples extracted from Vesely et al. (2006) of monitoring problems and approaches.

MONITORING CLONAL PLANTS

Given our lack of knowledge of the distribution of a clonal plant species, we are concerned that timber management plans could have a direct impact on remaining populations that have not yet been identified on our district. How will we know if a timber sale will impact this species?

    In this example, the plant species may have a geographic range extending well beyond the timber sale boundaries and may extend over multiple forest ownerships. The concern is that populations of this species are patchily distributed and poorly known. We may be concerned that population expansion and persistence may be highly dependent on mobility of propagules among population patches and that additional loss of existing patches may exacerbate loss of the population over a significant portion of its range. Consequently, the primary goal of a monitoring effort should be to identify the probability of occurrence of the species in a timber sale. A survey of all (or a random sample of) impending timber sales will provide the land manager with additional information with regard to the distribution of the species. Although information may be collected that is related to fitness of the clone (size, number of propagules, etc.), the primary information needed is an estimate of the probability of occurrence of the organism prior to and following management actions. Indeed, this survey approach also can lend itself well to development of a secondary monitoring approach that utilizes a manipulative experiment. Identification of sites where the species occurs can provide the opportunity for random assignment of manipulations and control areas to understand the effect of management on the persistence of the species.

MONITORING THE OCCURRENCE OF A SMALL-MAMMAL SPECIES

Given the uncertainty in the distribution of a species of small mammal over a forest ownership, will a planned timber harvest have an undue impact on a large proportion of individuals of this species on this owner's forest?

In this example, the species geographic range extends well beyond the boundaries of the owner-ship, but the manager needs a context within which to understand the potential for adverse effects on the species. Based on survey information it is clear that the species occurs in areas that are planned for harvest. But do they occur elsewhere in the ownership? We need to have an unbiased estimate of the abundance of the species over the entire planning area to understand if the proposed management activities indeed represent the potential to impact a significant portion of the population for this spe-cies. With an estimate of abundance that extends over the ownership (or forest, or watershed, etc.) one can estimate (with known levels of confidence) if the proposed management activities might affect 1% of the habitat or population for this species or 80% of the habitat or population.

MONITORING TRENDS IN A SALAMANDER SUBPOPULATION

Given the history of land management on a forest and the plans for future management, will these management actions be associated with the abundance and distribution of a subpopulation of a salamander species that we know occurs on our forest?

In this example, the species again has a geographic range that extends beyond the boundaries of the forest, but there is concern that a subpopulation of a relatively immobile species may occur on our forest. The concern is that the subpopulation may decline in abundance over time as a result of the past and projected management activities on the forest. The goal is to document trends in abundance over time. Changes in abundance or even occurrence may be difficult to detect at a local scale (timber harvest or road building) because individuals are patchily distributed, but cumulatively over space and time, impacts could become apparent. Consequently, this trends-monitoring approach should extend over that portion of the forest where the species is known or likely to occur and provide an estimate of abundance of the species at that scale over time.

MONITORING RESPONSE OF NEOTROPICAL MIGRANT BIRDS TO FOREST MANAGEMENT

Concern has been expressed for several species of neotropical migrant birds whose geographic range extends across the forest. Is the proposed stand management causing changes in the abundance of these species?

In this example, we are dealing with a species that is probably widely distributed, reasonably long lived, and spends only a portion of its life in the area affected by proposed management. One could develop a trends monitoring framework for this species, but the data resulting from that effort would only indicate an association (or not) with time. It would not allow the manager to understand the *cause-and-effect* relationship between populations and management actions. In this case, there are several strata that must be identified relative to the management actions. Can the forest be stratified into portions that will not receive management and others that will receive management? If so, then are the areas in each stratum sufficiently large to monitor abundance of portions of the populations over time? Monitoring populations in both strata prior to and following management actions imposed within one of the strata would allow the managers to understand if changes occur in abundance or reproductive output. For instance, if populations in both managed and unmanaged areas declined over time, then the managers might conclude that population change is independent of any management effects and some larger pervasive factor is leading to decline (e.g., changes in habitat on wintering grounds). On the other hand, should the population in the unmanaged stratum change at a rate different from that on the managed stratum, then the difference could be caused by management actions and lead managers to change their plan.

In Figure 17.6, one of three replicate areas is shown prior to and following the management actions that included clearcut with reserves, two-story, and group selection harvests. A central control area can also be seen in the posttreatment photo. The results from this effort produced predictable responses, but the responses could clearly be linked to the treatments. White-crowned sparrows were not present on any of the pretreatment sites, but were clearly abundant on the clearcut and two-story stands following treatment (Figure 17.7). The treatments caused a response in the abundance of this species.

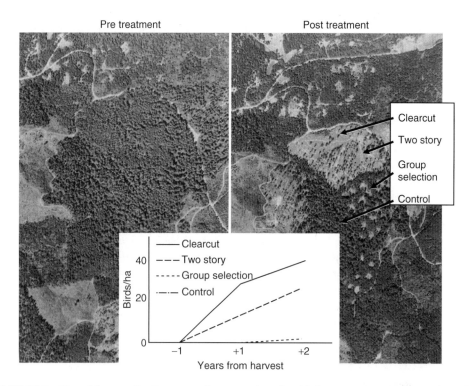

FIGURE 17.6   One of three replicate areas used to assess breeding bird response to treatments using a BACI experimental design and associated changes in white-crowned sparrow detections. Note that this species was absent from control areas both before and after treatments. (Redrafted from Chambers, C.L. et al. 1999. *Ecol. Applic.* 9: 171–185 and used with permission from Dr. Carol Chambers and the Ecological Society of America.)

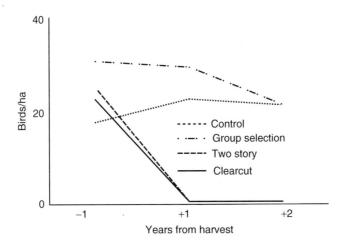

FIGURE 17.7   Changes in abundances of hermit warblers following three management actions. (Redrafted from Chambers, C.L. et al. 1999. *Ecol. Applic.* 9: 171–185 and used with permission from Dr. Carol Chambers and the Ecological Society of America.)

Conversely, hermit warblers declined in abundance on these two treatments, but remained fairly constant on the control and group selection treatments (Figure 17.7). Consequently, for this forest we could predict that future management actions such as these would produce comparable changes in the abundances of these two species with 95% confidence. To fully understand why these changes might have been observed, ancillary data on habitat elements important to these species also could be collected. If the treatments caused changes in important habitat elements, then the reasons for the effects become clearer (Chambers et al. 1999). These habitat relationships analyses can be particularly informative when developing predictive models of changes in abundance or fitness of an organism based on changes in habitat elements caused by management actions or natural disturbances.

MONITORING HABITAT ELEMENTS

Finally, say that we believe that the most likely factor affecting the change in populations of a wide-ranging raptor is change in nest site availability. Populations are low and the probability of detecting a change in abundance or fitness of this species over a large forest is very low. Managers may decide to monitor habitat elements that are associated with demographic characteristics of the species rather than try to monitor the population itself. So we might ask, "How is management affecting the abundance and distribution of potential nest trees for this species?"

Ideally, monitoring of the habitat elements and the associated demographic processes can be conducted to assess cause-and-effect relationships (see earlier text), but with rare or wide-ranging species this may not be possible. The monitoring data needed to develop *wildlife habitat relationships* includes an unbiased estimate of the availability of key habitat elements assumed to be associated with a demographic characteristic of the species and an estimate of the demographic characteristics assumed to be associated with each habitat element. It is important that the monitoring framework for the vegetation component of the habitat relationships must be implemented at spatial and temporal scales consistent with those used by the species of interest.

SUMMARY

Adaptive management is a formal process of treating management plans as hypotheses, implementing the plans and monitoring the implementation and effectiveness of the actions and validating key assumptions. The information gained from this process is then used to refine and improve future actions. The design of monitoring plans includes several key steps once goals have been identified: identify the appropriate response variables; identify the scope of inference; establish the experimental design; and estimate the appropriate sampling intensity, frequency, and duration. Monitoring of habitat elements usually includes use of both ground-based and remotely-sensed data. Habitat element and population monitoring may be conducted to survey the probability of occurrence of species or elements, examine trends, or establish cause-and-effect relationships. Deciding how to use the information to make changes in management direction should be a key part of the description of the DFC in the management plan.

REFERENCES

Barrett, R.H., and H. Salwasser. 1982. Adaptive management of timber and wildlife habitat using DYNAST and wildlife-habitat relationships models. Abstract of paper presented at the 1982 Joint Annual Conference of the Western Association of Fish and Wildlife Agencies and the Western Division of the American Fisheries Society, Las Vegas, NV.

Bruggink, J.G., and W.L. Kendall. 1995. *American Woodcock Harvest and Breeding Population Status, 1995.* U.S. Fish and Wildlife Service, Laurel, Maryland.

Chambers, C.L., W.C. McComb, and J.C. Tappeiner. 1999. Breeding bird responses to 3 silvicultural treatments in the Oregon Coast Range. *Ecol. Applic.* 9: 171–185.

Cody, M.L. (ed.). 1985. *Habitat Selection in Birds.* Academic Press, Orlando, Florida.

Foster, J.R. 2001. Statistical power in forest monitoring. *For. Ecol. Manage.* 151: 211–222.

Hurlbert, S.J. 1984. Pseudoreplication and the design of ecological field experiments. *Ecol. Monogr.* 54: 187–211.

Jones, K.B. 1986. Data types. In Cooperrider, A.Y., R.J. Boyd, and H.R. Stuart (eds.), *Inventorying and Monitoring Wildlife Habitat.* U.S. Department of the Interior, Bureau of Land Management. Denver, CO.

Kerr, R.M. 1986. Habitat mapping. In Cooperrider, A.Y., J.B. Raymond, and H.R. Stuart (eds.), *Inventory and Monitoring of Wildlife Habitat.* U.S. Department of the Interior, Bureau of Lands, Denver, CO.

Martin, K.J., and W.C. McComb. 2002. Small mammal habitat associations at patch and landscape scales in Oregon. *For. Sci.* 48: 255–266.

McGarigal, K., and W.C. McComb. 1995. Relationships between landscape structure and breeding birds in the Oregon Coast Range. *Ecol. Monogr.* 65: 235–260.

Ohmann, J.L., and M.J. Gregory. 2002. Predictive mapping of forest composition and structure with direct gradient analysis and nearest neighbor imputation in coastal Oregon, USA. *Can. J. For. Res.* 32: 725–741.

Resource Inventory Branch. 1998. Species inventory fundamentals. Version 2.0. Standards for components of British Columbia's Biodiversity No. 1. Resource Inventory Committee, Victoria, BC.

Richards, J.A., X. Jia, D.E. Ricken, and W. Gessner. 1999. *Remote Sensing Digital Image Analysis: An Introduction.* Springer-Verlag, New York, Inc. Secaucus, NJ.

Sauer, J.R., J.E. Hines, and J. Fallon. 2005. *The North American Breeding Bird Survey, Results and Analysis, 1966–2004.* Version 2005.2. USGS Patuxent Wildlife Research Center, Laurel, MD.

Vesely, D., B.C. McComb, C.D. Vojta, L.H. Suring, J. Halaj, R.S. Holthausen, B. Zuckerberg, and P.M. Manley. 2006. *Development of Protocols to Inventory or Monitor Wildlife, Fish, or Rare Plants.* USDA For. Serv. Gen. Tech. Rep. WO-72.

Walters, C. 1986. *Adaptive Management of Renewable Resources.* Macmillan, New York.

Walters, C.J., and C.S. Holling. 1990. Large-scale management experiments and learning by doing. *Ecology* 71: 2060–2068.

18 Forest Sustainability and Habitat Management

Sustainable. Does it mean leaving resources for the next generation that are equal to or greater than those that we enjoy? Or does it mean not losing any of Leopold's pieces? Or using resources wisely for the greatest good in the long run, as Pinchot suggested? Or restoring as much as possible to wilderness to be honored as a place of beauty and spirit as John Muir might have suggested? Clearly, sustainability literally is in the eye of the beholder. Equally clear is that society is demanding more services, *ecosystem services*, from our forests than it ever has.

I have heard some foresters say that they manage forests sustainably. They cut trees, they plant trees, the trees grow, and they cut them again. Trees are indeed a renewable resource, and theoretically production of wood fiber can continue for a very long period of time using currently accepted silvicultural practices. But in the past 20 years society has raised questions about what exactly is being sustained: 2×4s, plywood, and pulp? Water? Habitat for various wildlife species? Recreation? Aesthetics? Medicinal plants? Nontimber products? More? During the 1980s and 1990s, large areas of tropical forests were cleared for agriculture or to plant exotic tree species for timber production, and some forests were simply cut and abandoned. "Those who forget the past are doomed to repeat it," said Philosopher George Santayana. And indeed they have. This is a story that has repeated itself from the northeastern United States, through the Deep South and parts of the West, and in New Zealand, Australia, and many countries in South America, leading to an increasing number of questions regarding forest sustainability.

In 1992, 172 governments and 2400 representatives of nongovernmental organizations (NGOs) participated in a meeting in Rio de Janeiro, Brazil, in what became known as the Earth Summit. One outcome of that meeting was the endorsement of a set of 17 nonbinding forest principles that provided guidance for management and conservation of global forests. Also, during this meeting, an agreement on biodiversity conservation was reached, which stated that (United Nations Environment Program, convertion on biological diversity, Montreal, Quebec, Canada):

> ... the signatories must develop plans for protecting habitat and species; provide funds and technology to help developing countries provide protection; ensure commercial access to biological resources for development and share revenues fairly among source countries and developers; and establish safety regulations and accept liability for risks associated with biotechnology development.

One-hundred and fifty three nations endorsed this agreement. The U.S. did not.

The very next year, in 1993, another meeting was held in Montreal, Canada, that culminated in the Montreal Process. Participants included countries with temperate and boreal forests, including the United States. They developed a framework for measuring the progress of each country toward sustainable forest management. The framework included seven criteria and 67 indicators, known collectively as the Montreal Process Criteria & Indicators for the Conservation and Sustainable Management of Temperate and Boreal Forests (Washburn and Block 2001). The first criterion dealt with the conservation of biodiversity. This process and the discussions that led to it represented the social expectations, if not demands, that forests be managed in a way that ensured sustainability of a variety of resources, including especially biodiversity.

DEFINING THE RESOURCES TO BE SUSTAINED

Managing forests sustainably requires that those people responsible for forest management understand which resources need to be sustained. The Montreal Process provides an internationally accepted list of resources that should be considered during forest management (listed in order of criteria in the document; Washburn and Block 2001):

1. Conservation of biological diversity including ecosystem diversity (five indicators), species diversity (two indicators), and genetic diversity (two indicators).
2. Maintenance of productive capacity of forest ecosystems (five indicators).
3. Maintenance of forest ecosystem health and vitality (three indicators).
4. Conservation and maintenance of soil and water resources (eight indicators).
5. Maintenance of forest contribution to global carbon cycles (three indicators).
6. Maintenance and enhancement of long-term multiple socioeconomic benefits to meet the needs of societies, including production and consumption (six indicators), recreation and tourism (three indicators), investment in the forest sector (four indicators), cultural, social, and spiritual needs and values (two indicators), and employment and community needs (four indicators).
7. Legal, institutional, and economic framework for forest conservation and sustainable management including a legal framework (five indicators), institutional framework (five indicators), economic framework (five indicators), a means to measure and monitor changes (three indicators), and to conduct and apply research and development (five indicators).

Several of these criteria specifically identify areas of forest management where habitat must be explicitly considered. Clearly, criterion 1 identifies ecosystems, species, and genes that must be conserved. Managing forests in ways that achieve this goal is the motivation for writing this book. This indicator also ensures representation of forest types across a landscape and addresses levels of fragmentation of forests. Species richness, rare species protection, and population viability must be considered, as must species on the periphery of their geographic range. But other indicators also have implicit habitat management goals: plantations of native tree species, harvest of nontimber forest products (including berry-producing plants), management within the historical range of variability, maintaining ecological processes, providing coarse woody debris (as a carbon store), as well as socioeconomic goals such as employment and social stability. Many of the concepts addressed so far in this book relate directly to sustainable forest management as outlined in the Montreal Process. The devil is in the details, however, when trying to decide how to apply these concepts in forests at local, regional, and global scales.

SCALES OF SUSTAINABILITY

The Montreal Process framework is useful for assessing the status and trends of forest services, and is designed to enhance international communication about sustainable forest management (Washburn and Block 2001). With increased information within and among nations regarding the indicators identified in the Montreal Process, policy makers and stakeholders may be better able to make informed forest management decisions locally, regionally, and globally. But how do these indicators scale over space and time? Clearly, there is a temporal limit to sustainability. Societies change and so do their values. Climates change and hence the ability of ecosystems to provide services. Species go extinct, including humans. Stars die and planets grow cold. So the first step in developing a sustainable forest management plan is to define the scope of the plan. How far into the future will we strive to remain sustainable? As long as possible, recognizing that it is a continually moving target? For 100 years? 1000 years? Setting these goals may seem moot, but they do influence the actions that will be taken today to ensure that goals in the future are attainable.

We also need to define the size of the area over which we are striving to be sustainable. The Oregon Department of Forestry has used the Montreal Process criteria and indicators as a basis for developing a sustainable forest management plan for state forests. They defined their spatial scale as one ownership, theirs. But their ability to achieve their goals will be in part dependent on how their neighbors view forest management and how those landowner decisions aggregate to contribute to regional goals, and how regional contributions aggregate to achieve global goals. Think globally, act locally. So, although defining the temporal and spatial framework within which sustainable forest management will be conducted, it is equally important to define the context within which sustainable forest management is being conducted and how your actions on your lands contribute to some larger aggregate goal. The concept of explicitly considering the context for sustainable forest management is particularly important when addressing criterion 1. It should be clear by now that no single landowner and no set of reserves will meet the needs for all species and sustain global biodiversity along with other ecosystem services.

HUMANS ARE PART OF THE SYSTEM

The crux of the issue regarding sustainable forest management is how we can provide the depth and breadth of ecosystem services demanded by society that include economic products over space and time. Humans set the agenda, humans are part of the problem, and humans are part of the solution. It is all about us. Forest managers are granted a social license to practice forestry in a way that they choose so long as ecosystem services are provided. How does society grant that license? Unlike wildlife, which in the United States is a public resource that occurs on public and private lands, forests are considered private resources. This political dichotomy of ownership is a challenge when managing habitat on private lands to achieve public wildlife goals. But society benefits from forests regardless of whose land they are on. Policies and laws (see Chapter 19), the Montreal Process, and certification by a third party are mechanisms used to ensure that those benefits are sustained. Occasionally incentives are also used, such as selling carbon credits to forest owners. So, although one person or company may own forestland, what they are allowed to do on forestland is influenced by society both in their own country as well as in others.

FOREST CERTIFICATION

One international mechanism for ensuring that forests are managed sustainably and that biodiversity conservation is considered during management is third-party certification that the management is sustainable based on an audit by an objective organization. In this process, forest management practices are evaluated against a set of standards by an external certification organization. Certification benefits landowners economically because, at least in theory, certified forest products yield a higher return on an investment while assuring consumers that their purchase comes from a forest whose management meets certain standards (Washburn and Block 2001). Certification has two interconnected pieces: third-party certification of forest management activities that typically address some or all of the Montreal Process criteria and indicators and certification of chain of custody. *Chain of custody* is important because, as wood products pass from stump to mill and mill to wholesaler and wholesaler to retailer, there is assurance to the consumer that the wood purchased did indeed come from a certified forest.

There are a variety of third-party certification organizations in the world, and over 125 million ha (>3% of the world's forests) are certified by one or more organizations (Rametsteiner and Simula

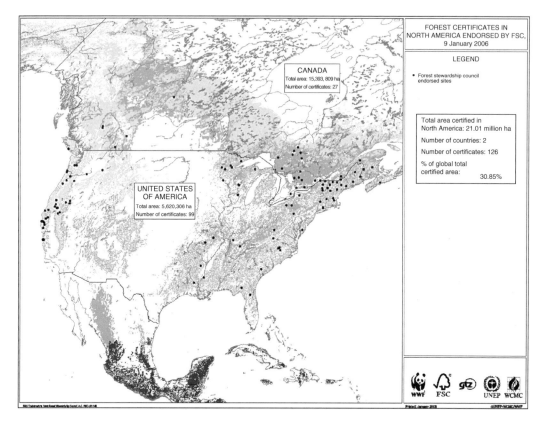

FIGURE 18.1   Map of certified forests in North America. (Map from United Nations Environment Programme, World Conservation Monitoring Centre; used with their permission.)

2003). Over 90% of these certified forests are in North America and Europe (Figure 18.1). Hence, the goal of conserving biodiversity in tropical forests using sustainable forest management principles has largely failed, but voluntary certification on private lands in temperate forests has been quite successful. The degree to which certification is effective in conserving biodiversity is unknown, but is assumed to be significant in temperate regions, though perhaps dependent on the certifying body (National Wildlife Federation et al. 2001). In the United States, there are two primary certification organizations: The Forest Stewardship Council (FSC) and Sustainable Forestry Initiative (SFI). Canada (e.g., Canadian Standards Association), Europe (e.g., Pan European Forest Certification [PEFC]), and other countries (e.g., the Keurhout Foundation in Holland and National Timber Certification Council in Malaysia) have additional certifying bodies (Rametsteiner and Simula 2003). The FSC originated in Europe and is active in 57 countries around the world, including the United States (Perera and Vlosky 2006). FSC is a nonprofit organization that evolved from the Earth Summit agreements. SFI was established by the American Forest and Paper Association (AF&PA). It is a comprehensive system of principles, objectives, and performance measures developed to integrate both responsible environmental practices and sound business practices (Perera and Vlosky 2006, SFI 2001). SFI certification includes first-party (self) and second-party (SFI board) audits as well as independent third-party audits of conformance to standards (Perera and Vlosky 2006). Since its establishment, over 124 million acres of forestland in North America has undergone third-party SFI certification (SFI 2006).

Consumers seem willing to support green certified products in the North America and Europe; so certification of forestlands in those countries continues to grow (Rametsteiner and Simula 2003). The cost to the landowner of being audited to receive certification is high, and cost-prohibitive for

FIGURE 18.2 Owners of small forest tracts in Massachusetts have developed an economy of scale through development of the Massachusetts Woodlands Cooperative. The cooperative is certified by the Forest Stewardship Council. (Photo by Susan Campbell and used with her permission.)

owners of small parcels. Indeed because most forestland in the eastern United States is in small parcels, most of it is not certified, unless innovative mechanisms such as woodland cooperatives are used to address an economy of scale (Barten et al. 2001; Figure 18.2).

When a landowner wishes to have his or her property certified, he or she makes a request from the certification body. The application for certification usually entails an application fee or membership in an organization, (i.e., AF&PA). Auditors then visit the property and compare what they see in owner records and on the ground with the standards for certification set by the certifying body. For third-party certification, the auditors are independent of the company or organization being certified so as to avoid conflict of interest. Certification is given when the auditors' report indicates that there are no noncompliances. If there are instances of noncompliance, then the landowner is given a fixed amount of time to address the issues before certification would then be granted. Periodic or annual audits are required by the certifying body. The landowner must not only pay the application fee but also the costs associated with the audit, and hence for small ownerships cost per certified forest acre can be prohibitively high.

Certification schemes can be broadly categorized into two groups: performance-based and process-based (Layton et al. 2002, Perera and Vlosky 2006). Performance-based standards define specific performance levels for various aspects of forest management; it is the framework used by FSC. Process-based schemes provide a systematic approach to developing, implementing, monitoring, and evaluating environmental policies, but they do not stipulate performance standards and are the framework used by SFI (Layton et al. 2002, Perera and Vlosky 2006). The two certification processes differ in the ways that the approach ensuring that biodiversity conservation and management of habitat for various wildlife species is considered (Table 18.1). In both approaches, the metrics used to define compliance or noncompliance may be areas of concern for landowners contemplating certification.

TABLE 18.1

Comparison of FSC and SFI Certification Approaches Dealing with Protection of Biological Diversity and Management of Habitat for Various Wildlife Species

Indicator	FSC	SFI
The status of forest-dependent species at risk of not maintaining viable breeding populations, as determined by legislation or scientific assessment	Safeguards shall exist that protect rare, threatened, and endangered species and their habitats. Conservation zones and protection areas shall be established... Ecological functions and values shall be maintained intact, enhanced, or restored, including . . . genetic, species, and ecosystem diversity . . .	Program participants shall apply knowledge gained through research, science, technology, and field experience to manage wildlife habitat and contribute to the conservation of biological diversity
Biological diversity	Landowners are required to maintain, enhance, or restore the long-term integrity of natural habitats, ecological processes, soil, water, and stand development	Each company develops its own policies, programs, and plans to contribute to the conservation of biological diversity and manage sites of ecological significance
Stand age-class distribution	Forest owners and managers maintain or restore portions of the forest to the range and distribution of age classes of trees that result from processes that would naturally occur on the site	Program participants shall have policies, programs, and plans to promote habitat diversity at the stand and landscape level
Habitat diversity	A diversity of habitats for native species is protected, maintained, and enhanced	Program participants shall apply knowledge gained through research, science, technology, and field experience to manage wildlife habitat and contribute to the conservation of biological diversity

Source: Based on Washburn, M., and N. Block. 2001. *Comparing Forest Management Certification Systems and the Montreal Process Criteria and Indicators.* Pinchot Institute, Washington, D.C.; and National Wildlife Federation et al. 2001. A Comparison of the American Forest & Paper Association's Sustainable Forestry Initiative and the Forest Stewardship Council's Certification System. National Wildlife Federation, Washington, D.C.

Layton et al. (2002) reviewed the metrics used and approaches taken to certification specifically dealing with wildlife and biodiversity evaluations. They concluded that auditors often indicate that professional judgment is used as the basis for determining if a criterion or indicator is met. Because audit teams vary from owner to owner, the consistency of evaluation is brought into question. Certification bodies have struggled with objective measures of performance especially when evaluating indicators of biodiversity. Lindenmayer et al. (2000) proposed the use of structure-based indicators such as stand structural complexity and plant species composition, connectivity, and heterogeneity in biodiversity conservation, and these may have utility in certification audits as well.

EFFECTIVENESS OF CERTIFICATION

How effective has certification been in conserving biodiversity? We do not know. There are indications that independent audits are an incentive for improving forest management consistent with the intent of the Montreal Process (Rametsteiner and Simula 2003). Gullison (2003) concluded that landowners who undergo FSC certification are more likely to improve management with respect to the value of managed forests to biodiversity conservation. But the incentives associated with certification are not sufficient to attract most producers to seek certification, particularly in tropical countries, where the costs of improving management to meet FSC guidelines are significantly greater than any market benefits they may receive (Gullison 2003). If FSC certification is to make greater inroads, particularly in tropical countries, significant investments will be needed both to increase the benefits and reduce the costs of certification (Gullison 2003). Conservation investors will need to carefully consider the biodiversity benefits that will be generated from such investments vs. the benefits generated from investing in more traditional approaches to biodiversity conservation (Gullison 2003). But a clear quantification of the impacts of green certification on conservation of biodiversity has not been completed. Nussbaum and Simula (2004) provided a series of case studies indicating potential benefits of certification to biodiversity conservation. They concluded that improved conservation of biodiversity appears to be a consistent benefit of certification as evidenced by increased protection of representative ecosystems and of rare, threatened, or endangered species. But few data are available on which to base rigorous assessment. Loehle et al. (2005) evaluated associations between a set of landscape structure metrics that could be used in certification programs and bird species richness. Their work suggested that indicators addressing aspects of landscape structure may be useful in evaluating potential contributions to biodiversity conservation. But the extent to which certification has led to improved ecosystem functions, reduced risk to loss of biodiversity, or greater probability of persistence of endangered species is still unclear (Nussbaum and Simula 2004). Nonetheless, certification represents a pragmatic international system of private governance that could lead to increased levels of biodiversity conservation. We should anticipate that the processes and standards will evolve and improve over time, and the disparate approaches currently used among certifying organizations will likely coalesce into a common framework. What is needed, however, is a monitoring system that clearly documents the gains and losses of biodiversity and ecosystem processes from certified and uncertified forests.

SUMMARY

Society demands more and more ecosystem services from forestlands, both public and private. International conferences and agreements have developed a set of principles defining sustainable forest management, and these principles have been included in certification protocols. Sustainable forest management certification is private governance of forest management and provides the social license for managers to continue to manage forests. Enrollment in certification programs has grown dramatically since the 1990s, especially in temperate forests. The incentives to landowners for certification in tropical forests are not apparent, especially given the high costs associated with certification. Although the benefits of certification to biodiversity conservation are not known, the generally accepted assumption is that certified forests should do more to conserve aspects of biodiversity than uncertified forests. Data are sorely needed to confirm or reject this assumption.

REFERENCES

Barten, P.K., D. Damery, P. Catanzaro, J. Fish, S. Campbell, A. Fabos, and L. Fish. 2001. Massachusetts family forests: birth of a landowner cooperative. *J. For.* 99: 23–30.

Gullison, R.E. 2003. Does forest certification conserve biodiversity? *Oryx* 37: 153–165.

Layton, P., S.T. Guynn, and D.C. Guynn. 2002. Wildlife and biodiversity metrics in forest certification systems. Final Report. National Council for Air and Stream Improvement, Research Triangle Park, NC.

Lindenmayer, D.B., C.R. Margules, and D.B. Botkin. 2000. Indicators of biodiversity for ecologically sustainable forest management. *Conserv. Biol.* 14: 1523–1739.

Loehle, C., T.B. Wigley, S. Rutzmoser, J.A. Gerwin, P.D. Keyser, R.A. Lancia, C.J. Reynolds, R.E. Thill, R. Weih, D. White Jr., and P.B. Wood. 2005. Managed forest landscape structure and avian species richness in the Southeastern US. *For. Ecol. Manage.* 214: 279–293.

National Wildlife Federation, Natural Resources Council of Maine, and Environmental Advocates. 2001. A Comparison of the American Forest & Paper Association's Sustainable Forestry Initiative and the Forest Stewardship Council's Certification System. National Wildlife Federation, Washington, D.C.

Nussbaum, R., and M. Simula. 2004. *Forest Certification: A Review of Impacts and Assessment Frameworks.* The Forests Dialogue Publications. Yale University School of Forestry & Environmental Studies, New Haven, CT.

Perera, P., and R.P. Vlosky. 2006. A History of forest certification. Louisiana Forest Products Development Center Working Paper No. 71. Louisiana State University, Baton Rouge, LA.

Rametsteiner, E., and M. Simula. 2003. Forest certification — an instrument to promote sustainable forest management? *J. Environ. Manage.* 67: 87–98.

Sustainable Forestry Initiative (SFI). 2006. *SFI Program Participants that have Completed 3rd Party Certification.* American Forest & Paper Association, Washington, D.C.

Washburn, M., and N. Block. 2001. Comparing Forest Management Certification Systems and the Montreal Process Criteria and Indicators. Pinchot Institute, Washington, D.C.

19 Regulatory and Legal Considerations

One way of influencing change in the way forestry is practiced to benefit habitat for animals or conservation of biodiversity is through incentive programs such as certification (see previous chapter), tax relief, or compensation for ecosystem services (e.g., easements, land purchases, or purchases of carbon credits; Pagiola et al. 2002). More typically, though, certain practices are prescribed by law. In the United States, laws are policies enacted by legislature and signed into law by an executive branch, and enforced through a judicial branch of government. There are certain things that society values strongly enough to prescribe it: keep species. Manage forests for sustained yield. Do not participate in trade of globally endangered species.

Some policies are set at local levels such as counties, towns, and even neighborhoods. Zoning laws, building permits, and noise limitations are all set and enforced locally. The layers of policies, laws, regulatory agencies, and responsibilities regarding forest management and wildlife conservation are at times overwhelming especially to private-forestland managers. And that is where the crux of the habitat management problem often lies. Wildlife are public resources whose habitat is most often controlled by private landowners with private property rights. It is relatively easy to envision an ecosystem management plan for a public property whereby the outcomes of implementing the plan are a set of ecosystem services valued by society, using public land for the public good (Thomas et al. 2006). But private property owners have property rights, and in some places they may also have water and mineral rights, restricting what society can demand from their land (Bliss et al. 1997). So although society may say they want active habitat management for a rare species on private lands, unless the land is deemed critical habitat for an endangered species, society cannot make the landowner do anything, unless there are laws.

INTERNATIONAL LAWS AND AGREEMENTS

The Convention on International Trade in Endangered Species of Wild Fauna and Flora (CITES) is an international agreement among governments designed to protect species worldwide. Its aim is to ensure that international trade in wild animals and plants does not threaten their survival (Mace and Lande 1991). CITES was drafted as a result of a resolution adopted in 1963 at a meeting of members of The World Conservation Union (IUCN) and signed in Washington, D.C., in 1973 (Hutton and Dickson 2000). There are now 169 signatories. Today, it provides varying degrees of protection to more than 30,000 species of animals and plants, whether they are sold alive or sold as animal or plant parts (e.g., ivory). But there are no habitat provisions. However, several U.S. acts authorize Congress to appropriate funds to aid in international efforts at habitat conservation. The International Environment Protection Act of 1983 and the Neotropical Migratory Bird Conservation Act of 2000 provide funds to other countries to benefit species of animals, particularly those that cross international boundaries (Elliott et al. 2005).

Each country has its own set of laws regarding conservation of wildlife, forests, and habitat. Laws in some countries are extensive and enforced rigorously. In some countries laws exist, but are not consistently enforced. And in all cases, political pressures can lead to variable interpretation of the laws. I use examples of the hierarchical layers of laws in the United States to illustrate the complexities facing habitat and forest managers, but similar layers and problems are found in other countries.

NATIONAL LAWS

Since the beginning of the conservation debates in the Unites States between Gifford Pinchot and John Muir, legislation has been proposed that would influence how public and private lands are managed to provide public goods and services (Table 19.1). Many of the earliest acts of legislation that influenced private landowners had to more to do with influencing harvest rates of commercially important species of trees and animals than with governing management on their lands (Bean and Rowland 1997). But as early as 1937 the Federal Aid in Wildlife Restoration Act (also known as the Pittman–Robertson Act) matched federal funds collected from a tax on firearms with state funds to allow management of habitat to benefit many wildlife species (Oehler 2003). Although the original focus of these purchases was for game species, many non-game species benefited indirectly. Use of these funds for habitat management is now viewed more broadly to benefit hunted and non-hunted species in many states (Oehler 2003).

During the 1960s and 1970s, a suite of environmental laws emerged reflecting concern over sustained production of goods and services from public lands. The Multiple Use Sustained Yield Act, Endangered Species Act (ESA), National Environmental Policy Act (NEPA), and National Forest Management Act represent pieces of legislation that continue to shape the way that federal lands are managed and continue to fuel the debates regarding priorities for federal lands (Hibbard and Madsen 2003). One of these, the ESA, also has far-reaching influence on private lands. When a species is listed as threatened or endangered by the appropriate federal agency (usually the U.S. Fish and Wildlife Service), "taking" of individuals of protected species by a private citizen constitutes a violation of federal law (Sagoff 1997). The ESA requires the federal government to designate "critical habitat" for any species it lists under the ESA. Once designated, then any alteration of critical habitat that emperils one or more individuals of any protected species constitutes a violation of the law (Sagoff 1997) and the landowner can be prosecuted unless she has an incidental take permit (Smallwood 2000). Habitat Conservation Plans are designed to provide no net loss of a species while allowing private landowners the opportunity to continue managing their lands (see Chapter 14). The issue of controlling private property rights through federal law has met with considerable resistance, but because most habitats for many protected species occur on private (not public) lands, the provision continues to represent a powerful tool for habitat protection. Unfortunately, there are no provisions for fair compensation of the property owner when ESA restricts harvest of trees for commercial gain (Innes et al. 1998). A federal compensation program similar to what is used under the Conservation Reserve Program or Wildlife Habitat Incentives Program (WHIP) in the Farm Bill would probably ease considerably the tension between private forest landowners and federal regulatory agencies.

TABLE 19.1
Examples of U.S. Laws That Influence the Ability of Public- and Private-Forestland Managers to Provide Habitat for Animals or to Conserve Biodiversity

Timber Protection Act of 1922	Protects timber on federal lands from fire, disease, and insects
Migratory Bird Conservation Act of 1929	Established procedures for acquisition by purchase, rental, or gift of areas for migratory birds
Federal Aid in Wildlife Restoration Act of 1937	Provides federal aid to states for management and restoration of wildlife including acquisition and improvement of wildlife habitat
Taylor Grazing Act of 1934	Governs grazing on public lands
Transfer of Certain Real Property for Wildlife Conservation Purposes Act of 1948	Real property no longer needed by a federal agency can be transferred to the Secretary of the Interior if the land has particular value for migratory birds or to a state agency for other wildlife conservation purposes
Multiple Use Sustained Yield Act of 1960	Established purposes for the National Forest System, including outdoor recreation, range, timber, watershed and fish and wildlife

(Continued)

TABLE 19.1
(Continued)

Sikes Act of 1960	Planning, development, and maintenance of fish and wildlife resources on military reservations
McIntire-Stennis Act of 1962	Authorized a formula fund for forest research in all states
Wilderness Act of 1964	Considers all interior roadless lands of area >5000 acres for inclusion into the National Wilderness Preservation System
National Environmental Policy Act of 1969	Ensures that environmental values are given appropriate consideration, along with economic and technical considerations
Alaska Native Claims Settlement Act of 1971	Authorized Alaska Natives to select and receive title to 44 million acres of public land in Alaska
Endangered Species Act of 1973	Provided for the conservation of ecosystems upon which threatened and endangered species of fish, wildlife, and plants depend and implemented the CITES agreement
Federal Land Policy and Management Act of 1976	Constitutes the "Organic Act" for the Bureau and governs most uses of the public lands
National Forest Management Act of 1976	Constitutes the "Organic Act" for the Forest Service
Public Rangeland Improvement Act of 1978	Improves conditions of public rangelands for grazing, wildlife habitat, and other uses
Cooperative Forestry Assistance Act of 1978	Provides assistance on forest management issues to nonfederal forest landowners
Renewable Resources Extension Act of 1978	Increases extension emphasis in renewable resources, including fish, wildlife, and water resources on private forest and rangelands
Alaska National Interest Lands Conservation Act of 1980	Provided 79.54 million acres of refuge land in Alaska
Fish and Wildlife Conservation Act 1980	Financial and technical assistance to the States for the development, revision, and implementation of conservation plans and programs for nongame fish and wildlife
Tax Deductions for Conservation Easements of 1980	A taxpayer may take a deduction for a "qualified real property interest" contributed to a charitable organization exclusively for conservation purposes protected in perpetuity
International Environment Protection Act of 1983	Assist other countries in wildlife and plant protection efforts in order to preserve biological diversity
Food Security Act of 1985	Contains several provisions that contribute to wetland conservation including the Swampbuster Conservation Reserve program Wetland Reserve Program (WRP)
Cave Resources Protection Act of 1988	Management and protection of caves and their resources on federal lands
Emergency Wetlands Resources Act of 1986	Authorized the purchase of wetlands from Land and Water Conservation Fund monies
Land Remote Sensing Policy Act of 1992	Directs that LANDSAT 7 acquire high-priority land remote-sensing data in order to meet the needs of the U.S. Global Change Research Program
Hawaii Tropical Forest Recovery Act of 1992	Provides grants, contracts, and cooperative agreements to promote sound management and conservation of tropical forests in the United States
Partnerships for Wildlife Act of 1992	Authorizes grants to the States for programs and projects to conserve nongame species
National Wildlife Refuge Acts	Various acts to establish specific refuges
Neotropical Migratory Bird Conservation Act of 2000	Provides grants to countries in Latin America and the Carribean, and the United States for the conservation of neotropical migratory birds that winter south of the border and summer in North America
State Wildlife Grants program of 2001	Supports cost-effective conservation aimed at preventing wildlife from becoming endangered
Healthy Forests Restoration Act of 2003	Reduces the risk of catastrophic fire to communities, helps save the lives of firefighters and citizens, and protects threatened and endangered species

Source: Extracted from the *U.S. Fish and Wildlife Service Laws Digest*, Washington, D.C.

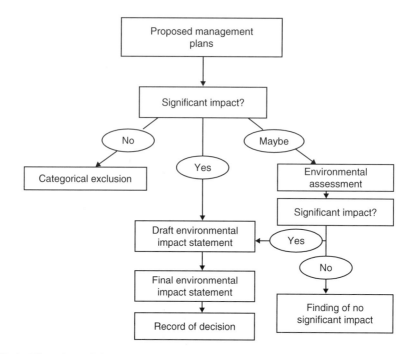

FIGURE 19.1   Flow chart of the NEPA process required for federal land managers and land managers using federal funds in the United States.

Another powerful federal law, the NEPA, is designed to document and prevent unwanted adverse environmental effects of federal actions. There are instances where this act can also influence management actions not only on federal lands but also on nonfederal lands as well (Moser 2000). For instance, if federal funds are used in a habitat improvement project on state or private lands, then NEPA processes would have to be followed (Figure 19.1). It is during this process that environmental impact statements are developed to compare the environmental consequences of proposed and alternative land-management actions. The process allows for public involvement and participation, creating a more transparent mechanism for ensuring that public concerns are addressed. The process is also costly both financially and in timing, often taking a year or more before a record of decision is made. Consequently, there have been efforts to change or "streamline" NEPA and allow management actions to proceed more rapidly where timing is critical (Smythe and Isber 2003). For instance the Healthy Forests Restoration Act included provisions to this effect, enabling a more rapid response with regard to salvage logging to reduce risks associated with reburning of forests. Such provisions meet opposition, however, when streamlined procedures are perceived to increase risk of environmental degradation.

A recent federal program has provided new opportunities for biodiversity conservation among the states in the United States. The State Wildlife Grants program of 2001 provides federal dollars to every state and territory to support cost-effective conservation aimed at preventing wildlife from becoming endangered. Funds are allocated to every state according to a formula based on each state's size and population. Each state and territory is responsible for developing a statewide wildlife action plan to identify species and their habitats of greatest conservation need and outline the steps that would be needed to conserve all species and natural areas. This program is supported by a coalition of more than 3000 organizations who are members of Teaming with Wildlife. Such broad-based grassroots support for this initiative has helped to ensure its success.

Federal funding is also provided to land-management agencies specifically to support habitat acquisition and management. National Wildlife Refuges represent the best example of a federal program of habitat management, but programs within the Natural Resources Conservation Service such as the WHIP provides cost-share habitat improvement support for private landowners as well.

Habitat management is also an integral part of land-management plans developed in the Forest Service, Bureau of Land Management, National Park Service, and military reservations. All tolled, many millions of federal dollars are spent annually on habitat management projects on public and private lands.

Federal funding also has been provided to support research and outreach activities to states. The McIntire-Stennis Act and Renewable Resources Extension Act provide formula funds to land-grant universities to support forest research and outreach activities, respectively (Thompson and Bullard 2004). These funds are often used to support research and outreach dealing with forest management, wildlife habitat management, and forest biodiversity conservation. This effort supplements efforts within the research programs of the various federal agencies (U.S. Fish and Wildlife Service, USGS Biological Resources Division, and U.S. EPA Office of Research and Development) and outreach arms of federal agencies (e.g., Forest Service State and Private Forestry program). Private industry also supports research into forest wildlife habitat management, largely through the forest industry-supported organization, National Council on Air and Stream Improvement (NCASI). In addition to these efforts, USDA supports a competitive grant program dealing with managed ecosystems that includes studies on forest–wildlife interactions. Information from all of these research efforts is extended to public- and private-land managers and to the general public to enable more effective management of land and water resources.

STATE LAWS

State laws generally have focused on traditional game management approaches with a lower priority given to protection of nongame species and biodiversity in general. But that trend seems to be changing. There is more and more pressure being placed on states to accept responsibility for biodiversity conservation and endangered species protection. But most states do not have adequate funding mechanisms to address these issues. Nonetheless, many states do have state ESAs and policies that influence management of habitat for these and other species (George et al. 1998). I provide a few examples of state wildlife laws that influence habitat management; these examples were largely extracted from Musgrave and Stein's (1993) overview of state wildlife laws.

Massachusetts identifies significant habitat for rare, threatened, and endangered species under its state ESA. Since most land in the state is privately owned, there are significant impacts on private forest land owners. Before a landowner can alter forest in a "significant habitat" (with some exceptions), the landowner must provide to the regulatory agency (MassWildlife) the following:

- A complete description of the project
- Alternatives to the proposed project
- Impacts of the proposed project on the subject species
- Plans for protection of the subject species and mitigation measures to be taken to offset these impacts
- A description of potential economic effects of the proposed project on the landowner and the community.

A permit to proceed with a project would be granted only upon a finding that the proposed action will not reduce the viability of the significant habitat to support the population of the subject species. Massachusetts also has developed a landowner incentives program for habitat improvement which is similar to the federal WHIP (Figure 19.2). So rather than simply regulating private landowner actions, they can encourage landowners to apply to participate in a cost-share program that would benefit certain species. Incentives and regulations represent the carrot and the stick approaches to maintaining biodiversity.

Louisiana also has policies regarding habitat improvement and protection for certain species. The state is authorized to contract with a private landowner for use of lands for at least 25 years to allow

FIGURE 19.2 Grasslands and early-successional woodlands managed for birds using Wildlife Habitat Incentives Program grant dollars, Arcadia Wildlife Sanctuary, Easthampton, Massachusetts.

establishment of wildlife management areas with a tax-free incentive to the landowner. In addition, the Louisiana Department of Wildlife and Fisheries established a "Louisiana Acres for Wildlife" program, which provides landowner assistance in habitat management. Biologists provide planting stock, guidance for habitat management actions, and habitat evaluation surveys. The program is voluntary and participants must have >0.4 ha of land and agree that management continue for at least 1 year. To date, the focus of this program has been on game species production, but clearly there are benefits for nongame species as well. Many other states, such as Oregon and Kentucky, have similar programs allowing the state to contract with landowners to establish wildlife refuges or management areas. Most states also are authorized to purchase lands to protect habitat, though until recently these purchases have been primarily to allow hunting of game species.

Minnesota's regulations include provisions for improvement or acquisition of "critical natural habitat." Identification of what constitutes a critical natural habitat is based on the following:

- Significance as existing or potential habitat for fish and wildlife and providing fish and wildlife oriented recreation.
- Significance to maintain or enhance native plant, fish, or wildlife species designated as endangered or threatened.
- Presence of native ecological communities that are now uncommon or diminishing.
- Significance to protect or enhance natural features within or contiguous to natural areas including fish spawning areas, wildlife management areas, scientific and natural areas, riparian habitat, and fish and wildlife management projects.

Purchase of lands for these purposes is restricted by availability of state funds so areas must be prioritized based on the above factors. States such as Massachusetts face similar decisions and where urbanization is an expanding land use causing purchase prices for important habitats to be excessively high. Consequently, some states are employing prioritization systems such as the Conservation Assessment and Prioritization System (CAPS, UMass–Amherst) and BIOMAP (Massachusetts Natural Heritage Program) to ensure that the purchases result in the greatest level of habitat protection for the most important species at the least cost.

The degree to which various states have begun to shift from a game management focus on their lands to an ecosystem management approach is highly variable. Because most state wildlife agencies are funded primarily from hunting and fishing license sales, expenditures on nongame species can easily be brought into question by their constituents. However, most state biologists have realized that "keeping all the pieces" is a correct course of action that will benefit both game and nongame species. An increasing number of states have developed alternative funding mechanisms for nongame species, including tax check-offs and vanity license plate sales, with proceeds going to biodiversity conservation programs.

Many states also have forest practices acts or policies restricting forest management actions to minimize the adverse effects of forest harvest and management on various ecosystem services. Initially, these were largely focussed on water quality and habitat for fish (e.g., Oregon and Washington Forest Practices Acts), but have expanded to include headwater and vernal pool amphibians (e.g., Massachusetts), and upland species (snag and green tree requirements in Oregon). Massachusetts, for instance, requires that a cutting plan be submitted to a state service forester who reviews it for compliance with state laws and also ensures that the plan is reviewed for presence of state-listed rare species by biologists in the Natural Heritage Program of MassWildlife. Such an approach may then lead to additional restrictions on activities based on the possible presence of a rare species and its habitat needs. The level of restrictions and level of review both prior to and following a harvest varies considerably from state to state. Often, separate state agencies within the same state are involved in the forest practices act enforcement and habitat management enforcement. Such an approach can at times pit one set of natural resources professionals against another. Approaches that allow both groups to identify a desired future condition (DFC) and mutually agree on approaches to work toward that DFC would often be more effective. In addition, increased levels of coordination of habitat management among states can increase effectiveness for many species because animals usually do not respond to political boundaries. Animals cross state lines during migration and dispersal. As climate change places additional pressures on geographic ranges, cross-boundary movements may become even more important. Interstate coordination of habitat acquisition, leasing, management, and prioritization may improve the effectiveness of approaches among participating states.

MUNICIPAL POLICIES

Habitat is also considered explicitly in the policies enacted at the county and town levels. These sorts of policies vary tremendously and often are associated with zoning laws, urban growth boundary designations, and building permit laws. For instance, Wisconsin County Forests are governed by the County Forest Law requiring that these forests be managed for multiple uses such as forest products, recreation, wildlife habitat, and watershed protection. Twenty-eight northern Wisconsin counties own and manage nearly 1 million ha of county forest lands. Counties are required to develop comprehensive land-use plans for each forest. Eau Claire, Wisconsin, manages a County Forest specifically as habitat for wildlife "common to Wisconsin." This plan recognizes that each species requires different forest conditions ranging from recently disturbed to old growth. They also pay particular attention to endangered species on the property (Karner Blue Butterfly, a federally listed endangered species) and the county has developed a habitat conservation plan for this species.

In the northeastern United States, county governance is not as important as town governance. Towns have policies that influence habitat management and availability and these vary widely from town to town. Few explicitly consider wildlife habitat. The town of Milan, New York, however, participates in an effort to "... protect the integrity and value of Milan's natural areas, and protect the Town's watershed and significant biological resources." The townspeople recognize that the diverse natural resources of the town are particularly vulnerable to adverse impacts associated with development and sprawl. They are conducting habitat assessments to provide baseline information and improve the town planners' ability to protect significant biological resources in the face of increasing development pressure. The habitat assessment program takes into consideration a number of environmental impacts associated with development including direct loss of habitat and reduced populations, habitat fragmentation and adverse edge effects, increasing effects of invasive species, degraded water quality, and increased pollution. Such a detailed approach to planning is becoming more and more common, especially in communities where land-use change is rapid and local communities become increasingly concerned about the long-term adverse effects of such changes.

POLICY ANALYSIS

The layers upon layers of policies that apply to forests across North America can easily overwhelm managers of forestlands. Policies are many, complex, and frequently changing. It is a challenge to develop forest management plans that follow current policies. By having foresters work with biologists to develop plans, it is more likely that pertinent forest practices rules and wildlife laws will be followed. But assuming that all policies are followed (clearly an incorrect assumption!), will they achieve society's goals? Will the application of the federal, state, and local policies to forest parcels across a mixed-ownership landscape lead to better biodiversity protection in 10 years? 100 years? Will forest management continue to be economically viable? In many instances the answer to these questions from policy makers is, "We think so." But in nearly every case, it is unclear how effective any single policy might be in achieving its goal and even less certain when considered in the presence of a set of additional policies, some with entirely different objectives.

Uncertainty of achieving desired results from policy implementation are related to the complexity or messiness of the policy problem. Lackey (2006) indicated that messy ecological policy problems often share several qualities:

1. Complexity — innumerable options and trade-offs.
2. Polarization — clashes between competing values.
3. Winners and losers — for each policy choice, some will clearly benefit, some will be harmed, and the consequences for others are uncertain.
4. Delayed consequences — no immediate "fix" and the benefits, if any, of painful concessions will often not be evident for decades.
5. Decision distortion — advocates often appeal to strongly held values and distort or hide the real policy choices and their consequences.
6. National vs. regional conflict — national (or international) priorities often differ substantially from those at the local or regional level.
7. Ambiguous role for science — science is often not pivotal in evaluating policy options, but science often ends up serving inappropriately as a surrogate for debates over values and preferences.

Lackey (2006) then proposed nine axioms that are typical of most current ecological policy problems that must be considered during policy development and debate:

1. The policy and political dynamic is a zero-sum game.
2. The distribution of benefits and costs is more important than the ratio of total benefits to total costs.

3. The most politically viable policy choice spreads the benefits to a broad majority with the costs limited to a narrow minority of the population.
4. Potential losers are usually more assertive and vocal than potential winners and are, therefore, disproportionately important in decision making.
5. Many advocates will cloak their arguments as science to mask their personal policy preferences.
6. Even with complete and accurate scientific information, most policy issues remain divisive.
7. Demonizing policy advocates supporting competing policy options is often more effective than presenting rigorous analytical arguments.
8. If something can be measured accurately and with confidence, it is probably not particularly relevant in decision making.
9. The meaning of words matters greatly and arguments over their precise meaning are often surrogates for debates over values.

Although scientific information is just one element of complex political deliberations in a democracy (Lackey 2006), it can and has influenced policy direction. One approach to considering the potential effectiveness of policies is to conduct a *policy analysis*, an organized projection of how implementation of the policy over space and time might possibly affect the resources valued by society. It is most useful to think of policies as hypotheses. The ones enacted now are being tested now, especially if we monitor the results. But there are two primary questions that should be asked by policy makers in particular and society in general: (1) are current policies achieving our goals? and (2) would an alternative policy be more effective in achieving our goals?

Spies et al. (2007) used the Oregon Coast Range as a case study to examine how forest policies might affect various measures of biodiversity over a multiownership region (Figure 15.8). The dominant federal policy, the Northwest Forest Plan, is designed to increase habitat for species associated with late-successional forests on federal lands (FEMAT 1993). Using projections of forest landscape development under pertinent federal and state policies, the forests of the region are expected to move toward the historical range of variation for most age classes of forest (Spies et al. 2007) and improve habitat for associated late successional species. But habitat recovery for some species may take >100 years; a very long-term policy would need to be in place! In addition, the current set of policies are expected to result in declines in diverse early successional forests and in hardwood forests in the region, with predicted declines in habitat availability for several species associated with these types of forests (Spies et al. 2007). These unanticipated changes in forest composition and structure may cause policy makers to develop alternative policies to ensure that such changes do not lead to a landscape in which additional species are placed at risk. Projections such as these represent "thought-experiments" that can provide policy makers insights into the possible outcomes of forest management policies (Oreskes 1997).

HOW DECISIONS IN THE UNITED STATES INFLUENCE HABITAT IN THE WORLD

Many nations of the world have complex forestland management policies that are enforced to ensure that ecosystem services demanded by society are provided into the foreseeable future. But the degree to which various countries have international policies addressing biodiversity conservation (the United States does), reducing the effect of climate change (the United States does not), or trade in endangered species varies tremendously. And policies enacted in the United States that conserve biodiversity can have significant impacts on biodiversity conservation in other countries.

The world has become a smaller place with global transportation and economies. All countries aspire for a high standard of living. They want a clean environment, a safe place to live, adequate health care, and to the degree possible stability and security in the lives of its citizens. Some countries can afford such a set of conditions (the "haves"), while others struggle to provide the basic services needed for human survival (the "have-nots"), and of course these extremes represent endpoints of a spectrum. The haves not only can afford to meet the expectations of most of its citizens most of the time, but also often have large appetites for resources such as energy, water, and timber. If the timber does not come from the haves, then it will come from the have-nots, the countries least able to afford enforcement of environmental laws. Where the have-not countries also represent areas of high biodiversity (e.g., some tropical countries) the haves countries conserve their biodiversity while threatening those in other countries — worldwide. How do we solve this problem? Altruism, a human quality of placing others before yourself is a human behavior that could be espoused by both citizens and the politicians who represent them. Realizing that, for the greater good of both humanity and biodiversity, some societies must have less so that others can have more, would begin to address these problems of inequity. Use less energy, less water, less wood, make fewer babies, accept stable economies and do not expect them to continue to grow, grow the food you eat, grow the wood that you use. These things have not happened yet. It is not clear what it might take for altruism to be more broadly expressed in the world.

SUMMARY

The myriad of interacting policies at the federal, state, and municipal levels influence the amount and distribution of habitat for various species from forest stands to global forests. Policies have begun to evolve at all levels from a strong focus on habitat for game species to one of biodiversity conservation. Increased levels of coordination in policy formulation across political lines (international, interstate, and cross-county) may be necessary to ensure biodiversity conservation in forested landscapes. Analyses of the future implications of forest policies can allow policy makers to better understand the time needed to see results of their policies if there are unanticipated consequences from their policies, and if alternative policies might better achieve their goals.

REFERENCES

Bean, M.J., and M.J. Rowland. 1997. *The evolution of national wildlife law*, 3rd edition. Praiger, Westport, CT.

Bliss, J.C., S. Nepal, R. Brooks Jr., and M.D. Larsen. 1997. In the mainstream: Environmental attitudes of mid-south forest owners. *So. J. Appl. For.* 21: 37–43.

Elliott, G., B. Altman, W. Easton, R. Estrella, G. Geupel, M. Chase, E. Cohen, and A. Chrisney. 2005. Integrated bird conservation along the Pacific Coast of North America: an action agenda. In Ralph, C.J., and T.D. Rich (eds.), *Bird Conservation Implementation and Integration in the Americas: Proceedings of the Third International Partners in Flight Conference*, Vol. 1. March 20–24, 2002; Asilomar, California. USDA For. Serv. Gen. Tech. Rep. PSW-GTR-191.

Forest Ecosystem Management Assessment Team [FEMAT]. 1993. *Forest Ecosystem Management: An Ecological, Economic, and Social Assessment*. Bureau of Land Management, Portland, OR. U.S. Fish & Wildlife Service, National Oceanic and Atmospheric Administration, U.S. Environmental Protection Agency, and National Park Service.

George, S.W., J. Snape III, and M. Senatore. 1998. *State Endangered Species Acts: Past, Present and Future*. Defenders of Wildlife, Washington, D.C.

Hibbard, M., and J. Madsen. 2003. Environmental resistance to place-based collaboration in the US West. *Soc. Natur. Resour.* 16: 703–718.

Hutton, J., and B. Dickson. 2000. Endangered species, threatened convention. The past, present and future of CITES, the convention on the international trade in endangered species of wild fauna and flora. Earthscan Publications Ltd., London.

Innes, R., S. Polasky, and J. Tschirhart. 1998. Takings, compensation and endangered species protection on private land. *J. Econ. Perspec.* 12: 35–52.

Lackey, R.T. 2006. Axioms of ecological policy. *Fisheries* 31: 286–290.

Mace, G.M., and R. Lande. 1991. Assessing extinction threats: toward a reevaluation of IUCN threatened species categories. *Conserv. Biol.* 5: 148–157.

Moser, D.E. 2000. Habitat conservation plans under the US endangered species act: the legal perspective. *Environ. Manage.* 26: S7–S13.

Musgrave, R.S., and M.A. Stein. 1993. *State Wildlife Laws Handbook*. Government Institutes, Rockville, MD.

Oehler, J.D. 2003. State efforts to promote early-successional habitats on public and private lands in the northeastern United States. *For. Ecol. Manage.* 185: 169–177.

Oreskes, N. 1997. Testing models of natural systems: can it be done? In Chiara, M.L. et al. (eds.), *Structures and Norms in Science*. Kluwer Academic Publishers. Amsterdam, pp. 207–217.

Pagiola, S., J. Bishop, and N. Landell-Mills (eds.). 2002. *Selling Forest Environmental Services*. Earthscan, London.

Sagoff, M. 1997. Muddle or muddle through?: takings jurisprudence meets the Endangered Species Act. *William and Mary Law Rev.* 38: 825–993.

Smallwood, K.S. 2000. A crosswalk from the Endangered Species Act to the HCP handbook and real HCPs. *Environ. Manage.* 26: 23–35.

Smythe, R., and C. Isber. 2003. NEPA in the agencies: a critique of current practices. *Environ. Pract.* 5: 290–297.

Spies, T.A., B.C. McComb, R. Kennedy, M. McGrath, K. Olsen, and R.J. Pabst. 2007. Habitat patterns and trends for focal species under current land policies in the Oregon Coast Range. *Ecol. Applic.* 17: 48–65.

Thomas, J.W., J.F. Franklin, J. Gordon, and K.N. Johnson. 2006. The Northwest Forest plan: origins, components, implementation experience, and suggestions for change. *Conserv. Biol.* 20: 277–287.

Thompson, D.H., and S.H. Bullard. 2004. *History and Evaluation of the McIntire-Stennis Cooperative Forestry Research Program*. Mississippi State Univ. Forestry and Wildlife Research Center, Starkville, Bull. FO249.

20 Should I Manage a Forest?

Most of this book has addressed active management — taking actions to achieve habitat goals for a species, a community or contributing to biodiversity conservation, while also considering the potential for providing wood and nontimber products for people. But many people do not feel compelled to manage their forests. The millions of small private landowners in the United States and Canada may own their forests for reasons other than timber, woodcock, or deer. They just like to have a forest. To walk through it, see it, sit in it, and listen to the birds in it (regardless of species). Except when there is a disturbance, forests change slowly. They provide a place that evokes stability, security, and spirituality. For people who view forests in this way, management is not only unnecessary, it is disruptive, and evokes instability, insecurity, and flies in the face of personal spirituality. And they may extend those feelings to all forests regardless of who owns them, because, after all, we are merely temporary tenants on earth, regardless of what we pay for the pieces we use. So, for many people, doing nothing is a perfectly acceptable management decision (Kittredge and Kittredge 1998). And doing nothing is indeed a management decision.

Other landowners wish to have certain animal species to hunt, wood to sell or burn, leaf colors to enjoy, as well as clean water to drink. They choose a more active management decision. Sometimes this involves regeneration methods that other landowners and neighbors do not find acceptable. NIMBY — not in my back yard — is a phrase that captures the essence of the disagreement between the two philosophies (Shindler et al. 2002). And some people want to restore a forest to a previous, "better" condition, with "better" being something valued by the landowner, or land manager, neighbor, or society. Restoration represents an example of a philosophy associated with active management, just as "doing nothing" and "commodity production" do. These philosophies are all points on a spectrum of values and behaviors. But restoration often represents the middle ground — the production of a desired future condition that is neither utilitarian nor "let nature take its course." So, I use restoration as an example of a management philosophy that we can examine more closely.

WHAT DOES RESTORATION MEAN?

Restoration is a noble goal — but restored to what? What is a reasonable target? What is our reference condition? Is it an ecological condition? A cultural one? Or both? And is it even possible to now "restore" a system that has been changed markedly because of recent intense human activities, climate change, invasive species, or toxic compounds. Higgs (2005) distinguished between the terms "ecological restoration" and "restoration ecology." *Restoration ecology* is the suite of scientific practices that constitute an emergent subdiscipline of ecology (Higgs 2005). *Ecological restoration* uses practices that represent restoration ecology as well as human and natural sciences, politics, technologies, economic factors, and cultural dimensions (Higgs 2005). Maintaining a broader approach to restoration requires respect for knowledge in addition to science, and especially the recognition of a morality, which is beyond the scope of science to be addressed fully. Some have made an analogy between ecological restoration and human health (Schaefer 2006). Schaefer (2006) suggested that portraying the human body as a metaphor of a natural ecosystem can be useful in identifying the breadth of strategies used to restore the "natural" environment. The use of science in restoration is analogous to the use of technology to address a disease through drugs or a surgery, while the cultural and ecological interactions are analogous to restoring health to the human body through a broader holistic/preventative approach to cultivate the mind–body connection (Schaefer 2006). So the integration of the tools, concepts, and examples described in this book in conjunction with cultural values and individual beliefs can be used to set restoration goals and objectives. It is not just about science but about the people who use science and other tools to restore a system to meet their needs. "Restoration" to one person may be quite different from what another person visualizes it as. Coming to an agreement on what "restoration" means to both the individuals and the community is central to setting restoration goals when many people are affected by a management decision. But agreement is difficult, especially when individual property rights enter the discussions. Rather, some proportion, at times a majority, of the citizens in an area (e.g., the watershed councils that have formed in Oregon) can agree to share a core set of goals and move the area toward that goal, recognizing that not all landowners will share those goals.

Human Requirements as Constraints on Goals

The historical range of variability (HRV) has often been proposed as a concept that can be used by forestland managers to guide conservation of ecosystem functions and biodiversity (Landres et al. 1999, Morgan et al. 1994, Swetnam et al. 1999). The rationale for use of the HRV in certain ecosystem properties is that biodiversity was assumed to persist, albeit with fluctuations in populations, over thousands of years of disturbance and recovery. Further, the concept assumes that as contemporary conditions depart from historical processes and states, the risk of losing species, both known and unknown, increases.

The following ideas reflect a conceptual framework developed by Dr. Sally Duncan, Dr. Norm Johnson, and myself for understanding the range of variability in a way that considers not only HRV as traditionally defined but also likely future ranges of variability given ecological processes and conditions that society finds acceptable. Authors have often used a probability distribution to reflect the probability of occurrence of certain ecological states or indicators over some reasonably long period of time (e.g., Wimberly and Ohmann 2004) in order to represent a HRV. The probabilities are derived from the outcome of physical disturbances that have occurred over this time frame, and typically included native American disturbances, but not usually contemporary disturbances produced since European colonization. Clearly, humans have been, are, and will be forces of disturbance and recovery in forested landscapes. The range of variability expressed at any time reflects human and biophysical drivers of landscape change while recognizing both temporal and spatial scale dependency. A key component of our understanding of the expression of variability in an ecosystem is the effect of a social range of variability (SRV). SRV reflects the cultural mores that collectively influence the range of conditions that society finds acceptable. Integration of the SRVs of the past, present, and future with disturbance and recovery during the past, present, and future interact to define both historical and future ranges of variability in ecosystem indicators: an ecological range of variability (ERV) and a SRV (Figure 20.1). The ERV is a product of ecosystem disturbance and recovery from any cause: physical, biological, or human. The patterns of disturbance and recovery over space and through time presumably produced a set of conditions that supported biodiversity historically and will influence biodiversity in the future. Societies have been more or less effective in influencing disturbance frequencies, intensities, sizes, and durations over time, with classic examples of recent short-term influences being flood control dams, fire control, and harvest of old-growth trees. But historical influences of humans on ecological conditions also are apparent. The Kalapuya Americans burned the Willamette Valley regularly to maintain a savannah system (Boag 1992). Native Americans also contributed to the fire frequencies of the longleaf pine savannahs of the Deep South (Denevan 1992).

The probability distribution of an ecological indicator arising from past disturbances (Figure 20.1) is directly influenced by the SRV, which can be described as a probability distribution of ecosystem conditions that are socially acceptable. For example, rivers channelized or dammed for irrigation and flood control (socially acceptable, at varying levels through time) no longer nourish wetlands during

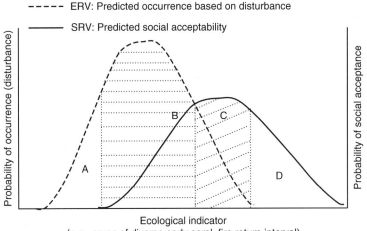

FIGURE 20.1 Conceptual diagram reflecting the interaction between the ERV resulting from disturbances and recovery in ecosystems, and the SRV reflecting social acceptability of ecological conditions. It is the tension between these two forces that produce the range of conditions that we see expressed across forested landscape over time. (Based on discussion with Duncan, S. and K.N. Johnson.)

and after flooding (a natural disturbance) (Shafroth et al. 2002). Suppressed wildfires (socially desirable) can no longer regulate levels of insect and disease populations (natural disturbance) (McCullough et al. 1998). Thus, ERV and SRV are inextricably linked. Even such disturbances as hurricanes and ice storms today may have their frequency affected by "desired" human activity across the planet, which is influencing our climate (Knutson and Tuleya 1999).

Historically, the full range of the probability distribution in an ecological indicator, including the part that may be socially unacceptable, may have been observed simply because human societies had insufficient influence on biophysical disturbance regimes. Earlier societies could not limit the occurrence of some unacceptable conditions driven largely by biophysical forces (e.g., stand-replacement wildfire). However, given the size of current human populations and their wealth and technological power, human societies today are capable of exerting significant pressure on the disturbance-based probability distribution, thereby altering possible future outcomes/trajectories. As the SRV departs from the ERV, social pressures can change drivers of ecosystem disturbance and limit the expression of the ERV (Figure 20.1). The more disparate the ERV and SRV, the greater is the potential for society to influence the expression of the ERV. This disparity is often reflected in the various management options considered for an area and the policies guiding the direction of land management. That disparity, or tension, between ERV and SRV will wax and wane depending on changes in social values as well as changes in nonhuman disturbance dynamics.

Given that human systems and ecosystems are inextricably linked, how can we be sure that decisions made today will not have unanticipated negative consequences in our future or in our children's futures? Science can provide the facts, but people also hold beliefs. Facts can be proven, beliefs cannot. It is through the dynamic combination of knowledge with evolving beliefs that change in societal goals occurs. Many events and experiences — both biophysical and social — affect beliefs and knowledge, and hence the ever-changing nature of societal goals.

DEVELOPING A PERSONAL MANAGEMENT PHILOSOPHY

As you go through your professional career you will use a variety of tools to make decisions: scientific data, belief, intuition, and stakeholder opinions, among others. All of these tools pass through the filter of your personal philosophy. Your core values and your beliefs give rise to your behaviors in using all of these tools. Two people using the same tools to make a decision but having different management philosophies will often arrive at different decisions. And these differences are not always apparent to those making the decisions. But the differences are often questioned by those holding different philosophies, and oftentimes in public settings. It is wise to be prepared to explain your personal philosophy to others, not to impose it on them but merely to explain it clearly and concisely. Although those with different philosophies may never agree with you, they might understand you, thereby making dialog possible. It is also important to keep in mind that our personal philosophies usually evolve throughout our lives. Checking in on your personal philosophy periodically through introspection is also prudent when faced with making daily management decisions. In addition, as you move into supervisory positions, you might consider encouraging management teams of all kinds and at all levels to similarly reassess their philosophies and share them with each other.

In the following sections I provide a glimpse into my personal philosophy, not to impose it on you, not to try to convince you that it is correct, but to make you think. You may not agree with some of the things that I have written, and some may make you angry. And if you do get angry, that is fine if it makes you think about why you have that reaction and how your philosophy differs from mine. Use my philosophy as a springboard to your own.

OUR PLACE ON EARTH

Humans are a species, and as such we have habitat needs to allow us to persist on this earth. We are simply one of millions of species on this earth, sharing space, energy, and time together. In my view, resources were not placed on this earth for us to use by some omniscient, omnipotent, or "omni-anything" being. We just are. Our life arose from the resources on this earth around our parents, the material bits of us will return to the resources of the earth, and our actions form the legacy that we leave to other humans and other species. Each of us is a blip in time and space. Or rather, we are 6 billion blips, each trying to survive, each using resources that could be used by other species on this planet — each arising from the earth, each returning to the earth. The earth is our source, and as far as we know the only source of any life for many light years in any direction. So it is prudent to treat it as we would our own home, and share it not only with those other individuals and species that are with us now but also with those that will come after us. In my view, we have a moral obligation to consider the effects of our 6 billion lives on the other species with which we share this planet, and to not pollute it, not overuse it, and not "poop in our own petri dish." But we have to recognize that we are a part of it, not separate, and how we make collective decisions affects the lives of other people and other species.

LIVING SIMPLY AND SUSTAINABLY

Energy can be changed from one form to another, but it cannot be created or destroyed. The total amount of energy in the universe is constant, merely changing from one form to another. The laws of thermodynamics apply to humans and all other life forms. From a practical standpoint, there is a solar constant. Allocating more energy to human needs and desires leaves less for other organisms. How do we redistribute energy on earth to allow coexistence with other organisms? Reducing population growth is an obvious first step. What is the carrying capacity of the earth for humans while ensuring the coexistence with other forms of life? Have we surpassed it already? Quite likely. Granted that technology and scientific advancement can provide us with a marginal gain in our ability to support humans and other species, but we are faced with the fact that there is only so much energy to be allocated to all species and organisms. Technology is simply fussing with the edges, fine-tuning the allocations. We will reach a point of diminishing returns on technological advancements to increasing human-carrying capacity (Czech 2003).

And if we were to seriously start now in reducing human populations, there will be lag time to population reduction. Consequently, reducing energy consumption is another step. Our lives are based on growth economies. Consumption is good — it spurs economic growth. But for how long? If money is indeed a surrogate for energy, and there is a solar constant, then how can energy consumption continually increase? It cannot. Accepting stable economies rather than continually growing economies is something that contemporary societies must begin to grapple with, or the differences in the quality of life between the "haves" and the "have-nots" will continue to increase, leading to greater and greater political instability.

And even if we reduce populations, control consumption, live more simply, and reduce carbon emissions and other pollutants on our earth, will it lead to continued coexistence with other species? Much depends on not only how much of the earth's renewable resources we consume, but also where and when. Even doing nothing comes at a cost. Location, frequency, and intensity of resource use matters to our own long-term well-being, and the well-being of other species. There are limits to the resistance and resilience of the ecosystems in which we all live. As we use the earth to meet our own needs, it is important to remember that it must be given the time to recover from those uses to allow ecosystems to express resilience to both human and natural disturbances.

LEAVING THE WORLD A BETTER PLACE

We all have our own ideas of "better." Much of how various people define "better" has to do with their time frames — better today, this week, next year, or next generation? Taking the long-term view, and assuming that most people most of the time will want to leave the world a better place for future generations, it becomes clearer that we cannot have it all now and in the future too. We could live in a way that allows us to be happy and prosper and also allow those who come after us to live lives as good or better as ours. We can provide that next generation with the management options they would need to develop and implement their own definitions for "sustainable." Leaving such a suitable legacy for our children takes commitment and discipline, and will mean that societies with the most will have to be satisfied with not having even more (Figure 20.2). The insatiable appetites for resources from rich countries pass the burden of production and loss of sustainability

FIGURE 20.2 Management philosophies often involve insurance that there are always intergenerational opportunities for use of wildlife and forest resources.

on to other societies. If we enter the global arena of natural resources management as equals, sharing the benefits and the costs with other societies, then I suspect that the image of the "haves" would improve significantly among the "have-nots." This would not only provide us with an opportunity to live in greater harmony than we do now, but would also provide all children of the future the options and opportunities that many do not enjoy now, including the opportunity to live on a diverse living planet.

SUMMARY

Management of the earth's resources to meet human needs and provide habitat for various species is dependant on the values and beliefs of people. We all approach management with the tools of science, beliefs, opinions, and culture, and pass those tools through the filter of our own philosophy. Defining a philosophy of forest and habitat management is important as you interact with others oftentimes holding different views. Developing and articulating your philosophy of management is something that should be done at the outset of your career and revisited periodically as your personal philosophy evolves throughout your life.

REFERENCES

Boag, P.G. 1992. *Environment and Experience: Settlement Culture in Nineteenth Century Oregon.* University of California Press, Berkeley, CA.

Czech, B. 2003. Technological progress and biodiversity conservation: a dollar spent, a dollar burned. *Conserv. Biol.* 17: 1455–1457.

Denevan, W.M. 1992. The pristine myth: the landscape of the Americas in 1492. *Ann. Assoc. Am. Geographers* 82: 369–385.

Higgs, E. 2005. The two-culture problem: ecological restoration and the integration of knowledge. *Restor. Ecol.* 13: 159–164.

Kittredge, D.B., and A.M. Kittredge. 1998. Doing nothing. *J. For.* 96: 64.

Knutson, T.R., and R.E. Tuleya. 1999. Increased hurricane intensities with CO_2-induced warming as simulated using the GFDL hurricane prediction system. *Clim. Dyn.* 15: 503–519.

Landres, P., P. Morgan, and F.J. Swanson. 1999. Overview of the use of natural variability in managing ecological systems. *Ecol. Appl.* 9: 1279–1288.

McCullough, D.G., R.A. Werner, and D. Neumann. 1998. Fire and insects in northern and boreal forest ecosystems of North America. *Annu. Rev. Entomol.* 43: 107–127.

Morgan, P., G. Aplet, J.B. Haufler, H. Humphries, M.M. Moore, and W.D. Wilson. 1994. Historical range of variability: a useful tool for evaluating ecosystem change. *J. Sustain. For.* 2: 87–111.

Schaefer, V. 2006. Science, stewardship, and spirituality: the human body as a model for ecological restoration. *Restor. Ecol.* 14: 1–3.

Shafroth, P.B., J.M. Friedman, G.T. Auble, M.L. Scott, and J.H. Braatne. 2002. Potential responses of riparian vegetation to dam removal. *BioScience* 52: 703–712.

Shindler, B., M. Brunson, and G.H. Stankey. 2002. *Social Acceptability of Forest Conditions and Management Practices: A Problem Analysis.* USDA For. Serv. PNW-GTR-537.

Swetnam, T.W., C.D. Allen, and J.L. Betancourt. 1999. Applied historical ecology using the past to manage for the future. *Ecol. Appl.* 9: 1189–1206.

Wimberly, M.C., and J.L. Ohmann. 2004. A multi-scale assessment of human and environmental constraints on forest land cover change on the Oregon (USA) coast range. *Lands. Ecol.* 19: 631–646.

Appendix 1:
Common and Scientific Names of Species Mentioned in the Text

MAMMALS

Alces alces	Moose
Aplodontia rufum	Mountain beaver
Arborimus albipes	White-footed vole
Arborimus longicaudus	Red tree vole
Bettongia pennicilata	Woylie
Bison bison	Bison
Blarina brevicauda	Short-tailed shrew
Canis lupus	Gray wolf
Castor canadensis	Beaver
Cervus elaphus	Elk
Clethrionomys californicus	Western (California) red-backed vole
Clethrionomys gapperi	Gapper's red-backed vole
Didelphis virginianus	Virginia opossum
Elaphus maximus	Asian elephant
Erithrizon dorsatum	Porcupine
Felis concolor	Florida panther, cougar
Felis lynx	Lynx
Glaucomys sabrinus	Northern flying squirrel
Glaucomys volans	Southern flying squirrel
Gulo gulo	Wolverine
Lepus americanus	Snowshoe hare
Lepus spp.	Hares
Lontra canadensis	River otter
Martes americana	American marten
Martes pennanti	Fisher
Mephitis mephitis	Striped skunk
Microtus oregoni	Creeping vole
Microtus spp.	Voles
Mustela vison	Mink
Myotis spp.	Myotis bats
Myotis volans	Long-legged bat
Neotoma spp.	Wood rats
Odocoileus hemionus	Mule deer
Odocoileus hemionus sitkensis	Sitka black-tailed deer

Odocoileus spp.	Deer
Odocoileus virginianus	White-tailed deer
Ondatra zibethicus	Muskrat
Peromyscus leucopus	White-footed mice
Peromyscus maniculatus	Deer mouse
Procyon lotor	Raccoon
Rangifer tarandus	Caribou
Rattus norvegicus	Norway rat
S. trowbridgii	Trowbridge's shrew
Sciurus carolinensis	Gray squirrel
Sciurus niger	Fox squirrel
Sorex pacificus	Pacific shrew
Sorex palustris	Water shrew
Sorex vagrans	Vagrant shrew
Sus scrofa	Domestic pig
Sylvilagus floridanus	Cottontail rabbit
Tamias striatus	Eastern chipmunk
Tamiasciurus douglasii	Douglas squirrels
Thomomys spp.	Pocket gophers
Trichosurus vulpecula	Brushtail possum
Ursus americanus	Black bear
Ursus arctos horribilis	Grizzly bear
Ursus spp.	Bears
Vulpes vulpes	Red fox
Zapus spp.	Jumping mice
Zapus trinotatus	Pacific jumping mouse

BIRDS

Accipiter gentilis	Northern goshawk
Actitis macularia	Spotted sandpiper
Aimophila aestivalis	Bachman's sparrow
Aix sponsa	Wood duck
Amazona vittata	Puerto Rican parrot
Anas rubripes	American black duck
Ardea herodias	Great blue heron
Bonasa umbellus	Ruffed grouse
Brachyramphus marmoratus	Marbled murrelet
Bubo virgianus	Great horned owl
Bucephala clangula	Common goldeneye
Buteo jamaciensis	Red-tailed hawk
Buteo swainsoni	Swainson's hawk
Campephilus principalis	Ivory-billed woodpecker
Catharus bicknelli	Bicknell's thrush
Catharus guttatus	Hermit thrush
Catharus ustulatus	Swainson's thrush
Certhia americana	Brown creeper
Ceryle alcyon	Belted kingfisher
Chaetura spp.	Swifts
Chaetura vauxi	Vauk's swift
Cinclus mexicanus	American dipper

Coccyzus americanus	Yellow-billed cuckoo
Colaptes auratus	Common flicker
Colinus virginianus	Bobwhite quail
Columba fasciata	Band-tailed pigeon
Contopus cooperi	Olive-sided flycatcher
Corvus corax	Common raven
Dendroica occidentalis	Hermit warbler
Dendroica pensylvanica	Chestnut-sided warbler
Dendroica petechia	Yellow warbler
Dendroica virens	Black-throated green warbler
Dolichonyx oryzivorus	Bobolink
Dryocopus pileatus	Pileated woodpecker
Ectopistes migratorius	Passenger pigeon
Empidonax hammondii	Hammond's flycatcher
Falcipennis canadensis	Spruce grouse
Falco peregrinus	Peregrine falcon
Gavia immer	Common loon
Haliaeetus leucocephalus	Bald eagle
Histrionicus histrionicus	Harlequin duck
Hylocichla mustelina	Wood thrush
Isoreus naevius	Varied thrush
Junco hyemalis	Dark-eyed junco
Melanerpes formicivorus	Acorn woodpecker
Mniotilta varia	Black-and-white warbler
Parus bicolor	Tufted titmouse
Passer domesticus	House sparrow
Phasianus colchicus	Ring-necked pheasant
Picoides borealis	Red-cockaded woodpecker
Picoides pusescens	Downy wood pecker
Pipilo maculatus	Spotted towhee
Poecile atricapilla	Black-capped chickadee
Scolopax minor	Woodcock
Seiurus aurocapillus	Ovenbird
Sialia sialis	Eastern bluebird
Sitta pusilla	Brown-headed nuthatch
Strix occidentalis caurina	Northern spotted owl
Strix varia	Barred owl
Sturnella magna	Eastern meadowlark
Sturnus vulgaris	European starling
Thryothorus ludovicianus	Carolina wren
Troglodytes aedon	House wren
Troglodytes troglodytes	Winter wren
Vermivora celata	Orange-crowned warbler
Vireo olivaceous	Red-eyed vireo
Wilsonia canadensis	Canada warbler
Zonotrichia leucophrys	White-crowned sparrow

AMPHIBIANS

Ambystoma maculatum	Spotted salamander
Ambystoma opacum	Marbled salamander

Aneides ferreus	Clouded salamander
Ascaphus truei	Tailed frog
Dicamptodon tenebrosus	Pacific giant salamander
Ensatina eschscholtzii	Ensatina salamander
Gyrinophilus porphyriticus	Spring salamander
Plethodon dunnii	Dunn's salamander
Plethodon spp.	Slimy salamander
Plethodon stormi	Siskyou mountain salamander
Plethodon welleri	Weller's salamander
Rana aurora draytonii	California red-legged frog
Rana cascadae	Cascades frog
Rana catesbiana	Bullfrog
Rana septentrionalis	Mink frog
Rana sylvatica	Wood frog
Rhyacotriton variegatus	Southern torrent salamander
Taricha granulosa	Rough-skinned newt

REPTILES

Agkistrodon contortrix	Northern copperhead
Agkistrodon piscivorous	Cottonmouth
Chelydra serpentine	Snapping turtle
Clemmys marmorata	Western pond turtle
Crotalus spp.	Rattlesnake
Crotalus viridis	Prairie rattlesnake
Elaphe obsoleta	Rat snake
Elgaria multicarinata	Southern Alligater Lizard
Gopherus polyphemus	Gopher tortoise
Sceloporus occidentalis	Western fence lizard
Terrapene carolina	Box turtle

INSECTS

Adelges tsugae	Hemlock wooly adelgid
Blatella spp.	Cockroaches
Camponotus spp.	Carpenter ant
Chrysomela confluens	Leaf beetles
Dendroctonus frontalis	Southern pine beetle
Dendroctonus spp.	Bark beetles
Lycaeides melissa samuelis	Karner blue butterfly
Lymantria dispar	Gypsy moth

FISH

Carassi carassius	Carp
Oncorhynchus clarki	Cutthroat trout
Oncorhynchus kisutch	Coho salmon
Salmo spp.	Trout
Salvelinus fontinalis	Brook trout

PLANTS

Abies balsamea	Balsam fir
Abies grandis	Grand fir
Acer macrophyllum	Bigleaf maple
Acer pensylvanicum	Striped maple
Acer platanoides	Norway maple
Acer rubrum	Red maple
Acer saccahrum	Sugar maple
Acer saccharinum	Silver maple
Acer spicatum	Mountain maple
Acer spp.	Maples
Ailanthus altissima	Tree-of-Heaven
Alnus rubra	Red alder
Amalanchier spp.	Serviceberries
Arbutus menziesii	Pacific madrone
Betula alleghaniensis	Yellow birch
Betula lenta	Black birch
Betula paperifera	White birch
Betula populifolia	Gray birch
Betula spp.	Birches
Brachypodium sylvaticum	False-brome
Carya glabra	pignut hickory
Carya ovata	Shagbark hickory
Carya spp.	Hickories
Castanea dentata	American chestnut
Chamaecyparis lawsoniana	Port-Orford cedar
Cornus florida	Flowering dogwood
Cornus spp.	Dogwoods
Corylus cornuta	Hazlenut
Crataegus spp.	Hawthorns
Cytisus scoparius	Scotch broom
Eleaganus umbellata	Autumn olive
Endothia parasitica	Chestnut blight fungus
Fagus grandifolia	American beech
Fomes pini	Red heart disease
Gallium spp.	Bedstraws
Ilex spp.	Hollies
Ilex verticillata	Winterberry
Ilex vomitoria	Yaupon
Kalmia latifolia	Mountain laurel
Juglans spp.	Walnuts
Liquidambar styraciflua	Sweetgum
Liriodendron tulipifera	Yellow-poplar
Magnolia fraseri	Fraser magnolia
Melaleuca quinquenervia	Australian paperbark tree
Nyssa aquatica	Water tupelo
Nyssa sylvatica	Blackgum
Phaeocryptopus gaeumannii	Swiss needle cast
Phytophthora lateralis	Root rot
Picea glauca	White spruce

Picea mariana	Black spruce
Picea rubens	Red spruce
Picea sitchensis	Sitka spruce
Picea spp.	Spruces
Pinus banksiana	Jack pine
Pinus contorta	Lodgepole pine
Pinus echinata	Shortleaf
Pinus elliottii	Slash pine
Pinus jeffreyi	Jeffrey pine
Pinus palustris	Longleaf pine
Pinus ponderosa	Ponderosa pine
Pinus resinosa	Red pine
Pinus spp.	Pines
Pinus taeda	Loblolly pine
Populus deltoides	Eastern cottonwood
Populus trichocarpa	Black cottonwood
Populus spp.	Aspens, Cottonwoods
Populus tremuloides	Quaking aspen
Prunus pensylvanica	Pin cherry
Prunus spp.	Cherries
Pseudotsuga menziesii	Douglas-fir
Quercus alba	White oak
Quercus falcata	Southern red oak
Quercus falcata var paegodifolia	Cherry bark oak
Quercus garryanna	Oregon white oak
Quercus lyrata	Overcup oak
Quercus marilandica	Blackjack oak
Quercus nigra	Water oak
Quercus palustris	Pin oak
Quercus phellos	Willow oak
Quercus spp.	Oaks
Quercus stellata	Post oak
Quercus velutina	Black oak
Quercus bicolor	Swamp white oak
Quercus coccinea	Scarlet oak
Quercus prinus	Chestnut oak
Quercus rubra	Northern red oak
Rosa multiflora	Multiflora rose
Rubus spectabilis	Salmonberry
Rubus spp.	Brambles, raspberries, blackberries
Salix spp.	Willows
Taxodium distichum	Baldcypress
Thuja plicata	Western redcedar
Tsuga canadensis	Eastern hemlock
Tsuga heterophylla	Western hemlock
Viburnum acerifolium	Maple-leaf viburnum
Viburnum alnifoium	Hobblebush
Viburnum spp.	Viburnums
Viburnum trilobum	High-bush cranberry

Appendix 2: Glossary

Active adaptive management   Management treated as a hypothesis to be tested using monitoring data

Adaptive management   A process of continual improvement in management using monitoring data to refine plans

Advance regeneration   Seedlings and saplings present in the stand prior to a disturbance which releases them

Aerial photographs   A capture of the reflectance of items on the earth's surface on a photographic film

Allele   Expression of a gene

Allochtonous material   Leaves, needles, and plant parts that fall into a water body

Alluvial   Downstream movement of sediments

Artificial regeneration   Planting seedlings or seeds usually at a particular spacing to establish a new stand

Barrier   An intervening patch type with a low probability of survival

Basal area   Cross-sectional area of trees on a hectare or acre at 1.4 m above ground

Broadcasting   Extrapolating data to other units of space outside of the scope of inference

Brood   A cohort of young birds

Brood parasite   Birds that reproduce by laying their eggs in the nests of other birds

Browse   Herbivore consumption of woody plants

Carrying capacity   A point in population growth where births equal deaths and further population growth is limited

Chain of custody   A process that assures the consumer that wood products came from a certified forest

Clearcut   A regeneration method in which all or most trees are removed to allow establishment of a new cohort of trees

Co-dominant   Trees in an even-aged stand receiving full sunlight from above and comprising the main canopy layer

Community   An assemblage of populations over space and through time

Composition   The types or classes of features in an areas, such as species of plants and types of soils

Connectivity   The degree to which the landscape facilitates or impedes movement among habitat patches

Context   An area beyond the landscape extent that we are not managing but which affects the function of the landscape

Core   The interior of a patch or home range

Corridor   An intervening patch type with a high probability of survival

Critical habitat   Specific areas and habitat elements essential to the conservation of species listed under the U.S. Endangered Species Act

Crown classes   Differentiation of trees into classes in response to growth rates and competition in an even-aged stand

Cutting cycle   Period of time between harvests when some trees of all tree diameters in an uneven-aged stand are cut

Decomposition pathway   An energy web passing through decomposers

Deferred rotation   Also known as a clearcut with reserves; retains some trees through two rotations

Demographic stochasticity   The variability represented in vital rates due to fluctuations in survival and reproduction

Desired future condition   A description of the structure and composition of a stand or landscape that you wish to achieve

Diameter-limit cutting   Cutting of all the trees above some minimum diameter during each cutting cycle in an uneven-aged stand

Digestible energy   That portion of food that can be used by an animal for energy and nutrients

Dominant crown class   Uppermost trees in an even-aged stand receiving sunlight from above and from the sides

Dynamic carrying capacity   Changing carrying capacity due to fluctuations in resource availability

Dynamic corridor   A corridors that "floats" across the landscape over time to provide connectivity at all times

Ecological restoration   Use of practices of restoration ecology as well as human and natural sciences, politics, technologies, economic factors, and cultural dimensions

Ecosystem management   A management approach designed to increase the likelihood that resources will be socially sustainable

Ecosystem services   Services provided by ecosystems to meet society's needs, including but not restricted to commodities

Ectotherm   A species that receives most of its body heat from the surrounding environment

Edge associates   Species that find the best quality habitat where there is access to resources in two or more vegetation patch types

Edge density   Edge length per unit area

Edge specialist   A species likely to only occur where edges between two or more vegetative patch types exist

Effect size   The difference (or slope) that you could detect given your sample size, sampling error, and the probability of making an error when rejecting a null hypothesis

Effectiveness monitoring   Monitoring designed to determine if habitat elements, populations, or processes are responding as expected and effectively achieving management goals

Endotherm   A species that generates its own body heat

Environmental stochasticity   Uncertain environmental events that influence population vital rates

Establishment cut   Second step in a shelterwood regeneration method to release trees to produce seeds and to provide growing space for regeneration

Eutrophic system   Nutrient-rich aquatic system

Extent   The outer bounds of the landscape over which we manage resources

Extinction vortex   Accelerated population declines irreversibly leading to extinction

Fecundity   Number of young produced per female over a given time period

Filter approach   An approach to biodiversity conservation that employs coarse-, meso-, and fine-filter management strategies

First-order selection   Selection of a geographic range by a species

Fledgling   A bird that successfully leaves the nest

Forecasting   Predicting trends into the future based on past trends

Forest interior species   A species that avoids edges and uses the core of a patch

Forest structure   The physical architecture of a forest in three dimensions

Forest type   Forest community dominated by representative tree species

Founder effect   Low genetic variation often seen in a newly established population

Fourth-order selection   Selection of specific food and cover resources acquired from the patches used by the individual within its home range

Genetic bottleneck   Dramatic decline in a population resulting in loss of alleles

Genetic drift   Some alleles may dominate in small populations by chance alone

Grain   The smallest unit of space in a landscape that we identify and use in an assessment or management plan

Grazing   Herbivore consumption of herbaceous plants

Group selection   Creation of small openings to establish patches of regeneration in an uneven-aged stand

Guild   A group of species that share common nesting or feeding resources

Habitat   The set of resources necessary to support a population over space and through time

Habitat conservation plan   A plan designed to offset any harmful effects of a proposed activity on endangered or threatened species allowing issuance of an incidental take permit

Habitat element   Piece of a forest important to many species such as vertical structure, dead wood, tree size, plant species, and forage

Habitat fragmentation   A process whereby a habitat for a species is progressively subdivided into smaller, geometrically more complex, and more isolated fragments

Habitat generalist   A species that can use a broad suite of food and cover resources

Habitat selection   A set of complex behaviors that each species has evolved to ensure fitness in a population

Habitat specialist   Species that use a narrow set of resources

Habitat types   Vegetation type or other discrete class of the environment that is associated with habitat for some species

Hard mast   Hard fruits such as nuts and acorns

Harvesting systems   The means of removing the trees from the site and to a landing during forest management

Heuristic   Use of models to teach us something about the system

Home range   Area that an individual (or pair of individuals) uses to acquire the resources that it needs to survive and reproduce

Human commensal   A species that typically is associated with humans

Hyporheic zone   Subsurface saturated sediments along the stream bottom

Ideal despotic distribution   A distribution of individuals reflecting high individual fitness in the highest-quality patches at lower-than-expected densities caused by territoriality

Ideal free distribution   A distribution of individuals reflecting the freedom of each individual to choose the patch that will provide the greatest energy or other required resources

Implementation monitoring   Measurements that document compliance with a stand prescription or management plan

Incidental take permit   A permit issued by the US Fish and Wildlife Service to allow activities that might incidentally harm (or "take") species listed as endangered or threatened under the Endangered Species Act

Indicator species   Species that are assumed to be surrogates for other species having similar resource needs

Individual tree selection   Removal of one or a few trees from a location in the stand to create a canopy gap to allow tree regeneration to occur

Induced edge   Edge between two patch types of different successional condition

Inherent edge   Edge formed by differences in the floristic composition of two patches

Intermediate crown class   Trees in an even-aged stand receiving partial sunlight from above

Intra-riparian gradients   Continuum of conditions from the headwaters to the confluence with larger water bodies

Intrinsic rate of natural increase   Each species' potential for population increase

Lambda   The population parameter used to estimate population change

Landscape   A complex mosaic of interacting patches

Logistic growth   As resources become limiting, population growth becomes asymptotic

Longevity   The age at death of the average animal in a population

Marsh   Wetlands dominated by nonwoody vegetation

Matrix   The landscape patch type within which focal patches are embedded

Matrix management   Managing the matrix condition to be made more permeable to dispersing organisms

Metapopulation   A population distributed among smaller, interacting subpopulations that contribute to overall population persistence

Mortality rate   The number of animals that die per unit of time (usually 1 year) divided by the number of animals alive at the beginning of the time period

Natality   The number of young individuals born or hatched per unit of time

Natural catastrophe   Extreme case of environmental uncertainty such as hurricanes, fires, and epizootics that can cause massive changes in vital rates

Natural cavity   Tree hole resulting from fungal decay

Natural regeneration   Stand regeneration from seedling establishment or sprouting following the disturbance

Neotropical migratory bird   Birds that nest in temperate latitudes but migrate to the tropics during the winter

Oligotrophic   Nutrient poor aquatic systems

Orographic effects   As air is moved over mountains, it increases in elevation, cools, and moisture precipitates

Orthophoto maps   Aerial photos corrected for distortion and usually with topographic information superimposed

Overwood removal   Final step in a shelterwood regeneration method to release newly established regeneration

Passive adaptive management   The "best" management option is identified, implemented, and monitored

Phreatophytic vegetation   Vegetation associated with high soil moisture or free water

Policy analysis   An organized projection of how implementation of the policy over space and time might affect the resources valued by society

Population   Self-sustaining assemblages of individuals of a species over space and through time

Population viability analysis   A structured approach to examine population performance based on demographic characteristics and habitat quantity and quality

Preparatory harvest   First step in a shelterwood regeneration method to encourage seed production

Prescriptions   Silvicultural management plans for stands

Primary cavity nester   A species that excavates a cavity in living or dead wood

Proximate cue   An element of structure and composition that an individual uses to predict resource availability

***Q*-factor**   The factor by which the number of trees in one diameter class is multiplied to get the number in the next smallest diameter class in an uneven-aged stand

Refereed journal   Scientific literature in which papers are reviewed and can be accepted or rejected based on review by peers

Response variable   Specific indicator or metric used to test a hypothesis

Restoration ecology   The suite of scientific practices that constitute an emergent subdiscipline of ecology designed to return functions to systems where they have been eliminated

Riparian area   The interface between the water and the land

Riparian associate   A species which tends to be found more commonly near water but does not require free water directly

Riparian obligate   A species that requires free water

Risk analysis   A structured way of analyzing the potential effects of decisions when outcomes are uncertain

Rotation   A complete growing cycle in an even-aged silvicultural system

Rotation age   The stand age when the stand is harvested and a new even-aged stand is regenerated

Satellite imagery   Reflectance values collected by satellites for discrete places on the earth

Scope of inference   The space and time over which data can be used to assess changes in a response variable

Secondary cavity user   Species that use natural cavities or ones created by primary cavity nesters

Second-order selection   Establishment of a home range

Seed bank   Seeds stored in the soil

Seedbed   Growing site for seedlings and sprouts

Seed-tree regeneration method   Natural regeneration is established by leaving some trees after harvest to provide a seed source

Serpentine soil   A soil enriched in toxic metals, including nickel, magnesium, barium, and chromium, and lacking in calcium

Shade intolerant   Plant species that do not survive under low light conditions and grow well only under full sunlight

Shade tolerant   Plant species that can survive under low light conditions

Shelterwood regeneration   Natural regeneration needs protection from sun or frost so a light canopy cover is maintained after harvest

Shifting-gap phase   Forests maintained by frequent small-scale gap disturbances

Silviculture   The art and practice of managing forest stands to achieve specific objectives

Sink habitat   Habitat patches in which populations are declining or are maintained by immigration

Site fidelity   A behavior in which an individual returns annually to the same location despite drastic changes in the habitat

Site index   Height of the dominant trees in an even-aged stand at a specified age

Soft mast   Soft fruits such as berries and drupes

Source habitat   Patches in which individuals are fit enough to support a stable or growing population

Source patch   During dispersal the patch that a disperser is leaving from

Stand   Unit of homogeneous forest vegetation used as the basis for management

Stand initiation   Early stage of stand development following a stand-replacement disturbance

Static corridors   Maintaining connectivity in a fixed location

Stepping stone   Small patches of habitat close to one another to enhance connectivity between high-quality patches

Stocking   The degree to which a site is occupied by trees of various sizes

Structure   Physical features of the environment such as vegetation, soils, and topography

Suppressed tree   Trees in an even-aged stand occurring below the main canopy in the stand

Survival   The number of animals that live through a time period, and is the converse of mortality

Survivorship functions   Types 1, 2, and 3 refer to high, medium, and low survival rates of juveniles, respectively

Swamp   A wetland dominated by woody vegetation

Target patch   During dispersal, the patch that a disperser is going to

Target tree size   The diameter class representing the largest harvestable trees in an uneven-aged stand

Territory   The space, usually around a nest, that an individual or pair defends from other individuals

Thermal neutral zone   The range of ambient temperatures in which an animal has to expend the least amount of energy maintaining a constant body temperature

Third-order selection   Use of patches within a home range where resources are available to meet an individual's needs

Transriparian gradients   Changes in conditions as you move from the edge of the stream into upslope forests

Trophic level   The feeding position in a food web

Ultimate resources   Food, cover, and other resources needed for survival

Validation monitoring   Measurements that provide the basis for testing assumptions during adaptive management

Vernal pools   Isolated ponds and wetlands that hold water for only a part of the year

Wolf trees   Large and often deformed legacy trees from the previous stand

Appendix 3: Measuring and Interpreting Habitat Elements

Basic to understanding current conditions and desired future conditions in stands and landscapes is measurement and interpretation of habitat elements. This field exercise introduces you to a few simple techniques for measuring the availability of key habitat elements. More comprehensive information on field sampling of habitat elements can be found in Bookhout (1994), James and Shugart (1970), Hays et al. (1981), and Noon (1981).

METHODS

Some habitat elements are particularly important to many species, depending on their size, distribution, and abundance. These include: percent cover, height, density, and biomass of trees, shrubs, grasses, forbs, and dead wood. Other habitat elements are associated with only a few species such as stream gradients (e.g., beaver; Allen 1983) and forest basal area (e.g., downy woodpecker; Schroeder 1982). Visit two areas with very different management histories such as a recent clearcut and an unmanaged forest. Then compare habitat elements between the two stand types and assess the relative habitat quality for a species between them using life history information, a habitat suitability index model, and a geographic information system.

RANDOM SAMPLING

Probably the most important part of sampling habitat is to sample randomly within the area of interest (stand, watershed, stream system, etc.). Systematic or subjective sampling can introduce bias into your estimates and lead to erroneous conclusions. In this example you will be sampling two stands. Within your stand you should collect a random sample of data describing the habitat elements. For the purposes of this exercise, you will collect data from three or more randomly located points in each stand.

1. Using a table of random numbers (nearly all statistics books have these) first select a three-digit number that is a bearing (in degrees) that will lead you into the stand. If the number that you select does not lead you into your stand, then select another number until you have a bearing that will work.
2. Select another three-digit number that is a distance in meters. Using your compass to establish the bearing and either a 30-m tape measure or pacing, measure along the assigned bearing the randomly selected distance and establish a sample point. You will collect habitat data at this point. Once you have completed collecting data at this point, repeat the process of random number selection three or more times in this stand and three or more times in another stand.

MEASURING DENSITY

One of the most common habitat elements that you will measure is density of items, usually trees, snags, logs, shrubs, or other plants. Density is simply a count of the elements over a specified area. When estimating the density of trees, you will usually count all the trees in a circular plot, usually 0.04 ha (0.1 acre) in size. Saplings and tall shrubs are usually measured in a 0.004-ha (0.01-acre) plot. Small shrubs and tree seedlings are usually measured in a 0.0004-ha (0.001-acre) plot.

1. From plot center, measure out in each cardinal direction (N, E, S, and W) 11.3 m (37.2 ft) (the radius of a 0.04-ha [0.1-acre] plot). Mark these places with flagging.
2. Using a diameter tape or a Biltmore stick, measure the diameter at 1.3 m (4.5 ft) above ground of all live trees in the plot that are more than 15 cm (6 in.) dbh and record the species of each tree. Repeat this procedure for all dead trees taller than 15 cm dbh. Expand this sample to a hectare (or acre) estimate by multiplying the estimates by 25 to convert to a per-hectare estimate (or multiply by 10 to get a per-acre estimate). This procedure can be repeated for smaller plot sizes to estimate seedling numbers, etc.

ESTIMATING PERCENT COVER

Using your four 11.3-m (37.2-ft) radii as transects, walk along each stopping at five equidistant points along each transect. At each of these points you will estimate canopy cover. There are a number of techniques available to estimate canopy cover including moosehorns (Garrison 1949) and densiometers (Lemmon 1957). A simple approach to estimating cover is to estimate the presence or absence of vegetation using a sighting tube (a piece of PVC pipe with crosshairs) (James and Shugart 1970). At each of the 20 points on your transects, look directly up and see if the crosshairs intersect vegetation (if so, record a 1) or sky (if so, record a 0). Repeat this at each of the five points on each of the four transects.

1. Tally the number of 1s recorded from these points.
2. Divide by 20 then multiply by 100 to estimate percent cover.
3. Now use this technique to measure understory herbaceous cover.

ESTIMATING HEIGHT

Use a clinometer with a percent scale (look through the view finder and you should see two scales, with units given on them, if you look straight up or straight down).

1. Measure 30 m (100 ft) from the base of the tree or other object that you wish to measure.
2. Looking through the view finder, align the horizontal line in the view finder with the top of the tree. Record the number on the percent scale (top).
3. Looking through the view finder, align the horizontal line in the view finder with the base of the tree. Record the number on the percent scale (bottom).
4. If the top number is positive and the bottom number is negative (<0), then add the absolute values of these two numbers together to estimate height in feet.
5. If the top number is positive and the bottom number is also positive (>0), then subtract the absolute value of the bottom number from the top number to estimate height in feet.

ESTIMATING BASAL AREA

Basal area is the cross-sectional area of all woody stems at 1.3 m (4.5 ft) above ground. It is a measure of dominance of a site by trees. The higher the basal area, the greater is the dominance

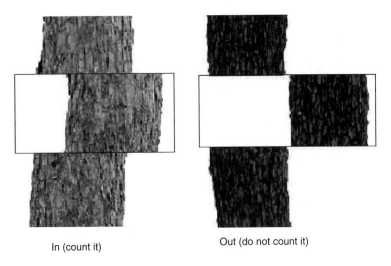

In (count it)   Out (do not count it)

FIGURE A3.1 When using a wedge prism, you have two images to compare — the one you see through the prism and the one above or below the prism. If they overlap, you count the tree as an "in" tree; if the images do not overlap, then the tree is not counted. Photo by Jesse Caputo and used with his permission.

by trees. There are two ways to estimate basal area. First, using your estimates of dbh from your sample of trees (see preceding section estimating density), you can calculate the area of each stem ($A = 3.1416 \times r^2$, where $r = $ dbh/2). By summing the areas on a 0.04-ha (0.1-acre) plot and then multiplying the total by 25, you can get an estimate of basal area per hectare (multiply by 10 to estimate basal area per acre).

Alternatively, you can use a wedge prism (Figure A3.1). Holding the prism over plot center, look at a tree through the prism. If the image that you see through the prism is connected to the image of the tree outside of the prism, then tally the tree and record its species. If the image that you see through the prism is disconnected from the image outside the prism, then do not record the tree. Moving in a circle around the prism that you continue to hold over the plot center, record all trees that have the prism image connected to the image outside of the prism regardless of whether they fall in the 0.1-acre plot or not. Tally up the number of trees that were recorded. Generally you will use a 10-factor prism, that is, each tallied tree represents 10 other trees per acre. Multiplying the number of trees tallied by 10 estimates the basal area in square feet per acre for this site.

ESTIMATING BIOMASS

Biomass of vegetation is usually estimated to provide information on food available for herbivores, typically in the winter when browse resources are essential to supporting herbivores (deer, moose, or hares). Herbivores will usually eat only woody growth from the most recent growing season, and during winter, this includes the twigs and buds, but not leaves (which will have fallen off).

Within a 1.1-m (3.7-ft) radius plot, using clippers, clip all of the twigs within the plot that have resulted from the most recent growing season. Remove and discard the leaves and place the twigs in a bag. Return to the lab and weigh the bag with the twigs. Remove the twigs and weigh the empty bag. Subtract the bag weight from the bag plus twigs weight to estimate biomass per 0.0004-ha (0.001-acre) plot. Multiply this number by 2500 (or 1000 in acres) to estimate biomass (kg) per hectare.

USING ESTIMATES OF HABITAT ELEMENTS TO ASSESS HABITAT PRESENCE

If you refer to Table A3.1 as an example (you will have your own numbers from your field samples), consider how you would interpret these data for a species of your choice, in this case

TABLE A3.1

Comparison of Average and Range of Habitat Elements between Clearcut (with a Legacy of Living and Dead Trees) and Uncut Forests, Cadwell Forest, Pelham, MA

	Clearcut	Mature forest
Trees > 15 cm/ha	3 (0–6)	308 (234–412)
Snags > 15 cm/ha	1 (0–2)	22 (4–43)
Basal area/ha	2.4 (0–3)	16 (12–18)
Canopy cover (%)	4 (0–7)	95 (90–100)
Canopy height (m)	23 (18–34)	27 (23–33)
Browse (kg/ha)	1234 (554–2600)	387(122–788)

downy woodpeckers. DeGraaf and Yamasaki (2001) describe habitat for downy woodpeckers as "...woodlands with living and dead trees from 25 to 60 cm dbh; some dead or living trees must be greater than 15 cm dbh for nesting." Although both sites contain trees and snags of sufficient size, the canopy cover data in Table A3.1 would suggest that the clearcut is not functioning as a woodland, and so we would probably not consider it suitable habitat for downy woodpeckers, though they certainly do use snags in openings at times.

USING ESTIMATES OF HABITAT ELEMENTS TO ASSESS HABITAT SUITABILITY

In addition to using your data to understand if a site might be used by a species, habitat suitability index models have been developed to understand if some sites might provide more suitable habitat than others (e.g., Schroeder 1982). Very few of these models have been validated especially not using fitness as a response variable. Nonetheless they do represent hypotheses based on the assumption that there is a positive relationship between the index and habitat-carrying capacity. If we take the example of the downy woodpecker then its habitat suitability is based on two indices: tree basal area (Figure A3.2) and density of snags >15 cm dbh (Figure A3.3). Consider first the uncut stand. Note that there is an average of 16 m²/ha of basal area and 22 snags/ha (8.8/0.4 ha). The corresponding suitability index score for each variable is 1.0 and the overall habitat suitability is calculated (in this case) as the minimum of the two values. Hence, this should be very good habitat for downy woodpeckers. In the recent clearcut, however, the suitability index for snags is approximately 0.1 and for basal area is approximately 0.2. Hence, the overall suitability in the recent clearcut for this species is 0.1 — not very good and certainly less than in the uncut stand. And in this case snag density is the factor most limiting habitat quality for downy woodpeckers in the recent clearcut. The best way to use these sorts of models is in a relative sense, to compare one site with another. If we were use this technique for snowshoe hares habitat assessment, then we might find the recent clearcut to be a much better habitat because of greater browse resources and hiding cover.

ASSESSING THE DISTRIBUTION OF HABITAT ACROSS A LANDSCAPE

It is often as important to know if stands are suitable habitat for a species and how they are arranged on a landscape. As shown in Figure A3.4, a 490-ha forest has been broken into habitat types based on overstory cover and stand structure. Field samples were taken at 117 points distributed across

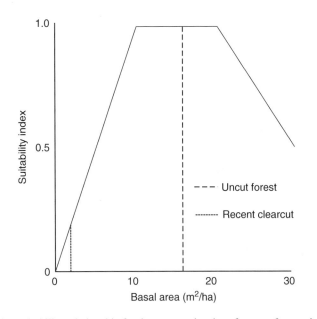

FIGURE A3.2   Habitat suitability relationship for downy woodpeckers for one of two suitability indices: basal area. (Redrafted from Schroeder, R.L. 1982. *Habitat Suitability Index Models: Downy Woodpecker.* U.S. Fish and Wildlife Service, FWS/OBS-82/10.38.)

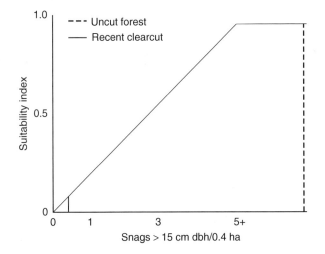

FIGURE A3.3   Habitat suitability relationship for downy woodpeckers for one of two suitability indices: snag density. (Redrafted from Schroeder, R.L. 1982. *Habitat Suitability Index Models: Downy Woodpecker.* U.S. Fish and Wildlife Service, FWS/OBS-82/10.38.)

the forest and habitat elements were sampled at each point. Habitat suitability index values are then calculated at each point and extrapolated to the habitat types as portrayed in this figure to illustrate how habitat availability for a species can be displayed over a landscape. A different pattern would emerge for other species using this same approach, and these would have to be overlain on stands used as the basis for management. In addition, these types of maps can guide harvest planning so as to achieve habitat patterns leading to a desired future condition for the landscape.

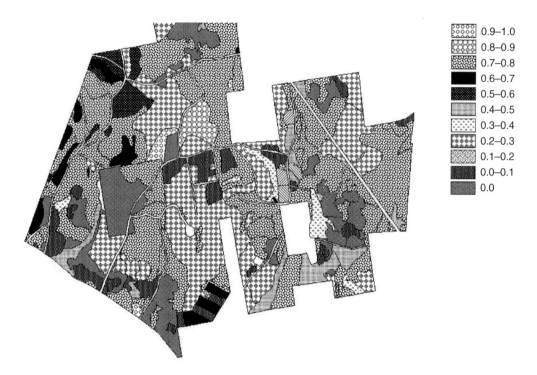

▓	0.9–1.0
▓	0.8–0.9
▓	0.7–0.8
■	0.6–0.7
▓	0.5–0.6
▓	0.4–0.5
▓	0.3–0.4
▓	0.2–0.3
▓	0.1–0.2
▓	0.0–0.1
▓	0.0

FIGURE A3.4   An example of a mosaic of habitat patches of varying suitability based on extrapolation of ground inventory data to digitized patches in Cadwell Memorial Forest, Pelham, Massachusetts.

REFERENCES

Allen, A.W. 1983. *Habitat Suitability Index Models: Beaver.* U.S. Department of the Interior, Fish and Wildlife Service, Washington, D.C., FWS/OBS-82/10.30 (revised).

Bookhout, T.A. 1994. *Research and Management Techniques for Wildlife and Habitats*, 5th edition. The Wildlife Society, Bethesda, MD.

DeGraaf, R.M., and M. Yamasaki. 2001. *New England Wildlife: Habitat, Natural History, and Distribution.* University Press of New England, Hanover, NH.

Garrison, G.A. 1949. Uses and modifications for the "moosehorn" crown closure estimator. *J. For.* 47: 733–735.

Hays, R.L., C. Summers, and W. Seitz. 1981. *Estimating Wildlife Habitat Variables.* USDI Fish and Wildlife Service, FWS/OBS-81/47.

James, F.C., and H.H. Shugart Jr. 1970. A quantitative method of habitat description. *Am. Birds* 24: 727–736.

Lemmon, P.E. 1957. A new instrument for measuring forest canopy overstory density. *J. For.* 55: 667–638.

Noon, B.R. 1981. Techniques for sampling avian habitats. In Capen D.E. (ed.), *The Use of Multivariate Statistics in Studies of Wildlife Habitat.* U.S. For. Serv. Gen. Tech. Rep. RM. 87: 41–52.

Schroeder, R.L. 1982. *Habitat Suitability Index Models: Downy Woodpecker.* U.S. Fish and Wildlife Service, FWS/OBS-82/10.38.

Subject Index

Note: Numbers in **bold font** indicate pages containing charts/diagrams/graphs/illustrations/tables.

Species Index

Note: Bolded pages contain photos or charts relevant to the species. See Appendix 1 for scientific names.